T0292117

Thermoluminescence of solids

Cambridge Solid State Science Series

EDITORS:

Professor R. W. Cahn
University of Cambridge

Professor E. A. Davis
Department of Physics, University of Leicester

Professor I. M. Ward
Department of Physics, University of Leeds

S.W.S. McKEEVER

Department of Physics, Oklahoma State University

Thermoluminescence of solids

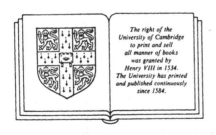

The right of the
University of Cambridge
to print and sell
all manner of books
was granted by
Henry VIII in 1534.
The University has printed
and published continuously
since 1584.

CAMBRIDGE UNIVERSITY PRESS
Cambridge
New York   New Rochelle
Melbourne   Sydney

CAMBRIDGE UNIVERSITY PRESS
Cambridge, New York, Melbourne, Madrid, Cape Town, Singapore,
São Paulo, Delhi, Dubai, Tokyo, Mexico City

Cambridge University Press
The Edinburgh Building, Cambridge CB2 8RU, UK

Published in the United States of America by
Cambridge University Press, New York

www.cambridge.org
Information on this title: www.cambridge.org/9780521368117

First published 1985
First paperback edition 1988

A catalogue record for this publication is available from the British Library

Library of Congress Cataloguing in Publication Data: 84 – 5028

ISBN 978-0-521-24520-3 Hardback
ISBN 978-0-521-36811-7 Paperback

To Mum, Dad and Joan

Contents

Contents   xi

Preface

Previous books dealing with thermoluminescence have tended to treat only specialist aspects of its usage – e.g., dating, dosimetry, geology or analysis of glow-curves. This is a reflection of the way in which thermoluminescence research has developed over the last decade; its study has become fragmented, each fragment being tailored to suit only the discipline to which it is being applied. As a result, the various disciplines have become somewhat insular, the advances in one arena not necessarily being transmitted to another. (For instance, one might find the occasional inappropiate application of different methods of analysis of glow-curves to problems associated with, say, geology or archaeology; or a lack of appreciation of the solid-state defect reactions which can take place in phosphors when thermoluminescence is applied to measure radiation dose.)

These facts do not appear to have imposed a limit on the rate of publication of papers dealing with thermoluminescence; several hundred articles on this topic appear each year. Considering this, perhaps the time is now appropriate for a pause in the accelerated use of thermoluminescence in order to reflect upon its capabilities and to become more aware of its limitations.

This book is intended as a step towards this goal, and towards the unification of the insular approaches by presenting the topic of thermoluminescence as a single subject. Extensive referencing of published works was felt to be an important requirement and as a result over 1000 references are listed. By fully realizing the limitations of the technique we will be able to make best use of its capabilities.

The book begins (chapter 1) by placing thermoluminescence in the context of luminescence phenomena in general and then traces the historical development of the subject, culminating in its modern-day usage in a wide variety of applications. Chapter 2 is primarily intended for those readers who have no knowledge at all of thermoluminescence. More experienced readers may wish to miss out sections 2.1 and 2.2 of this chapter and proceed directly with section 2.3.

Those readers primarily interested in the use of the technique as an investigative tool in trap-level spectroscopy will find chapter 3 to be the most useful of the chapters for their needs. Along with section 2.3, this chapter provides the reader with the necessary information to perform a variety of analyses on experimental glow-curves.

As a general information chapter covering several aspects of thermoluminescence production, chapter 4 deals with important phenomena such as supralinearity, sensitization, fading and optical and quenching effects. This is followed by the first of the 'applications' chapters (5) for which I chose, as illustrative material, to give a detailed account of thermoluminescence in alkali halides and SiO_2. This chapter is intended to present the reader with a view of how the technique can be used to infer information on a phosphor's solid-state properties. Alkali halides and SiO_2 were chosen because these materials, more than any other, have been subjected to such an intense examination over the decades that a wide variety of detailed information could be given, borrowing heavily from a rich literature on many other experimental techniques.

Chapters 6 and 7 deal with the most popular of the applications of thermoluminescence – namely, dosimetry and dating. The specialist reader may only need to read one or other of these chapters, but to obtain full usefulness from the book, it is recommended that both are read.

Chapter 8 examines the application of thermoluminescence to geology. Although much useful information has emerged in this context, the technique has tended to promise more than it can (so far) deliver.

Finally, chapter 9 deals with instrumentation. In some senses this is an isolated chapter in that it is not truly necessary to read the others before turning to it, if all that is required is information on experimental apparatus. Where reference to earlier material is desirable, appropriate cross-references are given.

It is a great pleasure to extend my warm thanks to those who have contributed so much to the writing of this book. Primarily my thanks go to Professor R.W. Cahn who first suggested the book and who has provided constant encouragement and helpful criticism throughout its preparation. I am indebted also to Dr P.D. Townsend who painstakingly read every page of the manuscript and whose constructive discussions, comments and criticisms have greatly improved the book's content. The useful remarks of Drs H.M. Rendell and L.E. Halliburton on various sections of the book are also gratefully acknowledged. Finally, I must extend a deep sense of gratitude to my wife, whose patience, support and typing skills played major roles in the book's production. (You can now have your husband back.)

S.W.S. McKeever
Oklahoma State University
July 1983

1   Introduction

1.1   What is thermoluminescence?

Volume 84 of the *Physics Abstracts Subject Index* for 1981 lists over 300 separate entries for papers concerning the topic of *thermoluminescence*. Bräunlich (1979) states that there are over 500 articles published per year on phenomena relating to thermoluminescence (generally termed *thermally stimulated relaxations*). Newcomers to this field and experienced research workers alike may find such a high rate of publication surprising. This response may be reinforced when they realize that, as an experimental technique, thermoluminescence finds favour in such diverse scientific disciplines as archaeology, geology, medicine, solid-state physics, biology and organic chemistry, to name just some of the mainstream areas of study. The reader may justifiably enquire what it is about thermoluminescence that makes it such a well-used experimental method, enjoying widespread popularity and displaying enormous versatility. The answer to this enquiry is the essential theme addressed in this book. To answer the question fully it will be necessary to illustrate how the thermoluminescence characteristics of a material relate directly to the material's solid-state properties and how these solid-state properties are being utilized in the diverse fields mentioned above. However, before delving into the detail necessary to answer the question in full an initial approach to the problem must be made by stating what thermoluminescence is.

Thermoluminescence is the emission of light from an insulator or semiconductor when it is heated. This is not to be confused with the light spontaneously emitted from a substance when it is heated to incandescence. Thermoluminescence is the thermally stimulated emission of light following the previous absorption of energy from radiation.

In this statement can be found the three essential ingredients necessary for the production of thermoluminescence. Firstly, the material must be an insulator or a semiconductor – metals do not exhibit luminescent properties. Secondly, the material must have at some time absorbed energy during exposure to radiation. Thirdly, the luminescence emission is triggered by heating the material. In addition, there is one important property of thermoluminescence which cannot be inferred from this statement as it stands at present. It is a particular characteristic of thermoluminescence that, once heated to excite the light emission, the material cannot be made to emit thermoluminescence again by simply cooling the specimen and reheating. In order to re-exhibit the luminescence the material has to be re-exposed to radiation, whereupon raising the temperature will once again produce light emission.

The fundamental principles which govern the production of thermoluminescence are essentially the same as those which govern all luminescence processes, and in this way thermoluminescence is merely one of a large family of luminescence phenomena. Therefore, before progressing further in order to examine the process of thermoluminescence in a more detailed manner, it is instructive to pause for a moment and to take a wider perspective by examining where thermoluminescence lies in the context of luminescence phenomena in general.

1.2   Luminescence

When radiation is incident on a material some of its energy may be absorbed and re-emitted as light of a *longer* wavelength (Stoke's Law). This is the process of luminescence. The wavelength of the emitted light is characteristic of the luminescent substance and not of the incident radiation. Usually, most studies of luminescence phenomena are concerned with the emission of visible light but other wavelengths can be emitted, such as ultra-violet or infra-red. It is only the emission of visible light which is dealt with in this book.

The various luminescence phenomena are given names which reflect the type of radiation used to excite the emission. Thus we have *photoluminescence* (excitation by optical or ultra-violet light), *radioluminescence* (nuclear radiations, i.e., γ-rays, β-particles, X-rays, etc.) and *cathodoluminescence* (electron beam). In addition to excitation by radiation, luminescence can also be generated by chemical energy (*chemiluminescence*), mechanical energy (*triboluminescence*), electrical energy (*electroluminescence*), biochemical energy (*bioluminescence*) and even sound waves (*sonoluminescence*). Unless otherwise stated, the luminescence referred to throughout this book will be concerned with excitation by radiation.

The emission of light takes place a characteristic time τ_c after the absorption of the radiation and this parameter allows us to sub classify the process of luminescence (see figure 1.1). Thus we can distinguish between *fluorescence* in which $\tau_c < 10^{-8}$ s and *phosphorescence* in which $\tau_c > 10^{-8}$ s (Garlick, 1949; Curie, 1960).

The value of $\tau_c < 10^{-8}$ s provides a definition for the essentially spontaneous process of fluorescence emission. Thus, in figure 1.2, the fluorescence emission is depicted as taking place simultaneously with the absorption of radiation and stopping immediately the radiation ceases.

Phosphorescence on the other hand is characterized by a delay between the radiation absorption and the time t_{max} to reach full intensity. Furthermore, the phosphorescence is seen to continue for some time after the excitation has been removed. Clearly, if the delay time τ_c is of the order of 1 s, it is easy to classify the emission as phosphorescence. However, for delays of a much shorter time it is more difficult to distinguish between fluorescence and phosphorescence. Phosphorescence itself may be conveniently subdivided into two main types (Garlick & Wilkins, 1945; Randall & Wilkins, 1945a,b; Curie, 1960) namely, short-period ($\tau_c < 10^{-4}$ s) and long-period ($\tau_c > 10^{-4}$ s) phosphorescence. From a practical viewpoint the only clear way to distinguish between fluorescence and phosphorescence is to study the

Figure 1.1. The 'family tree' of luminescence phenomena. The prefix to the term luminescence distinguishes between the modes of excitation, whilst the delay between excitation and emission τ_c distinguishes between fluorescence and phosphorescence.

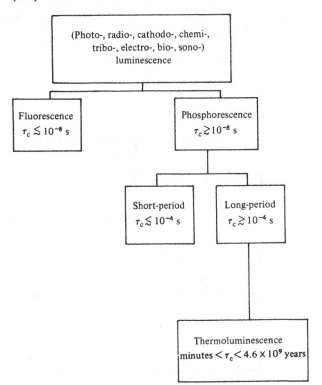

effect of temperature upon the decay of the luminescence. Fluorescence is essentially independent of temperature, whereas the decay of phosphorescence exhibits a strong temperature dependence.

In general, luminescence emission is explained by the transfer of energy from radiation to the electrons of the solid, thus exciting the electrons from a ground state g to an excited state e (transition (i) in figure 1.3a). The emission of a luminescence photon takes place when an excited electron returns to its ground state (transition (ii)). Thus, for fluorescence, the delay between transitions (i) and (ii) is less than 10^{-8} s and this process is temperature independent. According to Chen & Kirsh (1981), Jablonski (1935) gave the first explanation of temperature-dependent phosphorescence. Here (figure 1.3b) the energy level diagram is modified by the presence of a metastable level m in the 'forbidden' energy gap between e and g. An electron excited from g to e can now become trapped at m where it will remain until it is given enough energy E to return to e from where it can undergo a normal transition back to g,

Figure 1.2. Relationships between radiation absorption and the emissions of fluorescence, phosphorescence and thermoluminescence. T_0 is the temperature at which irradiation takes place; β is the heating rate; t_r is the time at which the irradiation ends and the decay of phosphorescence begins.

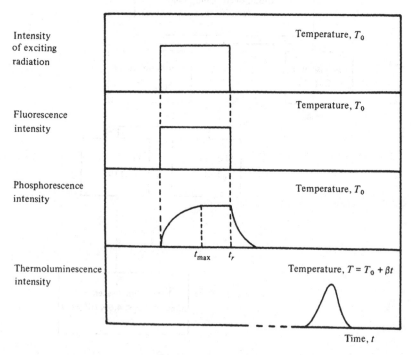

with the subsequent emission of light. Thus the delay observed in phosphorescence corresponds to the time the electron spends in the electron trap m. From thermodynamic arguments it can be shown that the mean time spent in the trap at temperature T is given by

$$\tau = s^{-1} \exp(E/kT), \tag{1.1}$$

where s is a constant and E is the energy difference between m and e (called the trap depth); k is Boltzmann's constant. Thus the phosphorescence process is exponentially dependent upon temperature.

This simple picture of phosphorescence based on the energy band theory of solids has been used with success to account for the luminescence properties of several phosphors (e.g., Johnson, 1939). However, the theory was not given a thorough formalism until the work of Randall & Wilkins (1945a). These authors assumed that once the electron had been freed from its trap (i.e., once it had made the transition m–e in figure 1.3b), the probability of it returning to m is much less than the probability of it returning to the ground state g. The intensity of phosphorescence emission at any instant $I(t)$ is proportional to the rate of recombination (i.e., the rate of e–g transitions). In this case, e–g transitions are simply governed by m–e transitions and thus $I(t)$ is proportional to the rate of release of electrons from the trap. Hence,

$$I(t) = -C\,dn/dt = Cn/\tau \tag{1.2}$$

where C is the constant of proportionality and n is the number of electrons trapped in m. Integrating (1.2) gives

$$I(t) = I_0 \exp(-t/\tau) \tag{1.3}$$

where τ is given by equation (1.1), t is time and I_0 is the intensity at $t = 0$. Equation (1.3) is the equation of the decay of phosphorescence at a constant temperature following the end of irradiation, i.e., for $t > t_r$ in

Figure 1.3. Energy transitions involved in the production of (a) fluorescence and (b) phosphorescence. Symbols defined in text.

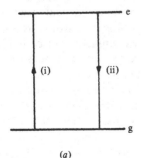

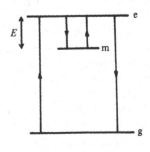

(a)   (b)

figure 1.2. At constant temperature, the decay is thus a simple exponential, or first-order decay.

It is often found in practice, however, that the decay of phosphorescence is non-exponential. One reason for this might be the overlap of several first-order processes so that at any one temperature several traps, each of different E values, are being sampled. A second reason, also dealt with by Randall & Wilkins (1945a), concerns the possibility that once released from its trap, the electron may either return to m, or recombine at g. In this case the recombination rate is proportional not just to the number of electrons but also to the number of available recombination sites. The assumption that these are equal (both equal to n) gives

$$I(t) = -C\,\mathrm{d}n/\mathrm{d}t = \alpha n^2 \tag{1.4}$$

where α is a constant at constant T.

Comparison of equation (1.4) with equation (1.2) shows that the phosphorescence intensity is now proportional to n^2 rather than n and integration gives

$$I(t) = I_0/(n_0\,\alpha t + 1)^2 \tag{1.5}$$

This kind of decay is termed second-order. The constant α is related to the mean lifetime τ and to a term which describes the relative possibilities of the retrapping of electrons at m and their recombination at g. This constant (α) will be discussed in more detail in a later section.

Equation (1.1) shows that the mean lifetime τ is exponentially dependent upon temperature. For phosphorescence the combined values of E and T are such that τ is very small and luminescence is observed easily at the temperature T_0 at which the irradiation takes place. However, if the trap is deep enough, then values of E and T_0 are such that $E \gg kT_0$ and thus τ is very large. In effect, this means that the electron will remain trapped in level m indefinitely, or rather the rate of release of trapped electrons $\mathrm{d}n/\mathrm{d}t = -n/\tau$ is very small at T_0. For example, for a trap depth $E = 1.5\,\mathrm{eV}$ and assuming that $s = 10^{12}\,\mathrm{s}^{-1}$, then $\tau = 7.3 \times 10^5$ years at $T = 298$ K. From a practical viewpoint this means that luminescence would never be observed from this trap at $T_0 = 298$ K, or less. However, luminescence emission can be induced by raising the temperature. For example, if the temperature is raised at a linear rate $\beta = \mathrm{d}T/\mathrm{d}t$, there will come a temperature at which $I(t) = n/\tau$ is large enough for the luminescence to be observed. As T rises, τ decreases and consequently the intensity increases as the electrons become freed from the trap and recombination takes place. Eventually, as the trap becomes depleted, $I(t)$ starts to decrease and the resultant intensity versus temperature curve is in the form of a peak. Because the luminescence has been stimulated by heating, it is thermoluminescence. A thermoluminescence peak is shown in figure 1.2 in comparison with the fluorescence and phosphorescence emissions. In this figure it is plotted against time, which is related to temperature by $\beta = dT/dt$. For a suitably deep trap the thermoluminescence can be triggered an indefinite time after the irradiation ceases. This time may range from minutes to years. A practical limit on the time between irradiation and thermoluminescence readout is set only by the age of the solar system at 4.6×10^9 years (figure 1.1).

The normal way of displaying thermoluminescence data is to plot luminescence intensity as a function of temperature – known as a 'glow-curve'. A typical glow-curve for LiF, one of the most studied thermoluminescent phosphors, is shown in figure 1.4. The temperature at which the peak maximum appears is related to the trap depth. In figure 1.4, four peaks are shown (2–5) indicating that four different species of trap are being activated within this particular temperature range, each with its own value of E and s. (In general, the larger the value of E, the higher the temperature at which the peaks will occur. However, this is by no means always true as will be discussed later.) The area under each peak is related to the number of filled traps, which, in turn is related to the amount of radiation initially imparted to the specimen.

In this simple description of the thermoluminescence process it is

Figure 1.4. A thermoluminescence 'glow-curve' from LiF, doped with Mg and Ti, following irradiation with 250 rad γ-rays at room temperature. Heating rate $\beta = 3\,^\circ\text{C s}^{-1}$.

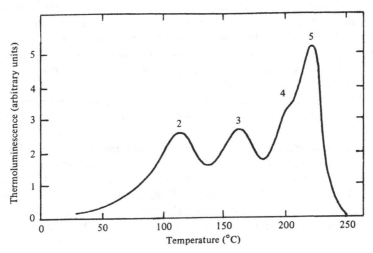

possible to observe all the essential elements of the initial statement on thermoluminescence made in the preceding section, namely, the energy band model associated with semiconductors and insulators; the absorption of energy from radiation building up a population of trapped electrons; and the thermally stimulated release of the electrons followed by recombination and luminescence emission. Furthermore, it is possible to see how a second irradiation is necessary if thermoluminescence is to be induced again after the first heating.

1.3   Early observations of thermoluminescence (pre-1948)

According to Becker (1973) medieval alchemists were aware that certain minerals glowed faintly when heated in the dark. However, possibly the first scientifically recorded observation of thermoluminescence was made in 1663 by Robert Boyle (Boyle, 1663) who noted a 'glimmering light' from a diamond by 'taking it to Bed with me, and holding it a good while upon a warm part of my Naked Body'. Boyle also stimulated the luminescence emission by more conventional means by using the heat from a hot iron, from friction and from a candle. In 1676 Elsholtz observed a similar effect from the mineral fluorspar (see Seeley, 1975). Early interpretations of the phenomena were that the heat itself was being directly converted into light. Oldenburg (1676), referring to the thermoluminescence of a phosphor called 'Phosphorus Smaragdinus', wrote that the material received its light 'from the Fire itself'. Most of the other observations at the time supported this or similar views. Du Fay (1726) thought that the luminescence was due to 'a sulphur' which actually burned on heating, but later he was to provide what was possibly the first clear evidence that the observed phenomenon (i.e., thermoluminescence) was nothing more than delayed phosphorescence (Du Fay, 1738). His experiments on natural quartz showed that the thermoluminescence could be reactivated by exposure of the sample to light. Heat only stimulated the emission, but was not its cause.

Déribéré (1936) reports that, in 1821, Calloud, a chemist from Annecy in France, discovered that heating sulphate of quinine produced an intense blue luminescence between 100 and 180 °C. This observation was confirmed shortly afterwards by Pelletier. In 1867, Becquerel also reported luminescence from fluorspar as it was heated (see Rutherford, 1913). Some years later, the first account of thermoluminescence from extraterrestrial minerals was written by Alexander Herschel. Herschel, grandson of the great astronomer William Herschel, observed that 'Some fine dust and grains obtained from the inner portion of the Middlesborough aerolite, when the meteorite was first being chemically and microscopically examined, were found, to my considerable surprise, to glow quite distinctly, though not very brightly, with yellowish-white light, when sprinkled in the usual way for these experiments on a piece of nearly red-heated iron in the dark.' (Herschel, 1889).

It is difficult to pinpoint exactly when the word thermoluminescence was first used in the published literature, but it is certainly used in 1895 by Wiedemann & Schmidt (1895) in what Becker (1974) describes as 'Probably the first careful investigation of experimentally radiation-induced TL (thermoluminescence) under its modern name....' A major difference between the work of Wiedemann & Schmidt and that of earlier investigators is that these authors induced the thermoluminescence themselves by irradiating the specimens with an electron beam in the laboratory. This type of thermoluminescence is sometimes referred to as 'artificial' whereas the earlier observations were on 'natural' thermoluminescence, i.e., that induced by natural radioactivity in the environment. Wiedemann & Schmidt studied a wide variety of synthetically produced phosphors. However, the fact that the thermoluminescence from natural specimens could be regenerated in the laboratory was first published by Trowbridge & Burbank (1898). These authors drained the natural thermoluminescence from fluorite by heating it and then re-excited the thermoluminescence by exposing the specimen to X-rays.

The connection between phosphorescence and radiation was the subject of extensive examinations in the late nineteenth century (see Rutherford, Chadwick & Ellis, 1957) and the study of radiation-induced thermoluminescence received a boost from Marie Curie in 1904 when she wrote in her doctoral thesis: 'Certain bodies, such as fluorite, became luminous when heated; they are thermo-luminescent. Their luminosity disappears after some time, but the capacity of becoming luminous afresh through heat is restored to them by the action of a spark and also by the action of radiation. Radium can thus restore to these bodies their thermo-luminescent property.' (Curie, 1904). In fact, the ability of β-rays from radium to restore thermoluminescence had also been observed by Wiedemann (see Rutherford, 1913).

At about this time the first thorough study of the spectra of the emitted thermoluminescence was made by Morse (1905). The mineral fluorspar was the subject of Morse's observations.

Lind & Bardwell (1923) continued the study of radiation-induced thermoluminescence by using radium to excite glow from various gems and transparent minerals. This was followed in 1924 by Wick, who made his own observation of natural thermoluminescence in fluorite (Wick, 1924). He and his colleagues proceeded to make a thorough study of X-ray- and electron-beam-induced thermoluminescence in a selection of natural minerals (Wick & Gleason, 1924; Wick, 1927) and synthetic phosphors (Wick & Slattery, 1927). In particular, Wick & Slattery (1928) made an extensive investigation of X-ray-induced thermoluminescence in a selection of the same synthetic phosphors as were studied by Wiedemann & Schmidt (1895). Included in these phosphors was $CaSO_4$ doped with manganese. Manganese became recognized as an especially good 'activator' of luminescence, and $CaSO_4:Mn$ in particular was an excellent thermoluminescent phosphor. This material was used by Lyman (1935) to detect ultra-violet radiation by monitoring the thermoluminescence induced by exposure of the phosphor to light from a spark. The degree of transparency of the air between the spark and the phosphor was deduced from the brightness of the thermoluminescence. In a sense Lyman's work was one of the earliest uses of thermoluminescence in dosimetry.

It has already been noted that the temperature at which the thermoluminescence peak maximum occurs is related to the electron trap depth. This was recognized by Urbach (1930) as the key to the use of thermoluminescence to study trap depth distributions. However, it was not until the now-famous papers by the research group at the University of Birmingham, England (Randall & Wilkins, 1945a,b; Garlick & Gibson, 1948), that any real progress was made in this regard. Randall & Wilkins formalized the theory of thermoluminescence by considering the first-order mechanism of electron detrapping, i.e., they assumed that once released from a trap the electron will undergo recombination rather than retrapping. Thus, the intensity of luminescence emission is proportional to the rate of release of trapped electrons (equation 1.2). Garlick & Gibson (1948) followed this by considering the second-order case. Here they assumed comparable retrapping and recombination probabilities and thus dealt with equation (1.4). By solving equations (1.2) and (1.4) for the non-isothermal case of a linear increase in temperature at rate $\beta = dT/dt$, these authors were able to relate, in a precise manner, the temperature, shape and size of a thermoluminescence peak to the trapping parameters E and s, and to the number of trapped electrons at the start of heating. Thus, Randall & Wilkins (1945a,b) arrived at the following equation for the shape of a first-order thermoluminescence glow peak:

$$I = n_0 s \exp(-E/kT) \exp\left[-(s/\beta) \int_{T_0}^{T} \exp(-E/kT)dT \right] \quad (1.6)$$

where n_0 is the number of trapped electrons at temperature T_0.

Similarly, the equation for a second-order glow peak as derived by Garlick & Gibson (1948) is:

$$I = n_0^2 s \exp(-E/kT)/N \left[1 + (n_0 s/N\beta) \int_{T_0}^{T} \exp(-E/kT) dT \right]^2 \quad (1.7)$$

where N is the number of *available* electron traps. From equations (1.6) and (1.7), simple expressions were produced from which the trapping parameters E and s could be derived from an experimental glow-curve. Thus, from this point on, thermoluminescence grew in popularity as a relatively simple technique for obtaining information on trap distributions.

1.4 Applications

1.4.1 Radiation dosimetry

The increased use of thermoluminescence which became evident in the late 1940s and early 1950s was only partly due to the work of Randall & Wilkins and of Garlick & Gibson. Although a quantification of the thermoluminescence mechanism was necessary in order to calculate the trapping parameters, several other applications of the phenomenon were beginning to emerge at this time. The prime motivators of this emergence of thermoluminescence as a practical research tool in several different fields of application were Farrington Daniels and his research group at the University of Wisconsin, USA, during the 1950s.

The absorption of radiation increases the level of thermoluminescence observed from a specimen by filling the localized energy levels with trapped electrons. The absorption of heat from the environment, on the other hand, tends to reduce the numbers of trapped electrons by thermally releasing them. Thus, the intensity of thermoluminescence from a specimen is the result of a competition between trap filling by radiation and trap emptying by thermal excitation. At a given temperature of irradiation, many materials display an intensity of thermoluminescence which is proportional (or nearly so) to the amount of radiation absorbed, and this led Daniels and colleagues to propose that thermoluminescence may be used as a means of radiation dosimetry. The first proper application of thermoluminescence to dosimetry was in 1953 when LiF was used to measure radiation following an atomic weapon test (quoted in Cameron, Suntharalingam & Kenney, 1968). LiF was found by Daniels to be a particularly good material for use in radiation dosimetry because of its high sensitivity, and small pellets of this material were used at the Hospital of the Oak Ridge Institute of Nuclear Studies to measure internal radiation doses received by cancer patients treated with radioactive isotopes (Daniels, Boyd & Saunders, 1953). By having the patients swallow small LiF pellets, which were recovered after passage through the digestive system, the accumulated dose received by the patients was obtained by measuring the thermoluminescence from the pellets and comparing it with that produced in similar crystals which had been irradiated with a known dose of radiation.

Following the pioneering work of Daniels and colleagues the application of thermoluminescence to radiation dosimetry has seen an immense escalation of effort and a vast literature now exists on this topic. (More than 50% of the papers published on thermoluminescence each year are concerned with dosimetry.) Most research has been devoted to the discovery and development of materials suitable for thermoluminescence dosemeters. As early as 1957, Daniels had despaired of the unpredictable properties of LiF and had favoured Al_2O_3 as a better dosemeter material (Rieke & Daniels, 1957). LiF did not re-emerge as a first-class dosemeter until the work of Cameron and his colleagues in the 1960s. It had been known since the work of Wiedemann & Schmidt (1895) that the presence of impurities within a crystal enhances the thermoluminescence response and so it was that the sensitivity of LiF was found to be improved by the incorporation of impurities (although at that time the exact nature of the impurity responsible for the enhanced luminescence in LiF was unknown).

It was also realized that some of the unpredictable behaviour of LiF could be offset by various pre- and post-irradiation annealing procedures (e.g., Zimmerman, Rhyner & Cameron, 1966). A commercial thermoluminescence dosemeter based on LiF became available shortly afterwards. In collaboration with Cameron's research group, the Harshaw Chemical Company, Ohio, USA, developed the dosemeter TLD-100 which consists of LiF doped with approximately 170 mole ppm Mg and 10 mole ppm Ti. This material remains the most popular thermoluminescent dosemeter in use today.

Much research was (and still is) carried out on the development of other dosemeters and for a detailed history of the development of thermoluminescence dosimetry in general the reader is referred to several recent books and review articles on the topic (Horowitz, 1981; Becker, 1973; McKinlay, 1981; Oberhofer & Scharmann, 1981). Since 1966 there have been seven international conferences on luminescence dosimetry, namely, Stanford (Attix, 1965); Gattlinburg (Auxier, Becker & Robinson, 1968); Risφ (Mejdahl, 1971); Krakow (Niewiadomski, 1974); Sao Paulo (Scharmann, 1977), Toulouse (Portal, 1980) and

Ottawa (Stoebe, 1984). The published proceedings from these symposia present a detailed picture of the development of thermoluminescence dosimetry into what is emerging as a preferred method of radiation dose measurement as it gradually ousts the more traditional techniques that use film badges and ionization chambers.

1.4.2 Age determination

Once the relationship between thermoluminescence intensity and absorbed radiation dose had been established it was only a short step to the use of thermoluminescence as a means of age determination. This application of thermoluminescence was also first suggested by Daniels and colleagues (Daniels et al., 1953) who offered the premise that the natural thermoluminescence from rocks is directly related to the radioactivity from uranium, thorium and potassium present within the material. This radioactivity results in the accumulation of a so-called 'geological' dose. If the rate of irradiation from the radioactive minerals is established, and if the rate of thermal release of the thermoluminescence during the rock's irradiation can be shown to be negligible, then the length of time over which the rock has been irradiated (i.e., its 'geological age') can be determined from:

$$age = absorbed\ dose/dose\ rate. \qquad (1.8)$$

Several attempts to determine the geological age of certain rock formations were carried out in the 1950s but the proper development of thermoluminescence as a means of age determination did not begin in earnest until the discovery of natural thermoluminescence from samples of ancient pottery (Kennedy & Knopff, 1960; see also Grögler, Houtermans & Stäuffer, 1960). This observation led to the immediate development of thermoluminescence as a means of archaeological dating. The basic premise is that the ingredients of the pottery (i.e., the clay) lose their accumulated geological dose when the pot is fired during its manufacture. Because of the high temperature experienced during firing the thermoluminescence level is re-set to zero. The newly formed pot is now subjected to natural radiation from the radioactive elements (chiefly uranium, thorium and potassium) naturally present in the clay and surroundings. Thus the pottery now accumulates an absorbed dose which is proportional to its 'archaeological age' – i.e., the time since firing. This age is also calculated by equation (1.8).

The method of thermoluminescence dating progressed through the 1960s and 70s via the development of improved methods of extracting the thermoluminescent grains from pottery in order to facilitate the calculation of the natural dose rate. Thus, the quartz-inclusion (Fleming, 1966) and fine-grain (Zimmerman, 1967) methods were introduced. A rather different method of dating, relying not on the thermolumines-cence due to the archaeological dose, D_a, but upon changes in the thermoluminescence sensitivity of the specimen induced by D_a, was introduced in 1971 (the so-called 'pre-dose' technique; Zimmerman, J., 1971a).

In addition to these major innovations, thermoluminescence dating has benefited from many valuable improvements in techni-que which have helped to overcome a plague of complexities (e.g., difficulties in dose rate evaluation, the problems of 'spurious' luminescence – i.e., not due to radiation – and of athermal fading). The result is that thermoluminescence dating is now establishing itself among archaeologists as a respectable method of age determination.

Several review articles on thermoluminescence dating have been written. Among them books by Aitken (1974) and Fleming (1976) deserve particular attention. Both of these books are concerned with methods of dating in archaeology and contain detailed chapters on thermoluminescence whilst Atiken (1984) and Fleming (1979a) devote complete books to the topic. In addition, a book by Fleming (1975a) on authenticity testing in art contains a detailed treatment of the use of thermoluminescence to detect forgeries of ceramic artefacts. Review articles by Seeley (1975), Aitken (1978a) and Wintle (1980) are also particularly helpful. Specialist seminars on thermoluminescence are now held regularly (Aitken, 1978b; Mejdahl & Aitken, 1982) with the most recent one having taken place at Helsingør, Denmark (Mejdahl *et al.*, 1983a).

1.4.3   Geology

The occurrence of natural thermoluminescence from rocks has long been known and several early observations have already been mentioned in this chapter – for example the study of thermolumines-cence in fluorite by Trowbridge & Burbank (1898). Calcite too was known to be a good thermoluminescent phosphor. According to Ellsworth (1932) the natural glow from this material was discovered accidentally by parties of prospectors in Ontario searching for radioactive minerals. When placed on a hot camp stove at night the calcite was observed to glow brightly. It would be a mistake, however, to think that fluorite and calcite were somehow unique among minerals. Daniels *et al.* (1953) stated that of over 3000 natural minerals studied (mostly granites and limestones), approximately 75% were found to exhibit natural thermolu-minescence of such an intensity that it was easily measured. The remaining 25% were also thought to emit thermoluminescence, but of lower intensity, and the sensitivity of the measuring apparatus at that time was not high enough to detect it. Some of the recorded natural thermoluminescence was bright enough to be 'sufficient for reading a newspaper'.

Daniels and his colleagues were instrumental in proving the relationship between the natural radioactivity and the natural thermoluminescence, although this connection had already been suggested earlier (Ellsworth, 1932). It was this relationship which led directly to the use of thermoluminescence as a means of arriving at the geological age of the specimen. The principle assumption is that the thermoluminescence is a measure of the absorbed radiation dose since the specimen was last heated, which, in the case of a rock, is its time of formation. Zeller (1968) provides a useful account of the principles involved.

However, age determination is not the only way that thermoluminescence is utilized in geology. In some instances thermoluminescence is more sensitive for detecting traces of radioactivity than more conventional means – e.g., a Geiger counter or a scintillation counter. Thus the technique has found widespread application in radioactive mineral prospecting. Some minerals are found to exhibit a particular glow-curve shape when extracted from one area, but the same mineral gives an entirely different glow-curve if extracted from another. Thus thermoluminescence has also found use in source identification. Many more specialized examples can be quoted.

One aspect of geology in which thermoluminescence is proving to be particularly useful is the study of meteorites and lunar material. Herschel's observation of thermoluminescence from the Middlesborough meteorite remained unexploited until the potential usefulness of the technique in meteorite research was recognized by Houtermans and his colleagues in the late 1950s (Grögler, Houtermans & Stäuffer, 1958). More recently, the balance between radiation filling of traps and the thermal emptying of them has been used to provide information relevant to the nearness of the meteorites to the sun whilst in space (i.e., their orbit) and to the time the meteorites have spent on earth (i.e., their 'terrestrial age'). A useful review is given by McKeever & Sears (1979). A classification scheme for certain rare meteorites (petrologic 'type 3') has also been devised on the basis of their thermoluminescence properties (Sears et al., 1980).

Early speculation on the thermoluminescence from lunar material (e.g., Sun & Gonzales, 1966; Sidran, 1968) was concerned primarily with a possible explanation for Transient Lunar Phenomena (anomalous brightness from specific regions of the moon's surface). However, the first measurements of the efficiency of thermoluminescence from the Apollo 11 specimens in 1969–70 ruled out this form of luminescence as an explanation of TLP. What did emerge though was the possibility of using thermoluminescence to establish the heat flow characteristics beneath the lunar surface, and to measure solar flare particle spectra averaged over the last 10^4 year. (See Hoyt, Miyajima & Walker, 1971a, for an early review.) These and other applications of thermoluminescence to lunar studies have been reported regularly since 1970 at the annual Lunar and Planetary Science Conference, Houston, Texas.

A detailed overview of the pre-1968 research into all the applications of thermoluminescence in geology can be found in a book edited by McDougall (1968a) and, more recently, in a report by Sankaran, Nambi & Sunta (1982).

1.4.4   Defects in solids

Our understanding of the nature of crystal defects has been increasing over the past 50 years or more owing to the development of refined experimental techniques. Defect concentrations, enthalpies and energies of formation and activation and other relevant parameters are generally determined from studies of ion movement, such as ionic conductivity, diffusion, defect reactions, etc. This type of measurement has been supplemented by many other techniques including optical absorption, electron spin resonance, X-ray diffraction, photoconductivity, optical scattering, impurity decoration and many more, all of which can serve for comparison with the other 'dynamic' methods.

In principle, experiments on thermoluminescence can be expected to yield useful information on the properties of the various types of defect present within an insulator or semiconductor. Since the work of Wiedemann & Schmidt (1895) it has been known that thermoluminescence is particularly sensitive to traces of impurities within the specimen. In most instances the exact role of the impurities is unknown but in most materials their presence is considered essential for thermoluminescence to occur. There have been countless studies of the effect of impurities on the thermoluminescence properties of various materials and notable early work includes that of Wick and colleagues (*op. cit.*) and of Daniels, Cameron and colleagues (for example, Rieke & Daniels, 1957). In general terms, it is believed that the impurities give rise to the localized energy levels within the forbidden energy gap (cf. figure 1.3) and that these are crucial to the thermoluminescence process. As a means of detecting the presence of these defect levels, the sensitivity of thermoluminescence is unrivalled. Townsend & Kelly (1973) estimate that the technique is capable of detecting as few as 10^9 defect levels in a specimen. When coupled with the ability to separate the energies of these levels (Chen & Kirsh (1981) is useful for a discussion of the analysis of glow-curves) thermoluminescence provides, in principle at least, a unique tool in the determination of the distribution of the defect energies. However, it is this very sensitivity of the method that makes interpretation of the results difficult and prevents its use in, say, chemical analysis where even the purest materials are 'impure' from the point of view of thermoluminescence.

In addition to defect levels caused by extrinsic defects (i.e., impurities) there are also those due to intrinsic defects, such as lattice vacancies and interstitials. The presence of this type of imperfection is also crucial to the thermoluminescence process in many materials. Defects produced by the radiation itself may also be important. Recent studies have shown that in some samples the production of one thermoluminescent photon may involve several defect levels in a kind of cooperative effort (Mayhugh, 1970). Conversely, one species of impurity may give rise to several thermoluminescence peaks.

Investigations which attempt to describe the defect state of a material by using thermoluminescence alone are of limited value. The most progress in recent years has been made when several different methods (e.g., optical absorption, photoconductivity, ionic conductivity, electron spin resonance, dielectric loss) have been used as an adjunct to thermoluminescence. The best example of this in recent years is the unravelling of the thermoluminescence process in the alkali halides where information from a wide variety of complementary experimental techniques has been brought together in an attempt to decipher the thermoluminescence signals. Even here, however, the thermoluminescence from LiF, possibly the most studied of all the alkali halides, is described by some three or four different models, each of which explains most of the observed experimental results, but none of which explains all of them.

Nevertheless, although it is fruitless to use thermoluminescence alone to describe the defect structure of a solid, it is a very useful technique when combined with other measurements. It is very rare that thermoluminescence provides no information at all.

1.4.5 *Other applications*

Most uses of thermoluminescence come under one of the applications mentioned so far, namely, dosimetry, age determination, geology or solid-state defect structure analysis. However, in recent years several novel and, at first sight, unlikely applications have been described in the literature. In biology the technique is finding increasing usage. Tatake (1975) summarizes the pre-1975 literature whilst more recently (Cooke *et al.*, 1980) thermoluminescence from DNA-base analogues has been observed. These materials act as tumour inhibitors by a mechanism of radical conversion and it is hoped that thermoluminescence will help us to understand the charge transfer processes involved. The technique has also been used to identify the dust particles responsible for lung disease in miners (Kriegseis & Scharmann, 1980) and the volcanic soil responsible for non-filarial elephantiasis (Townsend *et al.*, 1983*a*).

A full list of applications of thermoluminescence would also include the use of suitable phosphors for image storage devices, a test for fire damage in building materials, a quality control tool for ceramics and an aid in forensic science. Becker (1973) even notes that there have been attempts to identify the cause of ball lightning by examining the thermoluminescence from bricks close to where the phenomenon occurred.

Even without these minor applications thermoluminescence has proved itself to be a technique of immense versatility. Its accelerated usage over the last 30 years reflects credit on Farrington Daniels and his colleagues – in fact it would appear that most of the mainstream applications of thermoluminescence stem from the original suggestion made by him and his research group. The technique has come a long way since Boyle's observations on the 'glimmering light' from diamond.

1.5   This book

The aims of this book are twofold: firstly to illustrate how the solid-state properties of insulators and semiconductors determine their thermoluminescence characteristics; and secondly to show in detail how the technique is being utilized as a fundamental research tool in the fields already outlined. The purpose of this chapter has been to illustrate where thermoluminescence lies in the context of luminescence phenomena in general and, by tracing its approximate historical development, to illustrate the main fields in which it is proving to be particularly useful. The rest of the book is devoted to the details of how and why thermoluminescence is used in these different fields.

Despite the fact that each of the applications of thermoluminescence has a common root, the advances and problems met with in one discipline are not necessarily being communicated to another. Thus, I shall lay emphasis on an attempt to unify the different approaches to the subject.

The book will present the reader with those basic principles of the phenomenon which are required for a proper analysis of the data and will provide some practical direction in the use of the technique, together with a discussion of available apparatus and experimental methods.

In addition, it will give detailed examples of thorough studies and applications. In doing the latter, it will provide an up-to-date review of the most recent advances in the different spheres of utilization. Thus, the book is designed to be a reference work for thermoluminescence research, and to provide inexperienced research workers with a basic understanding of the phenomenon and how it can be used.

2 Theoretical background

2.1 Elementary concepts

The generally accepted picture of thermoluminescence outlined briefly in chapter 1 has its origins in the energy band theory of solids from which a simple explanation of the observed luminescence properties of various types of material can be obtained. It is, therefore, worthwhile to examine, in a rudimentary fashion, the energy band model of solids with a view to highlighting those characteristics which give rise to luminescence in general, and thermoluminescence in particular. Much of the modern-day theory on trapping, recombination and luminescence phenomena has remained unchanged since the original treatments provided by pioneers such as Shockley, Rose, Williams and others from the late 1940s to early 1960s. For purposes of explanation, I shall refer to several of these original works. More recent applications and confirmations of the theory will be treated in later sections and chapters of the book when specific examples are being discussed.

2.1.1 *Energy bands and localized levels: crystalline materials*

The solution of the Schrödinger equation for electrons subjected to a periodically varying potential reveals that the allowed energies for the electrons lie only in 'allowed zones'. Other possible energy values constitute 'forbidden zones' or 'band gaps'. This is the situation which arises in a solid where each atom is subject to a periodic array of potential wells, as in the Krönig–Penney model of a crystal.

The occupancy of each zone, or band, is described by the density of states function, namely,

$$N(E) = Z(E)f(E), \tag{2.1}$$

where $f(E)$ is the Fermi–Dirac distribution function given by

$$f(E) = 1/(\exp[(E - E_f)/kT] + 1). \tag{2.2}$$

In these equations, $N(E)$ is the density of occupied energy levels, $Z(E)$ is the density of available energy states, and E_f is the Fermi level, or chemical potential. At absolute zero, those energy levels below E_f are completely full, whilst those above E_f are completely empty. For semiconductors and insulators E_f lies above the uppermost valence energy with the result that the valence band is entirely full, which, in turn means that it is not possible to obtain a net transport of charge in the valence band in the direction of an applied electric field. Thus, at absolute zero, these materials are non-conductors. Electronic conduction will only result when the valence electrons are given enough energy to surmount an energy gap E_g in order to reach the next highest empty band, usually called the conduction band, and thus conductivity (σ) in these materials is characterized by a temperature dependence of the form $\exp(-E_g/kT)$ with $\sigma \to 0$ as $T \to 0$. Furthermore, for ideal materials, optical absorption only takes place for frequencies above E_g/h (h = Planck's constant). If, at zero temperature, E_f lies within the energy band for the valence electrons, then the band is only partially full and the material exhibits metallic conductivity, with σ tending towards a finite value as $T \to 0$. The situation is depicted schematically in figure 2.1.

Generally, for an ideal, crystalline semiconductor or insulator, $Z(E)$ $= 0$ when $E_c \rangle E \rangle E_v$. Here E_c is the bottom of the conduction band and E_v is the top of the valence band. However, whenever structural defects occur in a crystal, or if there are impurities within the lattice, there is a breakdown in the periodicity of the crystalline structure and it becomes possible for electrons to possess energies which are forbidden in the perfect crystal. One result of this is the generation of new optical absorption bands, sometimes giving the crystal a coloured appearance. The types of imperfection which may occur are many and varied and clearly the manufacturing or natural growth process of the crystal is

Figure 2.1. (a) The partially filled valence band for a metal at absolute zero. The shaded area represents the fully occupied states. This is compared with the completely filled valence bands for (b) a semiconductor and (c) an insulator. Here the filled bands are separated from the next highest empty band by a so-called 'forbidden' gap E_g. E_f is the position of the Fermi level and E_v is the top of the valence band.

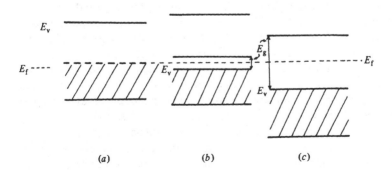

(a)   (b)   (c)

going to be an important factor in the determination of which type of defect predominates in any one material.

Important intrinsic point defects are Frenkel defects–i.e., interstitial molecules, atoms or ions (which are normally located on the lattice sites) along with the corresponding vacancies – and Schottky defects – i.e. lattice vacancies caused by the diffusion of the host ions to the surface of the crystal. These defects occur regularly in ionic crystals where one or the other type may dominate (Barr & Lidiard, 1970). In molecular crystals there is a predominance of Schottky defects owing to the difficulty of forming interstitials with large molecules. The incorporation of impurities into a material can give rise to lattice distortion, and may also result in the formation of point defects (extrinsic).

Line and planar defects such as dislocations (complex defects representing boundaries between slipped and unslipped lattice planes) and grain boundaries (interfaces of angular misfit between two sets of parallel arrays of lattice planes) may also give rise to allowed energy levels within the forbidden gap. The valence and conduction bands extend throughout the crystal whereas the defects states are centred upon the defects themselves and are thus termed 'localized energy levels'.

In some cases, for example when there is severe distortion around dislocations or boundaries, or when large clusters of point defects exist, or when impurities precipitate in a second phase, not only are localized energy levels created, but the width of the forbidden gap may widen or narrow owing to the local dilation and compression zones and/or subsequent change in lattice constant (Mataré, 1970). Such fluctuations in the energy gap are put to good effect in semiconductor heterojunction lasers of (Ga, Al) As-GaAs where the incorporation of Al into the GaAs structure by epitaxial growth widens the band gap with respect to that of the original material. For laser applications the band gap variation is optimized by controlling the Ga/Al mole ratio (Williams & Hall, 1978).

The energy levels introduced may be discrete, or they may be distributed, depending upon the exact nature of the defect and of the host lattice. In general terms, an understanding of how impurities and structural defects give rise to localized energy levels may be gained by using the example of an alkali halide crystal of the type M^+X^-. An electron freed from the valence band and wandering through the crystal may become attracted by the coulombic field of a vacant anion site (i.e., a missing X^- ion) and become 'trapped'; that is, no longer able to take part in conduction. The energy required to release the electron from the trap is less than that required to free a valence electron from an X^- ion and thus the anion vacancy has associated with it an energy level which lies somewhere between the valence and the conduction bands. A similar situation arises with cation vacancies where the missing M^+ ion results in a deficiency of positive charge which in turn results in a decrease in the energy required to free an electron from a neighbouring X^- ion. Once again a localized energy level within the forbidden gap is associated with this vacancy. The position of the localized level within the gap is determined by the decrease in the energy required to free the electron. Thus, for cation vacancies the energy level turns out to be below the equilibrium Fermi level and the centres are full of electrons and are thus potential hole traps. The anion vacancies, however, have an energy level above the Fermi energy meaning that the level is empty of electrons and that the defects are potential electron traps. Similar arguments apply to the incorporation of impurity ions (cation or anion) within the crystal lattice, either in substitutional or interstitial positions. The energy band diagram for an insulator or semiconductor containing electron and hole trapping levels is shown in figure 2.2.

One obvious cause of a breakdown in the periodicity of a perfect lattice is the presence of a surface. Surfaces cause interruptions in the periodic potential and can give rise to trapping levels within the forbidden gap in the surface regions. Usually, such levels are 'shallow', i.e., the energy required to free the electron or hole from the trap is small.

In addition to the atomic defects already mentioned, Seitz (1952; see also Mataré, 1970) lists electrons and holes, phonons, photons and charged and uncharged particles as primary defects in insulators and semiconductors. At absolute zero a non-equilibrium excess of electrons

Figure 2.2. Energy levels in an insulator in equilibrium at absolute zero. The levels below E_f are full of electrons, while those above are empty.

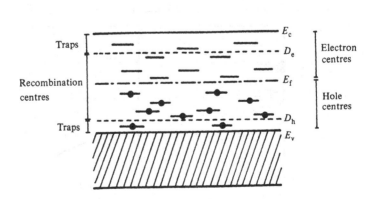

or holes may be created by, say, injection from electrodes. This excess concentration does not itself give rise to new localized levels, but simply alters the occupancy of those that already exist. Similarly, phonons (the quantum of energy associated with crystal vibrations) may also alter the localized level occupancies as the temperature is raised, but by themselves they do not introduce new levels. Photons and charged and uncharged particles may also alter the defect level occupancies by creating mobile electrons and holes via the process of ionization. These mobile charge carriers may then become trapped at the defect energy levels. Conversely, the absorption of photons may result in the photostimulation of previously trapped charge carriers back into the conduction or valence bands. The presence of photons and charged and uncharged particles does not by itself give rise to localized levels, but atomic defects may in fact be created by the interaction of the radiations within the crystal lattice and thereby localized energy states may be created indirectly.

2.1.2   *Non-crystalline materials*

The band theory for crystalline materials was developed using the concept of a semi-infinite periodic lattice giving rise to a periodically varying potential. In non-crystalline materials, however, such a concept clearly cannot be applied owing to the lack of long-range order resulting in random fluctuations in the potential. By placing a limit on the size of the fluctuations, solutions to the Schrödinger equation can be obtained which indicate that allowed zones do exist, each separated by an energy

Figure 2.3. (a) Density of states function $Z(E)$ for a non-crystalline semiconductor (or insulator) and (b) a crystalline semiconductor (or insulator). Localized states are shaded, E_c' and E_v' are the 'mobility edges', and $E_c' - E_v'$ is known as the 'mobility gap'.

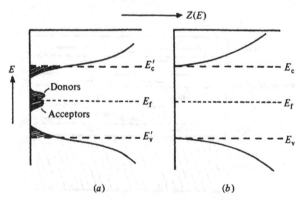

gap (see Mott, 1977, and Mott & Davies, 1979, for details). Furthermore, the calculations indicate that $Z(E)$ partially extends into the gap where it is finite (though small). Less restrictive approximations indicate that the allowed states may even extend all the way through the 'gap'. However, over a certain range of energies (the 'mobility gap', $E'_c > E > E'_v$) the energy states are localized and an activation energy ($E'_c - E$, or $E - E'_v$) is required to raise the electrons and holes from the localized states into the delocalized conduction and valence bands. In this sense this type of glassy material can still be regarded as a semiconductor or an insulator, with $\sigma \to 0$ as $T \to 0$, despite the fact that $Z(E) \neq 0$. In particular, with many intrinsic amorphous materials where E_f is located mid-way between the conduction and valence bands it is found that $Z(E_f) \neq 0$. A favoured explanation for this is that there is an overlap of intrinsic donor and acceptor states in the gap such that $Z(E_f)$ is finite (Mott, 1977). The density of states function for a typical amorphous semiconductor is shown in figure 2.3 where it is compared with the density of states for a crystalline material.

The donor and acceptor levels arise from the presence of point defects in the so-called 'continuous random network' of the material. In atomic materials they may be vacancies and clusters of vacancies. In compounds (oxides, chalcogenides) non-bridging atoms may be formed (e.g., oxygen bonded to only one Si atom in SiO_2).

The behaviour of impurities is more complex in amorphous materials than in crystals. The Fermi level always appears to be pinned near the centre of the gap, and it is possible that the impurities only produce deep-lying energy states (see Mott & Davies, 1979, for more details).

One consequence of the amorphous nature of the host lattice is that the localized energy levels, either intrinsic or extrinsic, tend to be distributed in energy rather than being discrete. However, the overall picture that emerges for non-crystalline materials is that they have conduction and valence bands with an energy gap (actually a 'mobility gap') containing localized states which may act as charge traps. In this sense they possess many features in common with their crystalline counterparts. For this reason, unless it introduces misunderstandings, I will refer to the normal crystalline energy band picture only.

2.1.3 *Traps and recombination centres*

An essential feature of all luminescence processes is the change in occupancy of the various localized energy states. These alterations in the population are implemented by electronic transitions from one energy state to another. Several kinds of transition are possible and some are shown, for both electrons and holes, in figure 2.4.

Transition (*a*) is the excitation of a valence electron from a host atom into the conduction band in which state it has enough energy to move freely through the lattice. Thus, transition (*a*) corresponds to the process of ionization and is a result of the absorption of energy from an external source, e.g., radiation. For every free electron in the conduction band a free hole is left behind in the valence band. Thus, ionization creates free electron–hole pairs which may wander through the crystal until such time as they each in turn arrive, and become localized, at defect centres. This results in the trapping of electrons (transition (*b*)) and/or of holes (transition (*e*)). The localized electrons and holes may be released from their traps by thermal or optical excitation (transitions (*c*) and (*f*)) whereupon they are once again free to move through the crystal.

A second option open to the free electrons and holes is that they may recombine with a charge carrier of opposite sign, either directly (transition (*h*)), or indirectly by recombining with a previously trapped carrier (transitions (*d*) and (*g*)). If either of these recombination mechanisms is accompanied by the emission of light (i.e., it is radiative) then luminescence results.

Thus, localized energy levels can act either as traps or as recombination centres and it becomes pertinent to determine what distinguishes a recombination centre from a simple trap. The classification

Figure 2.4. Common electronic transitions in (crystalline) semiconductors and insulators: (*a*) ionization; (*b*) and (*e*) electron and hole trapping respectively; (*c*) and (*f*) electron and hole release; (*d*) and (*g*) indirect recombination; (*h*) direct recombination. Electrons, solid circles; electron transitions, solid arrows; holes, open circles; hole transitions, open arrows.

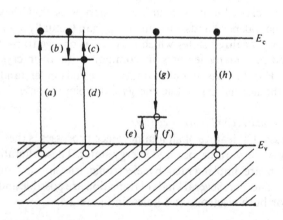

which is used to distinguish between the two types is based upon the relative probabilities of recombination and thermal excitation. For the electron trapping centre shown in figure 2.4, if transition (c) is more probable than transition (d), then the centre is classed as a trap. Conversely, if transition (d) is more probable than transition (c), then the energy level corresponds to a recombination centre. Similarly for the hole centre and transitions (g) and (f).

In equation (1.1) we saw that the probability of a charge carrier being thermally released from its trap is exponentially related to $-E/kT$, where E is the 'trap depth', i.e., the energy difference between the trap and the edge of the corresponding delocalized band. Thus, for a given temperature those centres of small E are more likely to be traps than centres of large E. For this reason, recombination centres are located towards the middle of the forbidden gap and traps are located towards the edges. Furthermore, it is possible to see from this how a centre which is a trap at one temperature may become a recombination centre at a lower temperature, and vice versa.

This distinction between a trap and a recombination centre, based on the relative probabilities of trapping and recombination, leads to the possibility that, at a given temperature, there will exist a defect level for which these transition probabilities are equal. Such a level, of depth D, would represent a demarcation between the traps and the recombination centres, such that a centre of energy depth E, where $E < D$, would be a trap, whereas if $E > D$, the site would be a recombination centre. Because the 'depth' of the trap refers to the energy difference between the localized level and the associated delocalized band (i.e., conduction band for electrons; valence band for holes) we have a demarcation level for electrons, D_e, and a corresponding one for holes, D_h. The demarcation levels are shown in figure 2.2.

The distinction between a trap and a recombination centre is not simply dependent upon the energy depth of the centre. From the definition of a demarcation level it is possible to write

$$s_e \exp(-D_e/kT) = n_r A_r^n \qquad (2.3)$$

for electrons, and

$$s_h \exp(-D_h/kT) = p_r A_r^p \qquad (2.4)$$

for holes. In these equations, the left-hand side represents the probability per unit time of thermal excitation of trapped electrons (equation 2.3) or holes (equation 2.4) from the electron and hole demarcation levels respectively. s_e and s_h are constants (cf. equation 1.1).

The right-hand sides of the equations represent the probability per unit time of recombination of a trapped charge carrier with a free carrier of opposite sign. Here, n_r and p_r are the densities of trapped charge, and A_r^n and A_r^p are the recombination transition coefficients (volume/unit time) for the electrons and holes respectively. Thus, the left-hand sides of equations (2.3) and (2.4) give the probabilities of transitions (c) and (f) in figure 2.4 and the right-hand sides give the probabilities of transitions (d) and (g). Adirovitch (1956) equates the transition coefficients A_r^n and A_r^p to the products $v\sigma_r^n$ and $v\sigma_r^p$ respectively. Here v is the thermal velocity of free carriers in conduction or valence bands and σ_r^n and σ_r^p are the cross-sections for the capture of the free charges (electrons or holes) by the trapped charges (holes or electrons). Thus, for a given concentration of trapped charge, it is the value of the capture cross-section which determines the probability of recombination.

The value of the capture cross-section depends upon the potential distribution in the region of the defect. Rose (1963) estimates typical capture cross-sections for various kinds of centre depending upon whether they are coulombic attractive, neutral or repulsive. The potential variations for the three cases are shown in figure 2.5. In the case of coulombic attraction (2.5a), Rose defines a critical radius for capture r_c at which the binding energy due to the coulombic force is equal to the kinetic energy θ of the carrier. Thus, at r_c

$$\theta = q^2/r_c\varepsilon \tag{2.5}$$

where q is the electronic charge and ε is the dielectric constant. The capture cross-section is thus given by

$$\sigma = \pi r_c^2 = \pi(q^2/\theta\varepsilon)^2. \tag{2.6}$$

Figure 2.5. Schematic representation of the potential distributions $\phi(r)$ around defect centres. (a) Coulombic attractive; (b) neutral; (c) coulombic repulsive. Typical values for the capture cross-sections are $10^{-16}\,m^2$, $10^{-19}\,m^2$ and $10^{-26}\,m^2$, respectively.

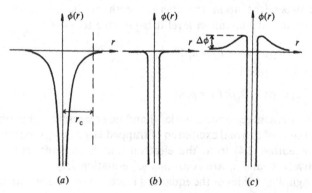

(a)   (b)   (c)

The maximum values of σ which have been observed for attractive centres are of the order of 10^{-16} m^2 and these have been termed 'giant traps' by Lax (1959) who proposes a method of cascade capture to account for them (Lax, 1960). Henry & Lang (1975), however, suggest that these large capture cross-sections result from a multi-phonon emission process. (See section 2.2.2 for further details.)

A consequence of Rose's analysis is that the capture cross-sections for coulombic attractive traps and recombination centres should be dependent upon temperature. The kinetic energy of free carriers is approximately equal to $3kT/2$[†] and thus $\sigma \propto T^{-2}$. In equations of the type (2.3) and (2.4) the T^{-2} dependence of σ is usually ignored compared with the exp$(-E/kT)$ dependence of the release probability. None the less, at very low temperatures, evidence does exist for a T^{-2} law for shallow traps in germanium. Rose (1963) compiles data from Ascarelli & Rodriguez (1961) to show an approximate T^{-2} dependence for σ in this material.

For carrier capture at a coulombic repulsive centre (figure 2.5c) a potential barrier $\Delta\phi$ must be overcome before capture can take place. Only small values of $\Delta\phi$ are necessary to reduce the capture cross-section by several orders of magnitude. Rose (1963) equates the factor by which the cross-section is reduced to the Boltzmann factor exp $(-q\Delta\phi/kT)$ and thus the capture cross-section for a repulsive centre is strongly dependent on temperature. A typical value for the capture cross-section of a repulsive centre is $\sim 10^{-26}$ m^2.

The capture cross-section for a neutral trap (figure 2.5b) is dealt with in a similar fashion to that of a coulombic attractive centre, except that the potential variation is no longer coulombic (i.e., proportional to r^{-2}) but follows a r^{-4} law. This kind of potential results from the interaction between the charge of the carrier and the polarization of the centre within the lattice (see Kao & Hwang, 1981). A typical value for a neutral centre capture cross-section is approximately 10^{-19} m^2.

Because a centre which is attractive for charge carriers of one sign is neutral or repulsive to those of opposite sign, each centre is characterized by two capture cross-sections, one for electrons and one for holes. It is the relative magnitudes of these cross-sections which determine the behaviour of the centre in trapping and recombination processes. The attractive or repulsive nature of a centre is related to its charge state.

[†] The most frequently quoted value for the average thermal velocity v of a charge carrier is the root mean square value, namely $(3kT/m^*)^{1/2}$; m^* is the carrier effective mass (Lax, 1960). Thus, $\theta = 3kT/2$. Milne (1973) suggests a value for v of $(4kT/m^*\pi)^{1/2}$, whereas Crowell (1976), assuming a Maxwellian distribution of thermal energies, suggests $v = (8kT/m^*\pi)^{1/2}$ for an isotropic system.

In an analysis of the trapping and recombination of electrons and holes in semiconductors, Shockley & Read (1952) assume that the charge of a centre will have two values, one for electrons and one for holes, differing by unity (i.e., one electronic charge q). Thus, in the Shockley–Read process of charge trapping and recombination, electron transitions between centres involve transitions between charge states. The trapping of an electron (or the release of a hole) means that the centre becomes more negative, and, therefore, less likely to attract an additional electron. Similarly, trapping a hole (or releasing an electron) makes the centre less negative (more positive) and reduces the probability of attracting another hole to the same site.

2.1.4   *Transitions not involving the delocalized bands*

The discussion so far has been concerned only with transitions involving the delocalized bands. For example, in figure 2.4 transition (c) involves excitation of an electron from a centre *to* the conduction band, whereas transitions (b) and (g) involve capture, by centres, of electrons *from* the conduction band. Transitions (d), (e) and (f) likewise involve hole transitions into and out of the valence band. However, in many materials electron and hole transitions can occur directly between centres without the carriers being raised into the conduction and valence bands. In many phosphors transitions of this type are important in luminescence processes (including thermoluminescence).

Figure 2.6 illustrates the type of centre-to-centre transition that can take place. In this figure only electron transitions are represented for simplicity. Similar transitions occur for holes. An electron trapped at level A may undergo recombination directly with the hole trapped at B (transition (i)). A recombination of this nature may take place if energy levels A and B are within the same atom. One of the earliest classical examples of this type of recombination is the luminescence from thallium in potassium chloride, when an electron from the $3P_1$ state in thallium decays to its ground state $1S_0$ (Williams, 1949; Johnson & Williams, 1952) with the emission of light of wavelength 305 nm.

The rare earths are particularly good examples of substances which exhibit this kind of recombination. When incorporated as dopant into many different kinds of lattice, luminescence can be induced which is primarily a characteristic of the particular rare earth used, and not of the host material (Dieke & Crosswhite, 1963). The incomplete 4f shell in the rare earth ions results in a large number of well-defined luminescence emission lines caused by transitions between the wide range of low-lying energy levels. The 4f electrons are effectively shielded from their surroundings by the $5s^2$ and $5p^6$ shells. Hence the 4f states are not affected by the surrounding host ions and therefore the spectra of the emissions remain essentially invariant in most host lattices. The transitions are normally radiative with the result that rare-earth-doped materials are finding increased popularity as luminescent phosphors. In some materials, when the coupling between the host and the dopant is large, the excited energy is dissipated as heat before luminescence can take place – most hydrated salts are examples of this. In other lattices the emission is strong (but not from all levels) – anhydrous halides and oxides provide good examples. Many examples of thermoluminescence from these materials have been reported in recent years (e.g., Becker, Kiessling & Scharmann, 1973; Nambi, Bapat & Ganguly, 1974; Kirsh & Kristianpoller, 1977; Sastry & Sapru, 1981).

Considering figure 2.6, if levels A and B are not within the same atom, a transition of type (*i*) can still take place by tunnelling, if the defects responsible for the levels are situated close to each other in the host lattice (i.e., a few lattice constants). This type of recombination is known to be important in a variety of organic and inorganic materials (e.g., Visocekas *et al.*, 1976; Hama *et al.* 1980; Charlesby, 1981).

An alternative possibility is that an electron has to be elevated to a higher energy level before recombination with a trapped hole can take place. Thus, in figure 2.6, an electron at level C has to be raised to A (transition (*j*)) before recombination (transition (*k*)) can take place. Again, an excellent example is provided by KCl:Tl (Johnson &

Figure 2.6. Electron transitions in a semiconductor or insulator not involving the conduction or valence bands. Electrons, solid circles; holes, open circles; electron transitions, arrows.

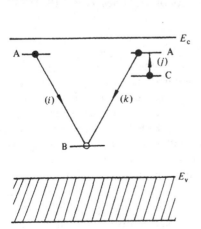

Williams, 1952). Here, an excited electron trapped in the $3P_0$ state of thallium is prevented from undergoing a transition to the ground state $1S_0$ by quantum mechanical selection rules. Thus, luminescence can only occur when the carrier is thermally excited from the $3P_0$ state into the $3P_1$ state, from where decay to the ground state is possible. A good example of several of these localized transitions taking place at once is suggested by Avouris & Morgan (1981) in Mn-doped zinc silicate. They suggest a model in which an electron is thermally excited from a deep level to a shallow, excited level, from where it can tunnel into an excited Mn^* state which then relaxes to a ground state with the emission of light.

One consequence of the fact that electron transitions of this sort do not involve excitation into the conduction band, is that there will not be any accompanying conductivity phenomena, whereas transitions of the type shown in figure 2.4 will be associated with changes in conductivity. (The situation is slightly more complex for some rare-earth-doped materials where detrapped carriers actually enter the delocalized band before recombining with oppositely charged carriers to produce the excited RE^{3+*} ion. Thereafter, decay of this excited ion to RE^{3+} produces the characteristic luminescence (Becker *et al.*, 1973; Kiessling & Scharmann, 1975; Kirsh & Kristianpoller, 1977; Sastry & Sapru, 1981; Bangert *et al.*, 1982*a*).)

In some rare earth systems the excitation energy from one type of rare earth can be absorbed by another type when the energy levels match. The emission occurs from the second rare earth and in some cases the emitted light can be seen to be of shorter wavelength than the absorbed light, in apparent contradiction with Stoke's Law. These systems are thus called Anti-Stokes phosphors (Davies *et al.*, 1974; Garlick & Richards, 1974*a*,*b*).

2.2   Recombination processes

All thermoluminescence phenomena are governed by the process of electron–hole recombination. From figure 2.4 and 2.6 three distinct types of recombination transition are possible, namely, band-to-band (transition (*h*)), band-to-centre (transitions (*d*) and (*g*)) and centre-to-centre (transitions (*i*) and (*k*)). The band-to-band recombination may be termed 'direct', whilst recombination involving localized levels, i.e., the centres, may be termed 'indirect'. Furthermore, in order for luminescence to occur, the recombination must be accompanied by the emission of a photon, i.e., it must be radiative. Thus, it becomes necessary to discuss those factors which govern the relative probabilities of direct and indirect, or radiative and non-radiative recombination processes.

2.2.1   *Direct and indirect recombination*

Luminescence emission arising from the recombination of free electrons and holes directly across the gap has been observed in a variety of materials. In CdS, for example, the electrical injection of free electrons and holes results in luminescence emission (electroluminescence) of wavelength ~ 520 nm at 300 K. The emission moves to shorter wavelengths as the temperature is reduced and in fact is seen to follow the absorption edge with temperature. The emission being observed in CdS is the result of a direct transition of type (h) (figure 2.4) and is seen to have the same temperature dependence as the process of absorption (a). Band-to-band recombination can take place in two ways: (i) if the minimum of the conduction band and the maximum of the valence band occur at the same value of the wave vector, then the transition can take place without momentum transfer and so the process has a relatively high probability; (ii) if the band maxima occur at different wave vectors then momentum transfer must take place which gives the band-to-band transition a low probability. Materials for which the energy band extremes coincide (type i) are called *direct-gap* materials, whilst those of type (ii), the more numerous, are called *indirect-gap* materials. An example of a technologically useful direct-gap material is GaAs, used in light-emitting diodes and lasers. (Other useful semiconductor examples can be found in Williams & Hall, 1978.) Further examples are some chalcogenide glasses and organic semiconductors (e.g., anthracene).

The temperature dependence of the lifetime τ_r of a free carrier for direct recombination (i.e., the mean time an electron spends in the conduction band before direct recombination with a free hole in the valence band) was determined by van Roosbroeck & Shockley (1954) and has been discussed several times since for both crystalline and amorphous materials (e.g., van Roosbroeck, 1973; van Roosbroeck & Casey, 1972). The lifetime is given by

$$\tau_r = n_i/2R \qquad (2.7)$$

where n_i is the intrinsic free carrier density and R is the temperature-dependent rate of direct recombination. Thus, we can write

$$R = \sigma n_i^2 v \qquad (2.8)$$

where σ is the direct-recombination cross-section. Values of τ_r consistent with equation (2.7) have been experimentally determined in several materials (see Bube, 1960; Williams & Hall, 1978). However, in many specimens recombination lifetimes are observed which are very much less than the lifetimes expected for band-to-band recombination (for examples see Kao & Hwang, 1981). This leads to the suggestion that a band-to-centre recombination mechanism may be dominant, particularly at low temperatures. At higher temperatures the measured lifetimes are often seen to approach those expected for band-to-band transitions.

The introduction of impurities and other lattice defects results in luminescence emission of wavelengths which are longer than those expected from conduction to valence band transitions, showing a shift from direct (band-to-band) to indirect (band-to-centre) recombination. Indeed, this general picture appears to be displayed by most materials, although narrow band-gap specimens show a greater tendency for direct recombination.

In general, impure insulators and wide band-gap semiconductors at normal temperatures will be characterized by a carrier lifetime which is limited predominantly by indirect (band-to-centre or centre-to-centre) transitions. That this should be so is not surprising when it is considered that, in band-to-band recombination, considerable momentum transfer as well as energy transfer is usually taking place. The free electrons and holes must meet with directly opposite velocities in order to conserve momentum.

The kinetics of indirect recombination involving free carriers (i.e., transitions (d) and (g) in figure 2.4) were first formulated by Shockley & Read (1952) and have been reviewed on several occasions since (Bräunlich, 1979; Kao & Hwang, 1981). The important feature of the Shockley–Read recombination process is that excited carriers may recombine with previously trapped carriers of the opposite sign. Thus, the recombination lifetime is dependent on the density of excited carriers, on temperature and on the density of recombination sites.

2.2.2   Radiative and non-radiative recombination

It is unlikely, in a direct transition, that the energy of the excited carrier can be totally dissipated by phonon interaction alone (which would result in a non-radiative transition). The excited electron must lose an amount of energy corresponding to the band gap and this would require the simultaneous creation of many phonons in order to dissipate the electron's energy. For this reason, direct transitions involve the emission of photons and are therefore radiative.

The energy dissipated in an indirect transition, however, is much less than the band-gap energy and may thus be dissipated either radiatively (via photons) or non-radiatively (via phonons). The energy may also be removed via Auger collisions, but although this is perfectly feasible theoretically, direct experimental confirmation is scarce. Auger processes have been observed to remove band-gap energy in InAs (Galkin,

Kharakhorin & Shatkovskii, 1971) and in Ge (Auston, Shank & Le Fur, 1975) but band-to-centre Auger processes have not been confirmed. In wider band gap materials, the role of Auger collisions is altogether uncertain (Landsberg, 1970; Henry & Lang, 1975) and no explicit experimental evidence exists to indicate that they might be important in thermoluminescence processes.

Whether or not a material will exhibit luminescence following irradiation and the absorption of energy, depends upon the relative probabilities of the radiative and non-radiative transitions. In general, the luminescence efficiency of a phosphor, η, is related to the probability of a luminescence transition, P_r, and the probability of a non-radiative transition, P_{nr} by

$$\eta = P_r/(P_r + P_{nr}). \qquad (2.9)$$

Several experimental measurements of η in a variety of materials have shown it to be strongly temperature dependent over given temperature ranges. Generally, the efficiency remains reasonably constant up to a critical temperature beyond which the efficiency decreases rapidly. The KCl: Tl system again provides a good example (figure 2.7a). As already discussed, the 305 nm emission from this material is due to indirect recombination of the type shown in figure 2.6. However, qualitatively similar behaviour is displayed by materials undergoing indirect recom-

Figure 2.7. Examples of thermal quenching in (a) KCl:Tl (after Johnson & Williams, 1952); and (b) ZnSe:Al (after Fillard, Gasiot & Manifacier, 1978).

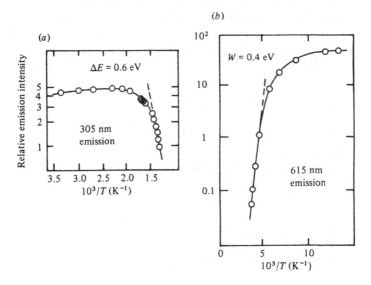

bination of the type represented in figure 2.4. A recent example is the luminescence from ZnSe:Al crystals, shown in figure 2.7*b*. Two explanations have been offered to explain the temperature dependence of η, depending upon the type of recombination under study. They are the Mott–Seitz and Schön–Klasens models and will be discussed here in a little detail.

The Mott–Seitz mechanism dates back to early work by Seitz (1940) and Mott & Gurney (1948) who considered electron transitions between an excited state and a ground state within the same atom in terms of a configurational coordinate diagram (figure 2.8). The configurational coordinate is the displacement of the atoms in the neighbourhood of the defect and, at equilibrium, in the ground state the electron will take on a minimum energy (at point A). Absorption of energy from radiation results in a transition to a higher excited state, B, without an adjustment of the configurational coordinate (Franck–Condon principle). However, in the excited state the configurational coordinate for the minimum energy is not the same as that for the ground state and thus the electron has to lose an amount of energy E_1, dissipated as heat, in order to reach a new energy minimum at C. Transition CD now results in luminescence emission, followed by a rearrangement of the configurational coordinate and the loss of energy E_2 as heat so that the electron returns finally to its original state, A. From this it can be seen that the luminescence energy CD is less than the absorbed energy AB by an amount $E_1 + E_2$. Thus, the emission bands lie on the long wavelength side of the corresponding absorption bands (the so-called Stoke's shift). Furthermore, the absorption bands have a finite width due to zero-point energy and thermal

Figure 2.8. Possible variations of electron energy with configurational coordinate for excited and ground states in a semiconductor or insulator.

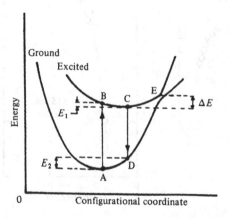

fluctuations causing the absorption transition to occur over a wide spread of energies. Likewise, the same fluctuations cause an even wider spread in emission energies so that the emission bands are wider than the original absorption bands.

An additional route by which the excited electron can return to its ground state is suggested in the configurational coordinate diagram. If the electron absorbs an amount of thermal energy ΔE whilst it is in the excited state, a transition from C to E can take place. The electron can now transfer easily to the ground state without the emission of radiation, but with just the dissipation of heat causing the return of the electron to the minimum energy at point A.

From this discussion it can be seen that the probability of a non-radiative transition P_{nr} (cf. equation (2.7)) is related to the temperature by a Boltzmann factor $\exp(-\Delta E/kT)$. The radiation probability P_r is unaffected by temperature, thus equation (2.9) may be rewritten as

$$\eta = 1/(1 + c\exp(-\Delta E/kT)) \qquad (2.10)$$

where c is a constant. This consideration thus provides an explanation for the temperature dependence of the luminescence efficiency typified in figure 2.7. From figure 2.7a, Johnson & Williams (1952) determined a value for ΔE of 0.6 eV for KCl:Tl.

An alternative mechanism, based directly on the energy band model, was advanced, independently, by Schön (1951a,b) and by Klasens (1946) for those materials in which the recombination involves displacement of the charge carriers through the crystal lattice (cf. figure 2.4). When electrons are excited into the conduction band, they may recombine with empty centres (i.e., trapped holes) to give luminescence. At high enough temperatures, however, electrons may be raised from the valence band into the empty levels (i.e., the holes may be thermally released) and the free holes may migrate and become trapped at other centres (termed 'killers'). Here the free electrons may recombine with the holes *without* the emission of light due to phonon interaction. An increase in the killer centre concentration and/or an increase in temperature can thus be seen to increase the probability of a non-radiative recombination. The mechanism is illustrated in figure 2.9.

The efficiency with which a free electron will recombine at the luminescence centre in figure 2.9 can be written

$$\eta = 1/(1 + c'\exp(-W/kT)), \qquad (2.11)$$

which is exactly the same form as equation (2.10) with c' a constant. The luminescence centre is acting as a recombination centre at low temperatures, but as a simple hole 'trap' at higher temperatures, in accordance with the definition of these centres given earlier in section 2.1.3. In order for the killer centre to be acting as a recombination level at the higher temperatures, its energy depth must be greater than W. From the data for ZnSe:Al given in figure 2.7b, Fillard, Gasiot & Manifacier (1978) determine W to be 0.4 eV.

In many materials, electron trapping is important in the luminescence process. The long luminescence lifetimes often observed imply that thermal activation is an integral part of the process. This in turn implies that the luminescence efficiency is expected to decrease if the temperature is decreased to a low enough value commensurate with the depths of the traps involved. The combined effect on the luminescence efficiency of electron trapping at low temperatures and of an activated competing radiationless process at high temperatures is shown in figure 2.10 for rare-earth-doped CaF_2 (Nambi, 1979).

Much effort has gone into describing the mechanism(s) by which the electron energy is dissipated during the non-radiative recombination of an electron and a hole. Besides the already mentioned Auger process (where the lost energy is used to excite another nearby electron) two mechanisms, both based on the transfer of energy from the electron to lattice phonons, have been suggested. Possibly the most important is the multi-phonon emission process, first investigated in detail by Huang & Rhys (1950). The essence of the multi-phonon emission mechanism is the existence of a strong coupling between the electron and the lattice vibrations. Huang & Rhys analyse the strength of this coupling and

Figure 2.9. The model of Schön and Klasens to explain thermal quenching of luminescence. 1 – radiative recombination; 2 – thermal release of hole; 3 – hole migration; 4 – hole trapping at a 'killer' centre; 5 – non-radiative recombination (energy dissipated by phonons). Electrons, solid circles; electron transitions, solid arrows; holes, open circles; hole transitions, open arrows.

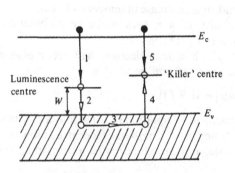

discuss the probability of an excited F-centre in an alkali halide returning to its ground state via a non-radiative transition. A similar analysis was developed by Richayzen (1957) who discussed non-radiative capture of excited electrons by localized states in the usual Shockley–Read recombination process. More detailed analyses of non-radiative Shockley–Read recombination, using a slightly different set of approximations, have been developed by Henry & Lang (1975) and by Pässler (1976, 1978). Non-radiative capture of free carriers takes place because the lattice vibrations cause the energy level to change its position in the forbidden energy gap. For large enough vibrations the empty energy level crosses into the conduction band and captures a free electron. The lattice relaxation which follows the capture lowers the position of the level back into the energy gap, the excess energy being propagated as lattice phonons. A similar process can be invoked for the non-radiative capture of the holes from the valence band.

Henry & Lang (1975) predict that the capture cross-section (for both free holes and electrons) is thermally activated, thus

$$\sigma = \sigma_{\infty} \exp(-E/kT) \qquad (2.12)$$

where experimental estimates of σ_{∞} and E range from $10^{-19}\,\mathrm{m}^2$ and

Figure 2.10. Thermal quenching effects in CaF_2 activated with various rare earths. At low temperatures, electron trapping effects dominate. At high temperatures, activated non-radiative processes dominate. (After Nambi, 1979.)

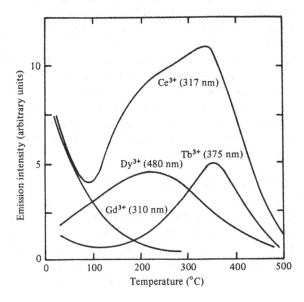

0 eV, to 10^{-18} m^2 and 0.57 eV, respectively. Here, E is the lattice energy necessary for a crossing of the energy level into the conduction band and in this sense it is equivalent to ΔE shown in figure 2.8. Pässler (1978) calculates the different capture probabilities and cross-sections that will arise depending on whether the centre is attractive, neutral or repulsive. At high temperatures in each case a thermal activation of the type $\exp(-E/kT)$ is dominant.

Thus, according to the multi-phonon emission mechanism, non-radiative electron–hole recombination can take place via a two-step sequence of, say, non-radiative electron capture (by the above mechanism) followed by non-radiative hole capture at the same centre (or vice versa). In this way, the energy $E_c - E_v$ (i.e., the energy gap) is transformed into heat in two stages ($E_c - E_t$ and $E_t - E_v$; E_t is the electron trap energy).

The work of Henry & Lang (1975) and of Pässler (1976, 1978) established that multi-phonon emission is a common process in many insulators and semiconductors and that this mechanism is able to explain the high capture cross-sections (i.e., up to 10^{-16} m^2) observed in some materials. However, the earlier treatment of multi-phonon emission processes by Huang & Rhys (1950) was unable to do so and this led Lax (1960) to propose an alternative mechanism for non-radiative recombination, which he termed 'cascade capture'. In this hypothesis the centre is presumed to possess a sequence of narrowly spaced excited states, the levels being close enough for a one-phonon transition between them to be possible. The electron energy is lost in a 'cascade' of these transitions, emitting one phonon with each step, until the electron reaches its ground state.

However, there seems to be little experimental evidence to support the existence of cascade capture mechanisms. Most centres appear not to have the necessary array of excited states. Henry & Lang (1975) studied non-radiative capture at several defect states in GaAs and GaP and concluded that the cascade capture cannot explain the observed values for the capture cross-sections, nor is it needed to explain the observed temperature dependence of these cross-sections.

2.3   Models for thermoluminescence

Many phosphors exhibit an increase in electronic conductivity during the absorption of energy from radiation. When excited by optical or ultra-violet light the induced decrease in the electrical resistivity is termed *photoconductivity*, but a more general term, which includes similar effects induced by nuclear radiations in wide band gap insulators, is *radiation-induced conductivity*. In sulphide phosphors (e.g., CdS, ZnS) it was discerned at the beginning of the present century that conductivity phenomena and luminescence were closely associated (see Garlick, 1958) and it became evident that photoconductivity arose from the liberation of charge carriers during luminescence. These ideas were fundamental in establishing the energy band model as a means of interpreting luminescence phenomena in many phosphors because this model is especially useful in providing an understanding of processes which involve transport of an electronic charge through the lattice. In this section the trapping and recombination phenomena which have been described so far will be used to discuss possible mechanisms for thermoluminescence.

2.3.1 Simple model
The earliest synthetic phosphors of the sulphide type were available only in powder form. Thus, experimental tests of the predicted interrelations between photoconductivity and luminescence could not begin in earnest until the advent of large single crystals (of CdS) in the late 1940s (Frerichs, 1947). Nevertheless, early theoretical treatments of luminescence were based on the energy band model (Johnson, 1939; Williams, 1949) and in fact it became well enough accepted to enable a quantitative theory of the kinetics of phosphorescence and thermoluminescence to be developed (Garlick & Wilkins, 1945; Randall & Wilkins, 1945a,b; Garlick & Gibson, 1948).

The simplest model of the type discussed by these authors is shown in figure 2.11. In this energy band scheme there are just two localized levels, one situated between the demarcation level and the delocalized band (i.e. D_e and E_c, or D_h and E_v in figure 2.11) and the other situated somewhere between D_e and D_h. Thus, one level acts as a trap (T) and the other acts as a recombination centre (R). In figure 2.11 the trap is situated above the equilibrium Fermi level E_f and thus is empty in the equilibrium state (i.e., before the absorption of radiation). It is therefore a potential electron trap. The recombination centre, on the other hand, is situated below the Fermi level and is thus full of electrons and is a potential hole trap.

The absorption of radiation of energy $(hv)_a > E_c - E_v$ (i.e., greater than the band-gap energy) results in the ionization of valence electrons, producing free electrons in the conduction band and free holes in the valence band (transition 1). The free carriers may either recombine with each other, become trapped or remain free in their respective delocalized bands. The last option would imply that, following the absorption of radiation, the sample would have a greater (stable) conductivity than before. This conflicts with observation and is not considered further.

As discussed in section 2.2.1 direct recombination of free electrons and holes across the band gap is a less likely process than indirect, Shockley–Read recombination at a localized defect state, especially in wide band gap semiconductors or insulators. Thus, in order for recombination to occur holes first become trapped at centres (R) (transition 5). Recombination takes place via the annihilation of the trapped holes by free electrons (transition 4). If, in this model, the recombination transition is assumed to be radiative, then luminescence will result.

The free electrons may also become trapped at level T (transition 2) in which case recombination can only take place if the trapped electrons absorb enough energy E to be released back into the conduction band, from where recombination is possible. Thus, the luminescence emission is delayed by an amount governed by the mean time τ spent by the electrons in the trap, given by the Arrhenius equation (1.1) and rewritten here as

$$p = \tau^{-1} = s \, \exp(-E/kT). \tag{2.13}$$

Here p is defined as the probability per unit time of the release of an electron from the trap. All other terms remain as previously defined.

If the trap depth E is such that at the temperature of irradiation, T_0, $E \gg kT_0$, then any electron which becomes trapped will remain so for a long period of time, so that even after removal of the irradiation there will exist a substantial population of trapped electrons. Furthermore, because the free electrons and holes are created in pairs and are annihilated in pairs, there must exist an equal population of trapped holes at level R. Because the normal equilibrium Fermi level E_f is situated below T and above R, these populations of trapped electrons

Figure 2.11. Simple two-level model for thermoluminescence. Allowed transitions: (1) ionization; (2) and (5) trapping; (3) thermal release; (4) radiative recombination and the emission of light. Electrons are the active carriers, but an exactly analogous situation arises for holes. Electrons, solid circles; electron transitions, solid arrows; holes, open circles; hole transitions, open arrows.

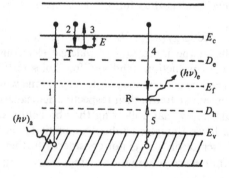

and holes represent a non-equilibrium state. The reaction path for the return to equilibrium is always open, but because the perturbation from equilibrium (i.e., the irradiation) was performed at low temperature (compared with E/k), the relaxation rate, as determined by equation (2.13), is very slow. Thus, the non-equilibrium state is metastable and will exist for an indefinite period, governed by the rate parameters E and s.

The return to equilibrium can be speeded up by raising the temperature of the specimen above T_0 such that $E \leq kT$. This in turn will increase the probability of detrapping p and the electrons will now be released from the trap into the conduction band. Thermoluminescence now results when the free electrons recombine with the trapped holes.

The intensity of the thermoluminescence $I(t)$ at any time during the heating is proportional to the rate of recombination of holes and electrons at level R. If n_h is the concentration of trapped holes then

$$I(t) = -\,dn_h/dt. \tag{2.14}$$

The relationship between $I(t)$ and n_h is shown schematically in figure 2.12. As the temperature rises the electrons are released and recombination takes place reducing the concentration of trapped holes and increasing the thermoluminescence intensity. As the electron traps are progressively emptied the rate of recombination decreases and thus

Figure 2.12. Relationship between thermoluminescence intensity $I(t)$ and the number of trapped holes n_h at the recombination centres. Also shown is the linear relation between time and temperature during heating.

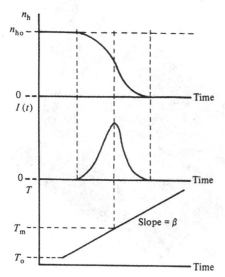

the thermoluminescence intensity decreases accordingly. This produces the characteristic thermoluminescence peak. In figure 2.12 the thermoluminescence is displayed as a function of time. Usually, in an experiment of this type, the temperature is raised as a linear function of time according to

$$T = T_0 + \beta t, \tag{2.15}$$

where β was defined in chapter 1 as the heating rate given by dT/dt. Equation (2.15) is displayed graphically in figure 2.12.

Because the probability of release of an electron from the trap is related to both the trap depth E and the temperature by equation (2.13) then the range of temperatures over which the thermoluminescence peak appears is related to the trap depth. In fact the position of the maximum in a thermoluminescence experiment is determined by the combination of E and s. For a given s, one would expect that the larger the value of E (i.e., the deeper the trap) the higher the temperature T_m at which the peak

Figure 2.13. Thermoluminescence of blue fluorite from Cave-in-Rock, Illinois, USA, after 10^6 rad ^{60}Co irradiation. The narrow emissions are due to rare earth impurities and the wider ones to defects or non-rare-earth impurities. Reproduced with permission of P.W. Levy, Brookhaven National Laboratory, New York.

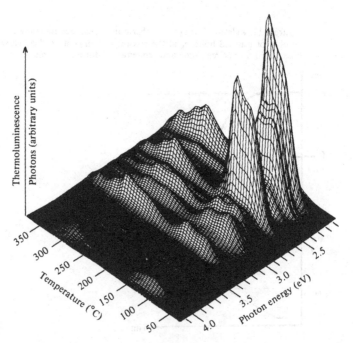

occurs. In this way a thermoluminescence experiment may, under favourable circumstances, provide pertinent information on the distribution of trapping states in a phosphor.

In order to obtain information on the position (in terms of energy) of the recombination centres, it is necessary to revert to measurements of the spectra of the emitted thermoluminescence. The energy of the emitted light $(hv)_e$ (figure 2.11) is governed by the energy difference between the excited state (in this case the conduction band) and the ground state (the recombination centre). The usual procedure is to perform a wavelength scan at a given glow-curve temperature (or over a narrow range of temperatures). The most informative measurement of this sort is obtained by simultaneously measuring the intensity of the thermoluminescence as functions of both sample temperature and emission wavelength (λ). Mattern *et al.* (1970) and Levy, Mattern & Lengweiller (1971) describe apparatus capable of performing such measurements and produce plots of $I-T-\lambda$. Figure 2.13 shows a typical example of the detail that can be obtained. An alternative way of plotting the same data is to produce a 'contour' plot, in which contours of equal intensity are plotted on a $T-\lambda$ grid. An example is given in figure 2.14a with the same data plotted in '3-D' in figure 2.14b.

The model outlined in figure 2.11 is the simplest that can be invoked to explain thermoluminescence when the process is accompanied by carrier transport through the lattice. Two energy levels is the minimum number needed in order to describe the thermoluminescence mechanism. The band model of an actual specimen may be much more complex than this but this simple picture can explain, at least qualitatively, all the fundamental features of thermoluminescence production.

The model may be analysed further by examining the rate equations which describe the flow of charge between the various energy levels during the stimulation of thermoluminescence. It becomes convenient to separate the phenomenon into two components – one describing trap filling during irradiation and one describing trap emptying during thermal excitation. Both trap filling and trap emptying will be analysed in greater detail in the next chapter but for the present it suffices to describe each process by a set of simultaneous differential equations which describe the traffic of charge carriers between conduction and valence bands, traps and recombination centres.

Following Chen, McKeever & Durrani (1981) the trap filling process may be described by the following four equations:

$$dn_c/dt = f - n_c A_r n_h - n_c(N - n)A, \qquad (2.16a)$$

$$dn/dt = n_c(N - n)A, \qquad (2.16b)$$

Figure 2.14. (*a*) Contour plot and (*b*) Isometric plot of thermoluminescence spectra from X-irradiated LiF TLD-600. (After Townsend *et al.*, 1983*b*.)

(*a*)

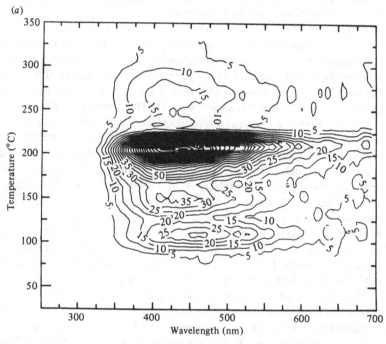

(*b*)

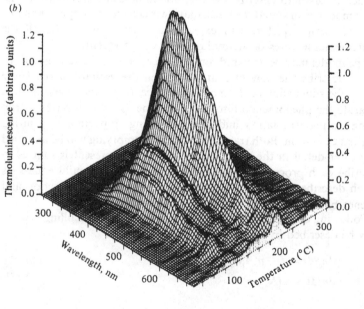

$$dn_v/dt = f - n_v(N_h - n_h)A_h, \qquad (2.16c)$$
$$dn_h/dt = n_v(N_h - n_h)A_h - n_c n_h A_r. \qquad (2.16d)$$

The terms involved are: n_c – concentration of electrons in the conduction band (per unit volume); n_v – concentration of holes in the valence band; n – concentration of electrons in traps; N –concentration of available electron traps (of depth E below the conduction band); n_h – concentration of holes in recombination centres; N_h – concentration of available hole centres; A – transition coefficient for electrons in the conduction band becoming trapped (volume/unit time; transition 2 in figure 2.11); A_h – transition coefficient for holes in the valence band becoming trapped in the hole centres (transition 5); A_r – recombination transition coefficient for electrons in the conduction band with holes in centres (transition 4) and f – the electron–hole generation rate (volume/unit time; transition 1).

In these equations, and in accordance with the charge transferences indicated in figure 2.11, direct band-to-band and indirect centre-to-centre transitions are forbidden (or, more accurately, their probabilities are assumed to be negligible). Importantly, transition 3 (i.e., thermal release of trapped electrons) is also ignored. This is easily ensured in a real situation by ensuring that the irradiation takes place at a sufficiently low temperature.

Each of the centres involved has a capture cross-section associated with it. It is possible, therefore, to relate the transition coefficients to the capture cross-sections by equations of the form $A = v\sigma$ where v and σ have already been discussed (p. 28). Additionally, for charge neutrality it can be seen that

$$n_c + n = n_v + n_h. \qquad (2.17)$$

The trap emptying process has also been dealt with by several authors and concerns the return from the non-equilibrium metastable charge state, induced by irradiation, to the equilibrium (pre-irradiation) condition. The irradiation is considered to have ceased and furthermore it is usually judged that at the start of the trap emptying period the free carrier concentrations in the valence and conduction bands are zero, i.e., $n_{c0} = n_{v0} = 0$. Thus, transitions 1 and 5 in figure 2.11 are not considered.

The rate equations describing the flow of charge between the various energy levels and bands during trap emptying have been described by Adirovitch (1956), Haering & Adams (1960) and Halperin & Braner (1960). The authors assume that once a trap is emptied, the freed electron can no longer distinguish between it and all other traps of the same type. Thus, it is possible to write

$$dn_c/dt = np - n_c(N - n)A - n_c n_h A_r, \qquad (2.18a)$$

$$dn/dt = n_c(N - n)A - np, \qquad (2.18b)$$

$$dn_h/dt = - n_c n_h A_r. \qquad (2.18c)$$

All the terms remain as previously defined. Transitions to or from the valence band are not considered in this case, so the neutrality condition now becomes

$$n_c + n = n_h. \qquad (2.19)$$

Additionally, for a thermoluminescence experiment the time dependence of the temperature, namely, equation (2.15), is also taken into account. Once again direct band-to-band and indirect centre-to-centre transitions are ignored.

The important difference between the trap filling and trap emptying processes described by equations (2.16a–d) and (2.18a–c) respectively is that the thermal release probability p is considered to be zero in equations (2.16a–d), but is large in equations (2.18a–c).

The temperature dependence of p is related to the values of both E and s. As already discussed, the energy term E is easily interpreted as the energy difference between the trap and the bottom of the conduction band – i.e., the trap depth. The Boltzmann factor $\exp(-E/kT)$ is the probability that the thermal energy imparted to the electron via phonon interaction is enough to free the carrier from the trap.

The term s in the Arrhenius equation (equation 1.1)) is commonly called the 'frequency factor' (Glasstone, Laidler & Eyring, 1941), although when applied to thermoluminescence it is often called the 'attempt-to-escape frequency' (e.g., Bube, 1960). The usual interpretation of s is that it represents the number of times per second, v, that a bound electron interacts with the lattice phonons times a transition 'probability', K (Glasstone *et al.*, 1941; Mott & Gurney, 1948; Bube, 1960; Curie, 1960). The *maximum* value expected for s is therefore the lattice vibration frequency, namely $10^{12}–10^{14} s^{-1}$. However, as used in equation (2.13), s also includes an entropy term. According to the thermodynamic expressions developed in absolute reaction rate theory (Eyring, 1936) and as applied to luminescence phenomena by Williams & Eyring (1947) (see also Curie, 1960), the rate of release of the trapped electrons can be written

$$p = vK \exp(-F/kT) \qquad (2.20)$$

where F is the Helmholtz free energy, given by $F = E - TS$, where S is the entropy change during the reaction. Thus, equation (2.20) becomes

$$p = vK \exp(S/k) \exp(-E/kT). \qquad (2.21)$$

By comparing equations (2.21) and (2.13), s can be identified with $\nu K \exp(S/k)$.

An examination of the conditions of applicability of the Arrhenius law by Gibbs (1972) suggests that s is essentially temperature independent. The absolute reaction rate theory of Eyring (1936), however, suggests that $s \propto T$. Keating (1961) has also examined the possibility that s is slightly temperature dependent. By application of the law of detailed balance (e.g., Bube, 1960), it is possible to show that

$$s = N_c v \sigma \qquad (2.22)$$

where N_c is the density of states in the conduction band. As already discussed, v is proportional to $T^{1/2}$ and σ can have several temperature dependencies, typically of the form T^{-a} (where $0 < a < 4$; Bemski, 1958; Lax, 1960; Rose, 1963). Keating (1961) includes the $T^{3/2}$ dependence of N_c to suggest, therefore, that s may vary with temperature like T^{-b}, with $-2 < b < 2$ (see also Chen, 1969b; Balarin, 1979a). In general, however, the weak temperature dependence of the frequency factor is of minor importance compared with the strong exponential temperature dependence of the Arrhenius term.

2.3.2   *Additions to the simple model*

Although it is possible to use the rudimentary two-level model to describe many of the essential features of thermoluminescence it is an unfortunate fact that there is not one known phosphor which is believed to possess such a simple energy band scheme. The best evidence for this comes in fact from thermoluminescence itself. For example, in this elementary, uncomplicated picture only one thermoluminescence peak should be observed on heating a specimen following its irradiation. Furthermore, the thermoluminescence emission should be of one basic colour. Already in this book two examples of actual glow-curves have been given in which this has not been observed experimentally. In figure 1.4 four thermoluminescence peaks are shown, overlapping each other to varying degrees, in the glow-curve for LiF between room temperature and 250 °C. If recombination takes place at a localized defect level, then the simplest interpretation of this is that there are four species of trap and only one recombination centre. If so, we should expect the colour of the emitted light for each glow-peak to be the same, and this, in broad terms, is what is observed, (Fairchild et al., 1978a, and figure 2.14). (The emission in fact is not spectrally identical for each peak. When analysed in detail some small differences are noted, but as yet these are not fully understood (Townsend et al., 1983b).) However, if the same sample of LiF were to be irradiated at low temperatures (say

− 150°C) even more glow peaks would be observed between − 150°C and room temperature, implying the presence of more traps. Furthermore, some of these would be observed to emit at an entirely different wavelength to those shown in figure 2.14, implying the presence of more than one recombination centre (e.g., Cooke & Rhodes, 1981). This increasingly more complex picture is well illustrated in figure 2.13 where the '3–D' thermoluminescence diagram for a sample of natural fluorite exhibits several glow peaks and several emission peaks.

It appears clear that, at the very least, there is a range of traps, or recombination centres, or both, which must be included in the energy level scheme. A more generalized band model may, therefore, be represented as in figure 2.15 (Hill & Schwed, 1955; Halperin & Braner, 1960). Here, there are a series of N_j electron traps and N_{hi} hole traps with n_j trapped electrons and n_{hi} trapped holes respectively. After irradiation and before heating the charge neutrality condition becomes

$$\Sigma n_j = \Sigma n_{hi}. \tag{2.23}$$

If we can consider the thermoluminescence from a specimen with an energy band picture like that in figure 2.15 to be simply the summation of several individual processes of the type discussed in figure 2.11, each taking place in its own temperature regime, then the complications introduced by the extra energy levels are trivial. However, if the production of thermoluminescence involves a complex interplay between the filling and emptying processes of all the energy levels, with several processes taking place at once, then the correct analysis and interpretation of the experimental results becomes much less straightforward.

Mention has already been made of the possibility of several traps and

Figure 2.15. Generalized energy level scheme for thermoluminescence.

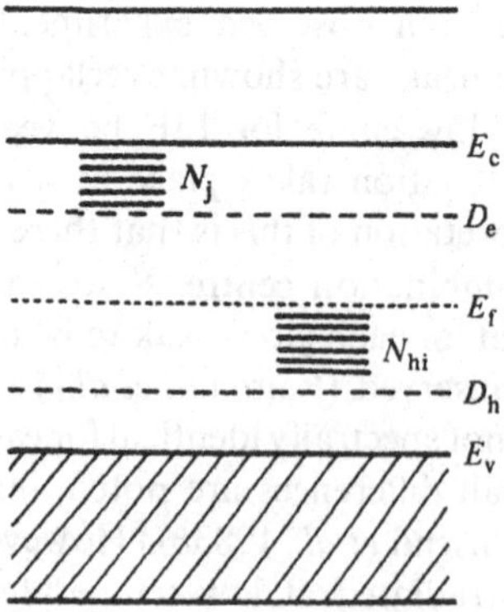

one recombination centre giving rise to several thermoluminescence peaks, each with the same emission wavelength but with different energy depths (activation energies). It ought to be mentioned that the converse is also a possibility, namely, one trap and several recombination centres, activated in different temperature regimes, giving several glow peaks, each with the same activation energy, but with different emission wavelengths. Such a possibility has been discussed by Hill & Schwed (1955) and by Bonfiglioli, Brovetto & Cortese (1956a, b) for thermoluminescence in NaCl and has been treated theoretically by Kivits (1978). In this mechanism the first glow peak corresponds to the exhaustion of the recombination centre with the largest capture cross-section; the next peak corresponds to the centre with the next largest cross-section, and so on. It must be stated, however, that this interpretation of the thermoluminescence in NaCl has been challenged by several later authors (e.g., Halperin *et al.*, 1960; plus other workers, for example those referenced by Alvarez Rivas, 1980). This mechanism, therefore, remains to be confirmed experimentally.

An obvious extension to the picture of several traps and recombination centres is to have a quasi-continuous or continuous distribution of trapping energies in which case distinct thermoluminescence peaks will not be observed. This situation may arise when dealing with polycrystalline or amorphous materials, either organic or inorganic. As was discussed in section 2.1.2, the 'normal' description of an energy band diagram consisting of conduction and valence bands separated by a forbidden gap is not fully applicable when dealing with amorphous specimens. However, if there exists a range of energies in which electrons are mobile and other energies in which they are not then the application of the energy band diagram to the analysis of thermoluminescence may be justified. In any case, amorphous and polycrystalline phosphors possess a short-range order which allows the energy band model to be applied at least over microscopic distances.

A distribution of trapping energies was first treated by Randall & Wilkins (1945b), and Shockley–Read recombination statistics were applied to this situation by Fowler (1956) and by Simmons & Taylor (1971). Broser & Broser-Warminsky (1954) discussed the rate equations governing the flow of carriers between the centres and bands for a generalized energy level distribution, but solutions of the equations can only be obtained by introducing specific and simplified cases. The thermoluminescence signal which results from a distribution of energy levels cannot be interpreted as a single peak, but rather must be viewed in the most simple case as the addition of an unknown number of closely spaced glow peaks resulting in a broad composite signal. The glow-curve is representative of the energy level distribution and individual processes cannot be separated. Analysis of such a glow-curve is difficult. Most analyses proceed by assuming a specific energy level distribution, such as linear, exponential or Gaussian (e.g., Roberts, Apsley & Munn, 1980) which extends over a limited energy range (e.g., Simmons & Taylor, 1971, 1972; Bosacchi, Franchi & Bosacchi, 1974; Pender & Fleming, 1975, 1977a).

An essential ingredient of the thermoluminescence mechanism discussed so far is that during heating, electrons are thermally released from traps into the conduction band before undergoing recombination with trapped holes. This being so an enhancement of the conductivity should be observed during the thermal stimulation. Thus, by subjecting the specimen to an electric field a *thermally stimulated conductivity* ought to

Figure 2.16. Examples of experimental thermoluminescence (TL) and thermally stimulated conductivity (TSC) in a variety of materials. (a) 365-nm-irradiated ZnS (Broser & Broser-Warminsky, 1954); (b) X-irradiated LiF (Böhm & Scharmann, 1971a); (c) X-irradiated quartz single crystal (Medlin, 1968).

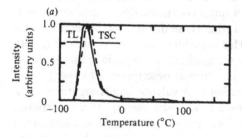

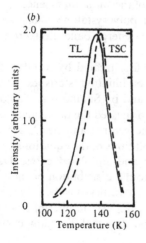

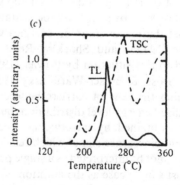

be seen. It is not surprising, therefore, that tests of the energy band model have been carried out by the simultaneous measurement of both thermoluminescence and thermally stimulated conductivity and indeed many such measurements have been reported in a variety of materials. Examples of some of the published curves obtained by simultaneous measurement of these phenomena are given in figure 2.16. It can be seen from this figure that both types of curve exhibit the same *general* shape, with each curve displaying a maximum. However, detailed examination reveals that the maxima do not occur at the same temperature and that there are important differences in shape between the two species of signal.

Several theoretical exercises have been undertaken to examine the differences that are expected between the thermoluminescence and conductivity curves based on the energy band model. Some of the early theories of thermally stimulated conductivity (e.g., Haering & Adams, 1960) dealt only with the simple energy level scheme of figure 2.11 but later theories introduced an extra energy level known as a 'thermally disconnected' trap (e.g., Dussel & Bube, 1967). The energy band diagram used by Dussel & Bube is shown in figure 2.17 in which the Fermi and demarcation levels have been left out for clarity. A thermally disconnected trap (concentration M in figure 2.17) is one which can be filled with electrons during irradiation but which has a trap depth which is much greater than the normal trapping level (concentration N) such that when the specimen is heated, only electrons trapped in the shallower level are freed. Electrons trapped in the deeper levels are unaffected, and thus this trap is said to be 'thermally disconnected'.

The existence of thermally disconnected traps is difficult to prove in any real situation, but generally it is thought to be likely. For any trap which is being activated in a certain temperature region, there is bound

Figure 2.17. Energy band scheme containing an additional energy level M corresponding to a concentration of 'thermally disconnected' traps.

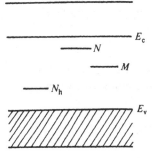

to be at least one deeper trap which is as yet unaffected. One might expect that eventually the temperature will be high enough to start draining even the deepest trap in the specimen; however, in most experiments on thermoluminescence this is unlikely owing to the experimental limitation imposed by the detection of background black-body radiation from the sample, heating strip and surroundings. This signal, which radiates in the infra-red, becomes large at high temperatures ($> 400°C$) and the sensitivity of the light detection equipment to it renders the measurement of thermoluminescence difficult, even with the use of filters to remove the unwanted signals. Thus, for a realistic assessment of thermoluminescence and its associated energy band diagram, thermally disconnected traps ought to be included.

The effect of incorporating this additional level into the energy band scheme is quite important when it comes to interpreting the observed thermoluminescence and thermally stimulated conductivity data. To begin with it becomes necessary to alter the rate equations describing trap filling and trap emptying, previously given in equations (2.16) and (2.18), respectively, for the simple energy level model. Most importantly we may write that, at the *start* of trap emptying, the charge neutrality condition becomes

$$n + m = n_h \tag{2.24}$$

which is simply a particular example of the more general equation (2.23). Here m is the concentration of trapped electrons in the disconnected trap. Dussel & Bube (1967) and most other theorists assume normally that $m = M$ (i.e., all the additional traps are full) but this is not a necessary condition. The important consideration is that m is a constant during the stimulation of electrons from the shallow trap.

In principle, figure 2.17 is a more satisfactory 'simple' energy level scheme than that shown in figure 2.11 to describe thermoluminescence and thermally stimulated conductivity. The rate equations (2.16*a–d*) which describe the filling of the traps during irradiation need only slight modification in order to incorporate the filling of M and it is not necessary to list them here. The main point to consider is that for charge neutrality during trap filling

$$n_c + n + m = n_v + n_h. \tag{2.25}$$

The equations for trap emptying are even simpler to deal with because the concentration m is unaffected, therefore, we may write

$$dm/dt = 0 \tag{2.26a}$$

along with

$$n_c + n + m = n_h. \tag{2.26b}$$

Several attempts have been made to solve equations (2.18a–c), with the addition of equations (2.26), in order to arrive at a description of both thermoluminescence and thermally stimulated conductivity. An exact solution using an analogue computer was obtained by Böhm & Scharmann (1971b) but the first numerical solution was that of Kemmey, Townsend & Levy (1967) who devised a system of 'book-keeping' to follow the time evolution of the various concentrations without the need for approximations. A notable publication was that of Kelly, Laubitz & Bräunlich (1971) who produced exact numerical solutions using a standard Runge–Kutta–Gill fourth-order program for a whole range of values of the relevant parameters (transition coefficients, trap and recombination centre densities, etc; (see also Shenker & Chen, 1972, and Hagebeuk & Kivits, 1976)). Earlier authors had attempted a solution of the equations by the introduction of simplifying approximations (e.g., Bräunlich & Kelly, 1970; Böhm & Scharmann, 1971a; Kelly & Bräunlich, 1970; Saunders, 1969). However, although these papers served as general pointers to the underlying trends, they were un- satisfactory because the approximations introduced placed restrictions on the generality of the kinetics of the trap emptying process. Thus approximate solutions only serve a limited purpose whereas the exact solutions given by Kelly *et al.* (1971) and others gave more detailed insights into the variety of the shapes that would be obtained for the luminescence and conductivity curves. By producing numerical values for the thermoluminescence intensity (given by equation (2.14)) versus temperature along with values of n_c against temperature, these authors were able to illustrate the variety of shapes that are possible and in particular the differences in shape and position between the thermolu- minescence and thermally stimulated conductivity curves that can be produced, depending upon the relative concentrations of N and M (or m; these authors also assume that $m = M$). (The thermally stimulated conductivity intensity is given by $n_c q\mu$, where q is the electronic charge and μ is the electron mobility which is probably temperature dependent. A value for the thermally stimulated conductivity intensity can only be obtained if μ and its temperature dependence are known.) Figure 2.18 shows the effect of the thermally disconnected traps on the curve shapes for some given values of the other parameters. A more thorough description of the analysis of the rate equations is reserved for the next chapter but this discussion is included here in order to demonstrate that the exact theoretical shape of thermoluminescence and TSC curves is critically related to the energy band model chosen and the relative transition probabilities between the various energy levels.

One important concept which is related to the comparison of

thermally stimulated luminescence and conductivity curves is that of the recombination lifetime of the free carriers, τ_r. The recombination lifetime is the mean time spent by an electron in the conduction band following its thermal release from a trap before recombination takes place and for free electrons recombining with trapped holes it is given by

$$\tau_r^{-1} = A_r n_h \tag{2.27}$$

where A_r and n_h have already been defined (p. 47).

For a given concentration of thermally disconnected traps we have already seen that

$$n_h = n + m \tag{2.24}$$

before the start of the heating, where m is a constant. If we make the additional stipulation that $m < n$ then n_h and, therefore τ_r are non-constant during thermoluminescence and TSC production. Using equation (2.14) and (2.18c) we have

$$I(t) = n_c n_h A_r.$$

Differentiating with respect to t gives

$$\begin{aligned} dI/dt &= dn_c/dt \times n_h A_r + dn_h/dt \times n_c A_r \\ &= dn_c/dt \times n_h A_r - n_c^2 n_h A_r^2. \end{aligned} \tag{2.28}$$

When $dI/dt = 0$, dn_c/dt must thus be positive. Similarly, when $dn_c/dt = 0$,

Figure 2.18. Selected curves from Kelly *et al.* (1971) to demonstrate the effect of thermally disconnected traps M. Parameter values are $N = 10^{18} \text{cm}^{-3}$; $E/k = 400 \text{ K}$; $s = 10^{11} \text{s}^{-1}$; $A/A_r = 10^6$. Curves A–C correspond to $M/N = 0, 10^{-2}$ and 1. Note that $\dot{n}_h$ is the temperature derivative of n_h and is related to the time derivative dn_h/dt by $T = T_0 + \beta t$.

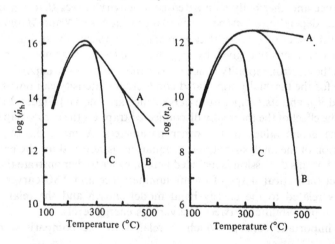

dI/dt must be negative. This can only mean that in general a thermoluminescence peak ($dI/dt = 0$) occurs at a lower temperature than a thermally stimulated conductivity peak ($dn_c/dt = 0$). Figure 2.16 gives good examples of where this is true. Occasionally, examples can be found where the luminescence and conductivity peaks for a given trap occur at the same temperature. This corresponds to the case where n_h is approximately constant during heating which in turn implies an approximately constant recombination lifetime. This will occur, for example, if $m \gg n$, whence $n_h \simeq m$ (a constant). As a rule, however, a displacement between the thermoluminescence and thermally stimulated conductivity peaks is observed (Böhm & Scharmann, 1971a; Chen, 1971). (Occasionally the opposite displacement may occur, i.e., a thermally stimulated conductivity peak appears at a lower temperature than the corresponding thermoluminescence peak. Chen & Fleming (1973) relate this type of temperature shift to the temperature dependencies of A_r and μ.)

2.3.3 An alternative model

An important feature of the models for thermoluminescence discussed so far is the fact that when the electron is raised from its ground state to its excited state during absorption of radiation a transition into the conduction band is involved. Similarly, when released from the trap by thermal excitation the electron is raised to the conduction band before undergoing recombination with the trapped hole. However, as already discussed in section 2.1.4 luminescence can result from indirect centre-to-centre transitions of the type shown in figure 2.6. In this case thermoluminescence will result when electrons trapped at level C are thermally raised to level A, from where recombination can take place with electrons in level B. Level A can simply be regarded as taking the place of the conduction band.

An obvious consequence of this type of transition is that the initial absorption of radiation will not be accompanied by radiation induced conductivity, and thermoluminescence will not be accompanied by thermally stimulated conductivity. One such phosphor which exhibits this property is KCl:Tl (Randall & Wilkins, 1945a) as has already been mentioned in section 2.1.4. Thermoluminescence is induced in this material in the region of 300–400 K and it is not accompanied by any detectable thermally stimulated conductivity signal. In fact, alkali halides in general, doped with many different species of atom, appear to be good examples of materials in which thermoluminescence is frequently not accompanied by thermally stimulated conductivity. KCl:In is an excellent example. Leiman (1973) observed two thermoluminescence peaks in this material, at 243 K and at 270 K. The higher temperature peak appears only after longer excitation times and appears to follow bimolecular kinetics (cf. Garlick & Gibson, 1948). The lower temperature peak on the other hand exhibits monomolecular kinetics (cf. Randall & Wilkins, 1945*a,b*). Leiman argues that the monomolecular peak is due to the recombination of electrons trapped (i.e. level C in figure 2.6) within 10–20 lattice constants of the ionized centre (level B) and no substantial electron displacement is involved. The 270 K bimolecular peak on the other hand is due to the recombination of freed electrons with recombination centres many lattice constants away –i.e., electron transport through the lattice is essential. To confirm this proposal Leiman notes that the 243 K peak is not accompanied by thermally stimulated conductivity, but that the 270 K peak is.

It should be noted that the complete lack of thermally stimulated conductivity or photoconductivity from a specimen does not automatically mean that only localized transitions are taking place. The induced conductivity may in fact be swamped by a high background current (e.g., ionic conductivity in alkali halides), or its detection may be complicated by the existence of space charges or adverse electrode effects. However, if at least one thermoluminescence peak is associated with a thermally stimulated conductivity peak then it is probable that any other thermoluminescence peaks which do not have thermally stimulated conductivity peaks are the result of localized transitions only.

Other proposals for localized trapping and recombination of a similar type in alkali halides have been made by Alvarez Rivas and colleagues for various crystals doped with divalent metals (e.g., Alvarez Rivas, 1980). These authors have studied KCl, KBr, KI, NaCl, NaF and LiF and in each case found examples of thermoluminescence peaks which were not accompanied by detectable TSC peaks, although some other glow peaks were accompanied by conductivity curves. Their proposal is that the irradiation creates F-centres and nearby interstitial atoms (i.e., interstitial halide ions with trapped holes). Thermal stimulation causes the correlated recombination of the interstitials and F-centre vacancies. The luminescence results from the recombination of the hole with the F-centre electron. This mechanism is similar to that discussed for KCl:In because neither the electron nor the hole enter their respective delocalized bands. Once again the thermoluminescence is characterized by monomolecular kinetics.

First-order, or monomolecular, kinetics is a natural consequence of this type of localized recombination. Because the excited electron does not reach the conduction band but remains in the 'field' of the defect centre, the rate of photon emission (i.e., recombination rate) depends only on the concentration of excited electrons. Thus a first-order expression of the type given in equation (1.2) is appropriate. Chen & Kirsh (1981) demonstrate that the first-order expression holds even if the retrapping transition (i.e., a transition from A to C in figure 2.6) is dominant over the recombination transition from A to B. Similar arguments are given by Hagekyriakou & Fleming (1982a). Conversely, for thermoluminescence which takes place via the excitation of the electron into the conduction band (figure 2.11) and the occurrence of macroscopic electron displacement, the recombination rate will depend not only on the number of free electrons, but also on the number of recombination sites available and a second-order expression, as in equation (1.4), becomes pertinent. This picture appears to be fully consistent with the results obtained by Leiman (1973) on KCl:In. As discussed by Randall & Wilkins (1945a) it is only in the special circumstances of the recombination probability being much larger than the retrapping probability (due to, say, A_r being much greater than A in equations (2.18a–c)) that first-order kinetics can emerge from the 'simple model' of figure 2.11.

The simultaneous measurement of thermoluminescence and TSC is a complex experimental procedure and in many instances it may be impossible to undertake. For example, in the alkali halides many thermoluminescence peaks occur above room temperature. However, because these materials are excellent ionic conductors they exhibit high electrical (ionic) conductivity in this temperature range and thus the measurement of TSC, which may be substantially weaker than the ionic conductivity, is extremely difficult. This means that the non-existence of an accompanying thermally stimulated conductivity peak may be difficult to establish and thus it is not necessarily easy to prove that the thermoluminescence mechanism involves only centre-to-centre transitions and not charge transport via the conduction band. None the less, in view of the large number of cases in which monomolecular kinetics have been established, it may be that the localized centre-to-centre mechanism may be more prevalent than is considered at present.

The rate equations which govern the production of thermoluminescence in this case have been examined by Halperin & Braner (1960). During trap emptying we may write:

$$dn/dt = sn_e - np, \qquad (2.29a)$$

$$dn_h/dt = -n_h n_e A_r \qquad (2.29b)$$

and

$$dn_e/dt = np - n_e(n_h A_r + s), \qquad (2.29c)$$

where n_e is the concentration of excited electrons in level A (figure 2.6).

The other terms retain their usual meaning. It is to be noted that s (the attempt-to-escape frequency) is identified with the probability of retrapping of the excited electrons by level C. This can be demonstrated by the principle of detailed balance (Halperin & Braner, 1960). Following Chen & Kirsh (1981) we can compare equations (2.29) and (2.18a,b) and note that the term sn_e replaces the term $n_c (N - n)A$ because the electron does not move through the lattice and can therefore only retrap into the same trap from which it was released.

Before leaving the discussion on localized centre-to-centre recombination it ought to be mentioned that it is possible for thermally stimulated conductivity to be observed from a trap (i.e., the electrons do enter the conduction band) but nevertheless the recombination centres and traps may still possess a strong spatial correlation such that the majority, or a substantial fraction, of the released electrons can combine only with the nearby recombination centre. For example when several point defects cluster together (in order to reduce the lattice strain) a local region of different lattice constant can be created which in turn may give rise to a narrowing of the energy gap at this location. Alternatively, dislocation strain may be responsible for an alteration in the band gap. In either of these situations, any electrons which are thermally released into the conduction band at or near the defect complex will preferentially drift back towards the defect and not throughout the lattice. Recently, a mechanism of this sort was suggested to account for the thermoluminescence behaviour of KCl doped with Tl (Toyotomi & Onaka, 1979a).

The trap and the recombination centre may themselves be different parts of the same complex, a suggestion which is becoming increasingly popular and will be more fully discussed in chapter 5. The result of this is that localized recombination will be preferred over general recombination throughout the lattice. Under these circumstances, the energy level model of figure 2.6 is not appropriate, but neither is the model of figure 2.11 because this model allows for migration of all freed carriers and their subsequent recombination with any of the trapped holes, not predominantly the nearby ones.

Evidence for 'preferred' recombination of this or a similar type has been presented by Fields & Moran (1974) in LiF:Mg who estimate the charge flow into the recombination centres during thermoluminescence and find it to be greater than the bulk carrier concentration measured by thermally stimulated conductivity. However, it must be stated that there is some doubt about this interpretation of these data. Fillard *et al.* (1978) suggest that the results stem from the inappropriate use of equation (2.14) (see the next section for details), whereas Chen & Kirsh (1981) question the coulombic model for the charge of the recombination centre, used by Fields & Moran. Another possible reason for the observed results is that the authors did not separate out the contribution to the measured current from the polarization of dipolar defects within the specimen (so-called *Thermally Stimulated Polarization Currents*) which have been shown to be prominent in the temperature regime of interest (McKeever & Hughes, 1975). Furthermore, it is a characteristic of these polarization currents that the conductivity peak is followed by a current reversal on the high temperature side (McKeever & Lilley, 1981). This may provide an explanation of the observed change in the sign of the 'charge ratio' $q/4\pi e$ (q = electronic charge; e = recombination centre charge) determined by Fields & Moran from the conductivity and thermoluminescence curves. Thus, although correlated recombination of this type remains an intriguing possibility, it has yet to be un-equivocally demonstrated.

2.3.4 More complex models

Recently a more complex model for the production of thermo-luminescence has been discussed. In this energy level model, introduced originally by Schön and colleagues (Riehl & Schön, 1939; Schön, 1942), not only is the thermal release of electrons and their recombination with trapped holes allowed, but also the thermal release of the trapped holes and their recombination with the trapped electrons is considered to occur at the same time. This energy level scheme is in fact that which was used by Klasens and others (Klasens 1946) to explain thermal quench-ing in sulphide phosphors (see section 2.2.2). The energy transitions considered in this model during the production of thermoluminescence are shown in figure 2.19, where once again the demarcation and Fermi-levels have been left out for clarity. Bräunlich & Scharmann (1966)

Figure 2.19. The model for thermoluminescence according to Schön (1942). Allowed transitions: (1) and (4) recombination; (2) and (5) thermal release; (3) and (6) retrapping. Electrons, solid circles; electron transitions, solid arrows; holes, open circles; hole transitions, open arrows.

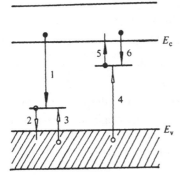

considered the rate equations for the flow of charge during trap emptying, thus:

$$dn_c/dt = np - n_c(N - n)A - n_c n_h A_r,$$ (2.30a)

$$dn/dt = n_c(N - n)A - np - n_v n A_r^*,$$ (2.30b)

$$dn_v/dt = p^* n_h - n_v(N_h - n_h)A_h - n_v n A_r^*,$$ (2.30c)

$$dn_h/dt = n_v(N_h - n_h)A_h - p^* n_h - n_c n_h A_r.$$ (2.30d)

In this set of equations the terms remain as defined in equations (2.16a–d) and (2.18a–c) with the addition of: p^* – probability per unit time of the release of a trapped hole into the valence band (transition 2 in figure 2.19); A_r^* – transition coefficient for the recombination of a free hole with a trapped electron (volume/unit time; transition 4). The hole release probability may be written

$$p^* = s^* \exp(-E^*/kT),$$ (2.31)

i.e., it is of identical form to equation (2.13) with E^* being the 'depth' of the hole trap above the valence band and s^* is the attempt-to-escape frequency. For electron and hole release to occur at the same temperature, it is clearly necessary that $p \approx p^*$. Additionally, the charge neutrality condition of equation (2.17) applies during the trap emptying period.

Evidence for this kind of mechanism has recently emerged from detailed measurements of simultaneous thermally stimulated conductivity and thermoluminescence in several different materials. An important feature of the Schön mechanism is that equation (2.14) will no longer apply. This is because holes are being removed from the hole trap by an additonal route to recombination. In this instance,

$$I(t) = n_c n_h A_r \neq dn_h/dt.$$ (2.32)

Remembering the equation for thermally stimulated conductivity (TSC), namely

$$TSC(t) = n_c q\mu$$

we may define a parameter R thus:

$$R = TSC(t)/I(t) = (q\mu/A_r)(1/n_h).$$

Differentiating with respect to time gives

$$dn_h/dt = -(q\mu/A_r)(d(1/R)/dt).$$ (2.33)

This analysis was developed by Gasiot & Fillard (1977) who then used equation (2.33) and the experimentally determined value of R to test the validity of equation (2.14). The results obtained by Gasiot & Fillard for stannic oxide clearly show that equation (2.14) is invalid.

Similar results are presented by Fillard *et al.* (1978) for ZnSe:Al and here, from the observed evolution of n_h, it is argued that the thermal release of holes must be taking place. Additional causes for the breakdown of equation (2.14) are suggested to be temperature dependence of the recombination centre capture cross-section in $CdF_2:Sm^{3+}$ (de Murcia *et al.* 1980) and refilling of recombination centres (i.e., n_h *increasing* during recombination) in AgBr (Fillard *et al.*, 1977).

3 Thermoluminescence analysis

3.1 Introduction

One of the prime objectives of a thermoluminescence experiment is to extract data from an experimental glow-curve, or a series of glow-curves, and to use these data to calculate values for the various parameters associated with the charge transfer process in the material under study. These parameters include the trap depths (E), the frequency factors (s), the capture cross-sections and the densities of the various traps and recombination centres taking part in the thermoluminescence emission. Of course, arriving at values for these parameters does not necessarily mean that we fully understand them, or that we are knowledgeable about the defect model with which they are associated. (Indeed, having calculated, say, E and s, there is often a temptation to ask 'so what?'.) Nevertheless, calculations of this sort are an important step in arriving at an acceptable level of understanding of the underlying processes and a great deal of effort has been directed towards the development of a reliable method of analysis. Unfortunately, this development is proving to be less straightforward than was at first imagined.

The most popular procedure begins by selecting the rate equations appropriate to a particular model (cf. chapter 2) and continues by introducing simplifying assumptions into these equations in order to arrive at an analytical expression which describes the variation in thermoluminescence intensity with temperature, in terms of the desired parameters. From these equations even simpler expressions are produced which relate the parameters directly to the data. However, this procedure firstly requires an assumption as to which model is used to describe the thermoluminescence production; and secondly, by introducing further assumptions into the chosen rate equations, additional restrictions are being brought into play which inevitably limit the generality, and possibly the validity, of the results obtained. Invariably, the model which is most often chosen for analysis is the simple two-level model of figure 2.11, sometimes with the addition of a third level (i.e., thermally disconnected traps, figure 2.17). However, more often than not, no tests are carried out to (a) check the validity of the simple model, and (b) confirm that the approximations introduced are indeed applicable. Some of these approximations place severe restrictions on the kinetics of the thermoluminescence emission. Fortunately, this is becoming recognized as an important factor in thermoluminescence analysis and many experimenters are making great efforts to try to be certain about the order of kinetics appropriate to their case. Nevertheless, other approximations (such as the constancy of the electron distribution in the conduction band) remain unverified, and the chosen energy level model stays unconfirmed. The reason for this lack of verification is quite straightforward – it is experimentally very difficult to do. Progress on this front has recently been made by the Montpellier (France) group who, by simultaneous measurement of thermally stimulated conductivity and thermoluminescence, have developed an experimental means of testing certain aspects of the two-level model (Fillard *et al.*, 1977, 1978; Gasiot & Fillard, 1977). Details of these tests were given in chapter 2, where we also saw that simultaneous thermally stimulated conductivity and thermoluminescence can show whether or not the released electron enters the conduction band. Additionally, theoretical exercises, such as those of Kelly *et al.* (1971), Shenker & Chen (1972) and Kivits and colleagues (Hagebeuk & Kivits, 1976; Kivits & Hagebeuk, 1977; Kivits, 1978), help to develop an understanding of the effects of the various assumptions on the analysis. However, all of these works deal with simultaneous measurements of thermally stimulated conductivity and thermoluminescence, and this is experimentally difficult to perform. Thus, with measurements of thermoluminescence alone (the vast majority!) one cannot carry out these checks with the result that the final results are often doubtful, or difficult to interpret.[†]

  However, the prospect of using thermoluminescence alone to arrive at values for the various trapping parameters is not quite as gloomy as this discussion implies so far. Recent analyses and arguments (e.g., Hagebeuk & Kivits, 1976; Kivits & Hagebeuk, 1977; Chen & Kirsh, 1981) indicate that certain methods of analysis can still be successfully applied despite the introduction of approximations in their development. Also, approximate solutions of the different rate equations (appropriate to a particular model) suggest that the actual model may also not be so important, in that trap depths and frequency factors may still be calculated (e.g., Halperin & Braner, 1960; Bräunlich & Scharmann, 1966). The problem then becomes one of interpreting the meaning of the

---

† It is an unfortunate fact that one often sees in published articles values for, say, E and s for complex glow-curves quoted without any detail of how they were calculated or any indication that the problems associated with their calculation and interpretation were properly considered. Regretably this kind of calculation is almost valueless.

results obtained. Although acceptable values of the parameters may be arrived at and these values may then be used to predict successfully the shape and behaviour of the thermoluminescence glow-curve, this unfortunately does not allow the experimenter to be definite about any one model.

In this chapter, these arguments will be developed in more detail. The simplified methods of analysis used to calculate the trapping parameters will be discussed and, in particular, it is intended to highlight the practical difficulties met with when attempting to extract the appropriate data from the actual glow-curves.

As with the discussion in section 2.3.1 it is again convenient to divide the analysis of thermoluminescence data into two parts – namely, trap filling and trap emptying. For didactic reasons the analysis of the trap emptying process will be given first.

3.2   Trap emptying

The proper analysis of the process of trap emptying during heating requires a separate treatment of each of the different models for thermoluminescence. These models have been subjected to varying degrees of detail over the years, but undoubtedly the one which has been most studied is the simple model, both with and without the addition of thermally disconnected traps.

3.2.1   *Equations for the simple model: order of kinetics*

The expressions which describe the rate processes taking place during thermal stimulation for the simple two-level model are given in equations (2.18*a–c*). If, for completeness, a thermally disconnected trap is added, equations (2.26*a,b*) must also be considered. Unfortunately, as they stand, these equations are intractable and it is necessary to resort to approximations in order to arrive at an analytical solution. The two assumptions normally employed are: (1) that the free carrier concentration in the conduction band is always very much less than the trapped carrier concentration, i.e.,

$$n_c \ll n; \tag{3.1}$$

and (2) that the rate of change of the free carrier concentration is always very much less than the rate of change of the trapped carrier concentration, i.e.,

$$dn_c/dt \ll dn/dt. \tag{3.2}$$

It is to be noted that these inequalities do not mean the same thing. Inequality (3.2) simply means that the free carrier concentration is approximately constant and does not require that $n_c \ll n$. Taken together they imply that the free carrier lifetime is much less than the trapped carrier lifetime, giving rise to a (small) quasi-stationary free carrier concentration. These assumptions have been widely employed in the solution of the differential equations, both with (Kelly & Bräunlich, 1970) and without (Halperin & Braner, 1960) the addition of thermally disconnected traps. Applying the inequalities to equations (2.18a–c), Halperin & Braner (1960) developed the equation

$$I(t) = - dn_h/dt = pnA_r n_h/(A_r n_h + A(N - n)), \qquad (3.3)$$

which, with $n_c \ll n$ (i.e., $n_h = n$, from equation (2.19)), becomes

$$I(t) = pn^2/(n + R(N - n)), \qquad (3.4)$$

where $R = A/A_r$.

Kelly & Bräunlich (1970) include thermally disconnected traps m in the analysis to yield

$$I(t) = pn(m + n)/((1 - R)n + m + RN). \qquad (3.5)$$

It is easy to show that when $m = 0$, equation (3.5) reduces to equation (3.4).

According to the criteria of Randall & Wilkins (1945a,b) the thermoluminescence models of figures 2.11 and 2.17 will yield first-order kinetics if the probability of retrapping is negligible compared with the probability of recombination. This may be expressed by the inequality

$$A_r n_h \gg A(N - n), \qquad (3.6)$$

or,

$$R \ll n/(N - n). \qquad (3.7)$$

Applying expression (3.7) to equation (3.4) yields

$$I(t) = pn = ns \exp(- E/kT). \qquad (3.8)$$

Because $dn_c/dt \ll dn/dt$, then

$$I(t) = - dn_h/dt = - dn/dt,$$

thus,

$$I(t) = - dn/dt = ns \exp(- E/kT). \qquad (3.9)$$

Integration from $t = t_0$, assuming a linear heating programme, gives

$$I(T) = n_0 s \exp(- E/kT) \exp\left[- (s/\beta) \int_{T_0}^{T} \exp(- E/kT) dT\right], \qquad (1.6)$$

which is simply the Randall & Wilkins expression for first-order (monomolecular) kinetics, introduced in chapter 1. (N.B. $T = T_0$ and $n = n_0$ at time $t = t_0$.) Böhm & Scharmann (1971a) show that exactly this equation can also be obtained when thermally disconnected traps are included (i.e., starting from equation (3.5) rather than equation (3.4)).

Garlick & Gibson (1948) preferred not to assume that retrapping was negligible, in which case the inequality expressed in (3.6) or (3.7) does not apply. Instead, if retrapping dominates

$$A_r n_h \ll A(N - n)$$ (3.10)

or,

$$R \gg n/(N - n),$$ (3.11)

which, when $n \ll N$, can be used with equation (3.4) to give

$$I(t) = n^2 s \exp(-E/kT)/RN.$$ (3.12)

This in fact is the second-order expression given in chapter 1 (equation (1.4)) with $\alpha = s \exp(-E/kT)/RN$. Under the special circumstances of $R = 1$ it becomes

$$I(t) = n^2 s \exp(-E/kT)/N.$$ (3.13)

Relationships (3.12) and (3.13) can be expressed generally as

$$I(t) = n^2 s' \exp(-E/kT),$$ (3.14)

with $s' = s/RN$ or $s' = s/N$, respectively, and the solution for a linear heating rate then becomes

$$I(T) = n_0^2 s' \exp(-E/kT)/\left[1 + (n_0 s'/\beta) \right.$$

$$\left. \times \int_{T_0}^{T} \exp(-E/kT) dT \right]^2$$ (1.7)

which is the second-order (bimolecular) expression of Garlick & Gibson (cf. chapter 1) with $R = 1$.

Figure 3.1. Computed thermoluminescence peaks of (I) first- and (II) second-order kinetics. $E = 0.42\,\text{eV}$; $s = 10^{10}\,\text{s}^{-1}$. (After Chen & Kirsh, 1981.)

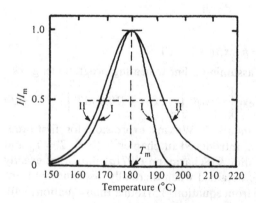

If it is remembered that a second-order glow-curve arises because of the increased probability of retrapping as compared to the first-order case (i.e., the light emission is 'delayed'), then it might reasonably be expected that a second-order curve will display more thermoluminescence during the second half of the peak than will a first-order curve. This can be seen in figure 3.1 in which computed glow peaks for first- and second-order kinetics are compared. Both curves have been normalized to give the same peak height. Examination of equations (1.6) and (1.7) will show that the increasing part of the peak is governed in both cases by terms of the form $a \times \exp(-E/kT)$. However, for a first-order peak, the decreasing part follows $\exp[-(a'/\beta)\int \exp(-E/kT)\mathrm{d}T]$, compared with the second-order case which follows $[a'' + (a'''/\beta)\int \exp(-E/kT)\mathrm{d}T]^{-2}$ (a', a'' and a''' are constants). This gives the first-order peak its characteristic asymmetry with most of the peak area being on the low temperature side of the maximum at T_m. The second-order curve is more symmetrical.

Further comparisons between the two types of glow peak can be made. Equation (3.14) might be rewritten as

$$I(t) = s''n \exp(-E/kT)$$

with $s'' = s'n$ being equivalent to s in the first-order case. Because the position of the peak (T_m) depends not only on E but also on s (or s'') then if n were to vary (by, say, initially exciting the trap to different levels of occupancy) it can be seen that a first-order peak will remain in the same position for different values of n, but the second-order peak will shift to higher temperatures as n decreases due to the variation in s''. This is illustrated in figure 3.2 where computed sets of first- and second-order glow-curves are shown for different levels of trap occupancy.

In several cases workers have reported that the shape of the actual glow peak does not conform to either the first- or second-order glow-curves expected from equations (3.8) and (3.14) respectively. This led May & Partridge (1964) to suggest a 'general-order' case for which we may write

$$I(t) = n^b s' \exp(E/kT), \tag{3.15}$$

where b is the kinetic order and is neither 1 nor 2. Development of this equation for $b \neq 1$ yields

$$I(T) = s'' n_0 \exp(-E/kT) \Bigg/ \left[1 + ((b-1)s''/\beta) \right.$$

$$\left. \times \int_{T_0}^{T} \exp(-E/kT)\mathrm{d}T \right]^{b/(b-1)} \tag{3.16}$$

where $s'' = s'n_0^{(b-1)}$. This, therefore, is the 'general-order' expression for thermoluminescence. Such a case may arise, for instance, if neither of the inequalities (3.7) or (3.11) apply. In fact there is a whole range of values of R, n_h, n and N which may not produce the classical first- or second-order glow-curves. Chen (1969b) develops this further by examining the differences in shape that exist between first-, second- and general-order

Figure 3.2. (a) First- and (b) second-order glow-curves computed with $E = 1.0\,\text{eV}$, $s = 10^{10}\,\text{s}^{-1}$ and $\beta = 10\,^\circ\text{C min}^{-1}$. The curves in (a) were computed with initial trapped charge concentrations of (1) 1.0, (2) 0.8, (3) 0.6, (4) 0.4, (5) 0.2 and (6) $0.05 \times 10^{22}\,\text{m}^{-3}$. The curves in (b) were calculated for the following fractions (1) 1.0, (2) 0.8, (3) 0.6, (4) 0.4, (5) 0.2 and (6) 0.05 of the traps filled prior to heating. The most intense first-order peak is shown for comparison. (After Levy, 1979.)

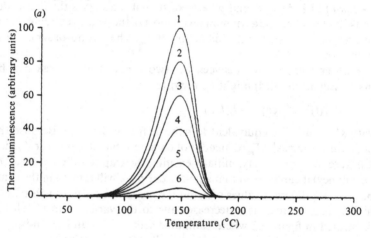

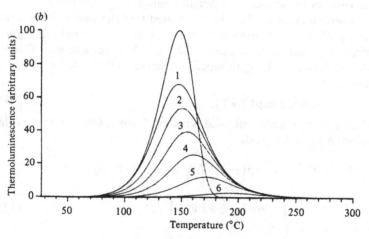

glow peaks. Chen uses a geometrical factor μ_g, defined by Halperin & Braner (1960) as $\mu_g = \delta/\omega$, where $\delta = T_2 - T_m$ and $\omega = T_2 - T_1$. The temperatures T_1, T_2 and T_m are defined in figure 3.3. By computing curves of $I(T)$ versus T for different values of b, Chen (1969b) determined that μ_g varies from 0.42 to 0.52 as b varies from 1 to 2. In principle, values of b outside this range are also possible but it is more difficult to give a detailed physical interpretation of such situations. In fact relating the curve shape to just one parameter, b, for a given E and s, is to oversimplify the situation because the glow-curve actually depends upon all of the parameters involved in equations (3.4) or (3.5).

An examination of the kinetics when thermally disconnected traps are included has recently been developed by Chen *et al.*, (1981a). In this case inequality (3.11) now becomes

$$R \gg (n + m)/(N - n) \tag{3.17}$$

and applying this to equation (3.5) with $N \gg n$ gives

$$I(t) = -\,\mathrm{d}n/\mathrm{d}t = n(n + m)s \exp(-E/kT)/RN, \tag{3.18}$$

or, when $R = 1$, from equation (3.5),

$$I(t) = -\,\mathrm{d}n/\mathrm{d}t = n(n + m)s \exp(-E/kT)/(N + m). \tag{3.19}$$

Clearly, both of these equations can be written in the form

$$I(t) = -\,\mathrm{d}n/\mathrm{d}t = s'n(n + m)\exp(-E/kT) \tag{3.20}$$

with $s' = s/RN$, or $s' = s/(N + m)$, respectively. Comparing this expression with equations (3.8) and (3.14) shows that it tends to the first-

Figure 3.3. Definition of the geometrical factor $\mu_g = \delta/\omega$ for an isolated glow peak. $\delta = T_2 - T_m$ and $\omega = T_2 - T_1$.

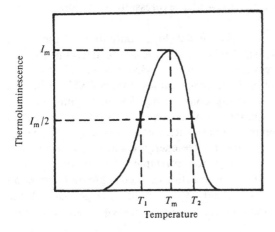

order case when $m \gg n$, and to the second-order case when $m \ll n$. More specifically, equation (3.20) can be expanded thus,

$$I(t) = -\,\mathrm{d}n/\mathrm{d}t = s'mn\exp(-E/kT) + s'n^2\exp(-E/kT), \quad (3.21)$$

which, as pointed out by Chen *et al.*, looks like the addition of the first- and second-order expressions. For this reason it has been termed 'mixed-order' kinetics. The solution of the equation is now

$$I(T) = \frac{s'm^2\alpha\exp\left[(ms'/\beta)\int_{T_0}^{T}\exp(-E/kT)\mathrm{d}T\right]\exp(-E/kT)}{\left\{\exp\left[(ms'/\beta)\int_{T_0}^{T}\exp(-E/kT)\mathrm{d}T\right] - \alpha\right\}^2} \quad (3.22)$$

where $\alpha = n_0/(n_0 + m)$. Once again a linear heating rate is assumed. The factor μ_g has been shown to vary between 0.42 (first-order) when $\alpha = 0$ (i.e., $n_0 \ll m$) and 0.52 (second-order) when $\alpha = 1$ ($n_0 \gg m$).

Thus, by introducing assumptions (3.1) and (3.2) into the rate equations for the simple model of thermoluminescence it is possible to derive simple expressions for first-, second- or general-order glow-curves when thermally disconnected traps are ignored; or expressions for first-, second- or mixed-order glow-curves when thermally disconnected traps are included. In both cases the shape of the glow-curve has been reduced from its dependence on seven parameters (E, s, N, n_h, m, A and A_r) to just three (E, s and b, or E, s and α) although it must be remembered that both b and α depend upon N, n_h, m, A and A_r. Thus the analysis of thermoluminescence glow-curves can only yield values for three parameters.

Values for the other five are very difficult to extract, because a wide range of combinations of these parameters will still give the same geometric factor μ_g. (It is to be noted that the mixed-order expression, equation (3.20), has been developed directly from the original rate equation, whereas the general-order expression, equation (3.15), is purely empirical.)

Since the preceding analysis was developed only through the use of approximations (3.1) and (3.2) it becomes necessary to determine if in fact these assumptions are valid, and, if not, what is their effect on the interpretation of thermoluminescence curves. Maxia (1978, 1980) has examined this theoretically by considering the thermoluminescence emission as two non-equilibrium transformations – (1) release of electrons to the conduction band and (2) their radiative recombination with trapped holes. By calculating the free energy changes associated with these processes Maxia applies the principle of minimum entropy production to show that the concentration of electrons in the conduction band does indeed remain steady throughout a large part of the thermoluminescence emission. However, the non-equilibrium thermodynamics as used by Maxia assumes a linear relationship between the entropy production rate and the transformation velocities, and this approximation is shown to be unreliable initially and only becomes reliable towards the end of the process – i.e., when equilibrium is almost restored (Maxia, 1980).

Because of the unreliability of the linear approximation of non-equilibrium thermodynamics as applied to thermoluminescence emission, probably the best way to test the applicability of inequalities (3.1) and (3.2) is to solve the rate equations numerically without the approximations and to compare the numerical results with those obtained when the approximations are included. This has been done by Kelly *et al.* (1971), Shenker & Chen (1972) and Hagebeuk & Kivits (1976). (The methods differ primarily in the numerical procedures used to arrive at the solutions.) Shenker & Chen (1972) find that, in the sets of parameter values used by them, the ratio of dn_c/dt to dn/dt ranges from 5×10^{-5} to 1.0, with, generally, the higher values of the ratio occurring towards the end of the glow peak – in contrast with the predictions of

Figure 3.4. Comparison of solutions of the exact and 'approximate' equations for $R = 1$, $m = 0$, $s = 10^6 \text{K}^{-1}$. Solid lines: solutions of exact equations (2.18*a*–*b*); dash-dot lines: solutions including approximations (3.1) and (3.2), dashed lines: solutions including approximation (3.2) only. A–C, $N = 10^{18} \text{m}^{-3}$; D–F, $N = 10^{20} \text{m}^{-3}$; G, $N = 10^{22} \text{m}^{-3}$, where all solutions agree. (After Kelly *et al.*, 1971.)

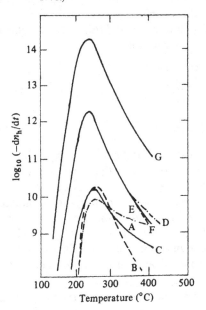

$\log_{10}(-dn_h/dt)$

Temperature (°C)

Maxia (1978, 1980). This finding is broadly confirmed by Kelly *et al.* (1971) who find that the approximations are valid only for a part of the range of acceptable parameter values. An example is given in figure 3.4 where comparison is made of the numerical solutions obtained both with and without the approximations. It can be seen that the disagreement between the solutions is most evident on the high temperature side. Thus, for those situations in which inequalities (3.1) and (3.2) do not apply, the equations for first-, second-, general- or mixed-order kinetics may not be used with confidence. For other sets of parameter values, however, the said equations may in fact be useful, simplified descriptions of thermoluminescence glow-curves.

3.2.2   *Equations for other models*

The rate equations describing the production of thermoluminescence when a transition to the conduction band is not involved were written by Halperin & Braner (1960) and have already been given in chapter 2 (i.e., equations (2.29a–c)). Because the released electron does not enter the conduction band (see figure 2.6) it cannot move through the lattice and is therefore only free to recombine with the centre from which it came. This removes the concentration dependence of the recombination rate and led Chen & Kirsh (1981) to replace the term $n_h A_r$ by γ, where γ is defined as a 'recombination probability', in s^{-1}, and is considered to be independent of n_h (the concentration of trapped holes at recombination sites throughout the sample).

Applying the inequalities $n_e \ll n$ and $dn_e/dt \ll dn/dt$ (i.e., the equivalent to inequalities (3.1) and (3.2)) we find that

$$I(t) = -dn_h/dt = -dn/dt = np\gamma/(\gamma + s)$$
$$= (ns\gamma/(\gamma + s))\exp(-E/kT). \tag{3.23}$$

This may be rewritten

$$I(t) = -dn/dt = n\bar{s}\exp(-E/kT), \tag{3.24}$$

where $\bar{s} = s\gamma/(\gamma + s)$. This, once again, is the first-order expression of Randall & Wilkins. Chen & Kirsh (1981) show that when the recombination transition dominates over the retrapping transition ($\gamma \gg s$), $\bar{s}$ reduces to s (i.e., first-order). However, when retrapping dominates ($\gamma \ll s$), $\bar{s}$ becomes equal to γ, and *the first-order form of the equation still remains*. This is consistent with the qualitative arguments of section 2.3.3.

Unfortunately there is no exact numerical solution to these equations which can be used to test the validity, and effect, of the inequalities concerning the concentration and rate of change of n_e.

The rate equations which describe the simultaneous release of holes during the thermal stimulation of the trapped electrons were given in equations (2.30a–d). These expressions are based on the Schön–Klasens model for thermoluminescence (figure 2.19) and were given approximate solutions by Bräunlich & Scharmann (1966). These authors considered four extreme cases, involving the rates of electron and hole retrapping and their comparison with the corresponding recombination rates. Referring to equations (2.30a–d), Bräunlich & Scharmann defined $R = A/A_r$ and $R^* = A_h/A_r^*$ and considered (i) $R \simeq 0$, $R^* \simeq 0$; (ii) $R \gg 1$, $R^* \gg 1$; (iii) $R \simeq 0$, $R^* \gg 1$; and (iv) $R \gg 1$, $R^* \simeq 0$. By applying the approximations $n_c \ll n$ and $n_v \ll n_h$, along with $dn_c/dt \ll dn/dt$ and $dn_v/dt \ll dn_h/dt$, they were able to show that cases (i), (iii) and (iv) each produced a glow-curve shape equivalent to the first-order glow peak of Randall & Wilkins, whilst case (ii) produced a shape of the second-order type of Garlick & Gibson. Because two activation energies (E and E^*) are involved in the production of thermoluminescence in this model, careful consideration must be given to the interpretation of the results obtained. This, in turn, depends strongly on the actual method used to calculate the trapping parameters (see next section).

Once again, exact numerical solutions of the rate equations for this model have not been published and it is not yet possible to discuss further the precise effect of the various assumptions introduced into the analysis.

So far, the message from the analysis of the thermoluminescence rate equations appears to be that no matter what model applies, if inequalities of the type (3.1) and (3.2) are valid, then the simplified expressions for thermoluminescence (i.e., first-, second-, general- or mixed-order) may be used to calculate values of E, s and b (or α). The other parameters (N, n_h, m, A and A_r) appear to be unobtainable because many combinations of these will still give the same value of b (or α). However, even if it was theoretically possible to obtain values for the other parameters, it is impossible from thermoluminescence alone to determine exactly what particular model appertains and, therefore, interpretation of the results must be made with extreme caution.

What becomes important now is to devise ways of extracting the values of E, s and b (or α) from the glow-curve, and then to test the suitability of the extraction method over a wide range of possible parameter values.

3.3 Methods of analysis

Since the pioneering work of Randall & Wilkins and Garlick & Gibson there has been a plethora of published papers dealing with methods by which the trapping parameters (mainly E and s) can be obtained from a glow-curve. Some utilize simple formulae, others consist of elaborate curve-fitting procedures. Although any treatise on thermo-luminescence would be incomplete without a discussion of such methods it is not the main purpose of this book to compile a lengthy list of analytical procedures, along with their advantages and disadvantages, simply for the sake of completeness. Instead, I shall lay emphasis on those methods which over the years have proved to be popular or reliable (not necessarily the same thing) and, where possible, promising new techniques will be included. In addition, special mention shall be made of the practical difficulties which are often met with when attempting to apply a given method to a real glow-curve. During the course of this section I shall make full use of several articles which have compiled, compared and contrasted many of the methods (Kivits & Hagebeuk, 1977, is a notable recent example) and in particular a review article by Chen (1976) and a book on analysis of thermally stimulated processes by Chen & Kirsh (1981) are recommended to the reader who may wish to refer to them for more detailed information.

As with most articles on methods of analysis I shall classify them according to the way in which the glow-curve data are handled – i.e., partial or whole curve analysis, peak shape methods, peak position methods and curve fitting. Additionally, one other classification will be included in which the data retrieval is generically quite different – namely, isothermal analysis. Finally, a section dealing with the special methods which have been devised to determine energy distributions will be included. These are used when, instead of discrete trapping levels, a distribution of traps and/or recombination centres is being examined (cf. section 2.3.2).

3.3.1   *Partial and whole curve analyses*

This class of methods stems from the recognition by Garlick & Gibson (1948) that the initial rise part of a thermoluminescence curve is exponentially dependent on temperature according to

$$I(t) = \text{constant} \times \exp(-E/kT). \tag{3.25}$$

This expression can be understood by examination of any of equations (3.9), (3.14), (3.15) or (3.21) in which it can be seen that if the temperature to which the specimen is heated is low enough for n to be approximately constant (i.e., for very little detrapping to have taken place) then equation (3.25) naturally emerges, independent of the order of kinetics. Clearly then, if a plot of $\ln(I)$ versus $1/T$ is made over this initial rise region, then a straight line of slope $-E/k$ is obtained, from which the activation energy E is easily found. This procedure is commonly termed the 'initial rise' method.

The important requirement in this analysis is that n remains approximately constant. Only upon an increase in temperature beyond a critical value, T_c, does this assumption become invalid. Haake (1957) has analysed the method and concluded that a straight line on a ln (I) versus $1/T$ plot is not the criterion to use for the applicability of the technique. He demonstrates that T_c is in fact considerably less than the peak temperature T_m. Kivits & Hagebeuk (1977) go further to demonstrate that the temperature T_c must not correspond to an intensity which is more than approximately 10–15% of the maximum intensity.

Clearly then, the initial rise technique can only be used when the glow peak is well defined and clearly separated from the other peaks. To overcome this problem in a real glow-curve, where the occurrence of a single, isolated peak is a fortunate and unusual occurrence, several methods of peak separation have been devised. Possibly the most popular of these separation procedures is the so-called 'thermal cleaning' technique described by Nicholas & Woods (1964). In this method the specimen is heated to a temperature just beyond the maximum of the first peak in the glow-curve, thereby substantially emptying the traps responsible for this peak. The sample is then rapidly cooled and reheated to a temperature just beyond the maximum of the next peak, and so on throughout the whole glow-curve. Thus, by removing in stages the lower temperature peaks, an essentially 'clean' initial rise part of the next peak in the series is obtained. With closely overlapping peaks, however, there is always the danger that the preceding peak will not have been completely removed. In this case, as discussed in detail by Haake (1957),

Figure 3.5. Change in the apparent activation energy with increasing partial heating temperature for three diamonds. The circles, triangles and squares represent different samples. (After Nahum & Halperin, 1963.)

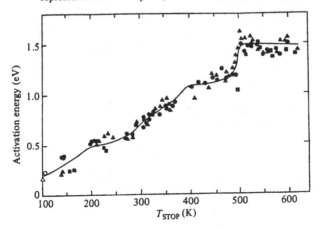

although a straight line ln (I) versus $1/T$ plot may be obtained, the value of E calculated will not be an accurate reflection of the activation energy for the second peak. In fact, the E value calculated will vary depending upon the degree of overlap and the temperature to which the sample is heated. Nahum & Halperin (1963) and Creswell & Perlman (1970) (the latter authors studying *Thermally Stimulated Discharge Currents*) have developed this principle so that many thermal cleaning cycles are made, each time using a slightly higher temperature, and a value of E is calculated for each heating from the initial part of the remaining glow-curve. The activation energies thus found are plotted against the temperature to which the sample was heated (T_{STOP}) and a 'staircase' shaped $E-T_{STOP}$ curve often results. An example is given in figure 3.5 for thermoluminescence in diamond (Nahum & Halperin, 1963).

However, there are some problems with this method. The errors involved in the activation energy calculations are often greater than the differences in the E values for the individual peaks. This is especially true for very narrowly separated energies, or if there is a continuum of trap depths. The 'staircase' shaped $E-T_{STOP}$ curve will not then be obtained. In any case, a stepwise increase in E with temperature will only be obtained when one is dealing with a glow-curve which has increasing activation energies for the peaks – a situation which is not always true. These problems will be heightened when the sample under study is a mixture of different phosphors, which may be the case with some complex geological specimens.

Figure 3.6. Schematic representation of the T_m-T_{STOP} procedure. The specimen is heated to a temperature T_{STOP} (curve A) before being cooled to room temperature. The sample is then reheated at the same rate in order to record all of the remaining glow-curve (curve B). The position of the first maximum in the glow-curve is recorded.

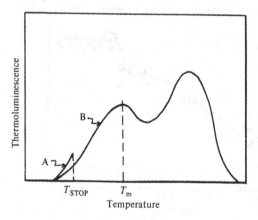

An alternative method of peak resolution was suggested by McKeever (1980a) in which the *position* of the thermoluminescence peak is monitored as T_{STOP} is increased. It was seen in figure 3.2 that the value of T_m for a first-order peak is independent of the initial concentration of trapped charge. Complications arise with non-first-order kinetics when a shift in peak position (to higher temperatures) is observed as the initial trapped charge concentration is reduced. Nevertheless it is a characteristic of the process that only a slight change in T_m is observed for the initial changes in trap occupation. Larger changes in T_m result only when the trap population is substantially altered (Garlick & Gibson, 1948). McKeever (1980a) uses this insensitivity of T_m to small changes in trap population by following the change in the position of the *first* maximum in the glow-curve as T_{STOP} is increased.

The technique begins by heating a previously irradiated sample at a linear rate to a temperature T_{STOP} corresponding to a position on the low temperature tail of the first glow peak (figure 3.6). The sample is then cooled rapidly to room temperature and reheated, at the same linear rate, to record all of the remaining glow-curve. The position of the first maximum is noted. The whole process is then repeated (on a freshly irradiated sample) using a new value of T_{STOP}. (T_{STOP} is increased in small steps of, say, 2 or 3 degrees.) By noting the value of T_m[†] a plot of T_m versus T_{STOP} is made. Figure 3.7 shows (schematically) the expected shape of the $T_m - T_{STOP}$ curve, for first- and second-order peaks and for different glow-curve shapes. For a glow-curve containing separated thermoluminescence peaks, a 'staircase' shaped $T_m - T_{STOP}$ curve will result. Each flat region in the curve is due to the presence of an individual peak. McKeever (1980b, c) applied the method to first-order peaks in LiF:Mg, and to complex, second-order glow-curves from meteorite samples. Here the technique proved to be particularly useful in pinpointing the individual peaks which were not obvious in the original glow-curve (figure 3.8).

The $T_m - T_{STOP}$ curves only reveal the most prominent of the peaks – smaller peaks remain hidden by their neighbours. Furthermore, only estimates of the true peak positions are given – the T_m values are usually higher than the actual positions because of peak overlap. The resolution of the technique is estimated to be approximately 5 °C (McKeever, 1980a).

Gobrecht & Hofmann (1966) extended the idea of repeated thermal

---

† The T_m value noted is the first maximum in the glow-curve. This is not the same as the true maximum of the first peak. This latter value will be slightly less due to peak overlap.

80   *Thermoluminescence analysis*

cleaning cycles to produce a method of analysis which enables the trap
energy distribution to be determined and, at the same time, checks that
the important ingredient of the initial rise analysis (namely, that n is
approximately constant) is satisfied. If the temperature is raised by only a
few degrees there will be very little change in n during heating. Thus, if
the thermoluminescence intensity is monitored during the cooling as
well, then there will be little difference in intensity between the heating
and cooling cycles. The whole thermoluminescence may be excited by
repeatedly heating and cooling the sample in small cycles, but many
times, and, from the slope of the heating and cooling curves, the average
depth of the traps being emptied during each temperature cycle may be
determined. Obviously, the smaller the temperature rise on each heating,
the closer the two slopes will be and so the more precise will be the
calculated value of E.

The procedure is illustrated schematically in figure 3.9 where the glow-
curve is represented as $\ln(I)$ versus $1/T$. Each temperature oscillation

Figure 3.7. Schematic $T_m - T_{STOP}$ curves for first-order (line B) and second-order
(line C) thermoluminescence. Column (a) shows a single glow peak and the
corresponding $T_m - T_{STOP}$ curves for first- and second-order kinetics. Column (b)
shows the 'staircase' shaped $T_m - T_{STOP}$ curves which result from overlapping
peaks. Closely overlapping, or a quasi-continuous distribution of peaks
(Column c) produces a straight line of slope ~ 1.0. (After McKeever, 1980a.)

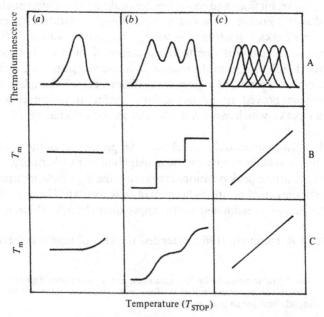

causes only a fraction of the total glow to be emitted and, therefore, Gobrecht & Hofmann called this the 'fractional glow technique'.

The number of trapped electrons released during each cycle is related to the partial light sum emitted during the same period. Thus, by adding each of those light sums which have the same average E value, the energy spectrum of the traps may be produced. An example is given in figure 3.10 for ZnS doped with Cu, In and Cu, Cl. Rudlof, Becherer & Glaefeke (1978, 1979) suggest improvements to the fractional glow technique by

Figure 3.8. Glow-curve (*a*) and associated T_m–T_{STOP} curve (*b*) for the natural thermoluminescence from the Soko Banja meteorite. The flat regions in the T_m–T_{STOP} curve represent individual peaks which are not apparent in the original glow-curve. (After McKeever, 1980*a*.)

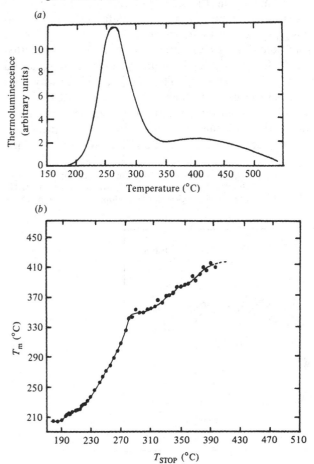

adjusting the calculated E values until a good fit to the experimental glow-curve is obtained. Further modifications are discussed by Tale (1981).

The main problems with the fractional glow technique are purely experimental. The heating rate should be quite high to give a short cycle interval (thereby ensuring that only a few traps are affected). This in turn requires excellent thermal contact between the sample and the heating strip. (Usually in thermoluminescence experiments the temperature actually measured is that of the strip, not the sample.) Also, at higher heating rates, a thermal gradient may be induced across the specimen. This in fact is a general problem in all thermoluminescence experiments and warrants further discussion in a later section (in chapter 9).

Additional problems lie in the data processing. The technique necessarily generates enormous amounts of data for which computerized handling procedures are desirable. This is especially so if the improvements of Rudlof *et al.* (1979) are incorporated. Thus, as admitted by Gobrecht & Hofmann, the expenditure of time, technique and equipment for this method is very high. This possibly explains why it has not found the widespread use that might have been expected.

One final difficulty suffered by the initial rise method ought to be mentioned; namely, the problems associated with thermal quenching, discussed in detail in section 2.2.2. The equations for first-, second-,

Figure 3.9. 'Fractional glow-curves' (schematic). The average slope $(m_1 + m_2)/2$ is determined for each cycle. Several hundred oscillations may be used. The dashed line shows the common glow-curve. (Adopted from Gobrecht & Hofmann, 1966.)

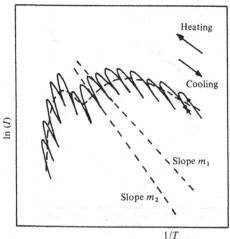

general- or mixed-order kinetics assume that the luminescence efficiency is unity, i.e., $\eta = 1$. However, a more general expression for the efficiency was given in equation (2.9), namely,

$$\eta = P_r/(P_r + P_{nr}), \tag{2.9}$$

where P_r and P_{nr} are the probabilities of radiative and non-radiative recombination transitions. Thus, a more exact version of, say, the general-order expression (equation(3.16)) would be:

Figure 3.10. Trap spectrum of ZnS activated with (a) Cu and In; (b) Cu and Cl. (After Gobrecht & Hofmann, 1966.) H_E is the relative trapped charge concentration.

(a)

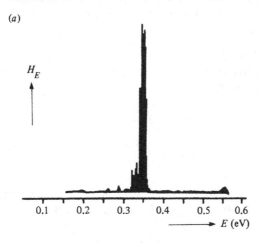

(b)

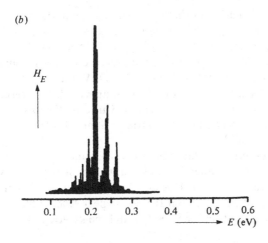

$$I(T) = \eta s'' n_0 \exp(-E/kT) / \left[1 + ((b-1)s''/\beta) \right.$$

$$\left. \times \int_{T_0}^{T} \exp(-E/kT) dT \right]^{b/(b-1)} \quad (3.16a)$$

From a consideration of the means by which non-radiative recombination can take place, it was shown that the efficiency can be described by an expression of the type

$$\eta = 1/(1 + c \exp(-W/kT)). \quad (2.10) \text{ or } (2.11)$$

The value of W can be evaluated from a plot of the luminescence emission intensity against reciprocal temperature, examples of which were given in figure 2.7. For kT large compared to W, the above expression reduces to

$$\eta = c^{-1} \exp(W/kT)$$

which when introduced into the new equation of general-order kinetics (equation (3.16a)) yields

$$I(T) = c^{-1} s'' n_0 \exp((W-E)/kT) / \left[1 + ((b-1)s''/\beta) \right.$$

$$\left. \times \int_{T_0}^{T} \exp(-E/kT) dT \right]^{b/(b-1)}.$$

During the initial rise period, the integral in the above expression is approximately zero, thus the initial rise expression (i.e., equivalent to equation (3.25)) now becomes

$$I(T) = \text{constant} \times \exp((W-E)/kT). \quad (3.25a)$$

Clearly then, a calculation of the trap depth by the initial rise method when thermal quenching is taking place calculates $(E - W)$, not E. This problem was extensively discussed by Wintle (1975a) who observed the phenomenon occurring over certain temperature ranges in quartz, in which she calculates $W = 0.64 \, \text{eV}$ and $E = 1.69 \, \text{eV}$ (for the 325 °C glow peak). Similar effects have been observed in many other materials (for example ZnSe (Fillard *et al.*, 1978; Hitier, Curie & Visocekas, 1981); AgBr (Fillard *et al.*, 1977); LiF (Kathuria & Sunta, 1981); and CaSO$_4$ (Sunta & Bapat, 1982)).

The methods described so far utilize only the initial rising portion of a glow peak, thus the number of data points which may be used to calculate E is severely limited. This may be overcome by using not just one small section of the curve but the whole peak (the whole or 'total' glow peak method) and this enables E, s and the kinetic order, b, to be calculated. The method stems from the general-order expression, equation (3.15), along with the linear temperature rise, equation (2.15), which may be combined to give

$$I(T) = (s'/\beta)n^b \exp(-E/kT).$$

From this, a plot of $\ln(I(T)/n^b)$ versus $1/T$ will give a straight line of slope equal to $(-E/k)$, and an intercept of $\ln(s'/\beta)$. The straight line will only emerge when the correct value for b is inserted (Halperin *et al.*, 1960).

The value for n is found from the glow peak by using $I(t) = -\mathrm{d}n/\mathrm{d}t$, which means that $n = (1/\beta)\int_T^\infty I(T)\mathrm{d}T$. The integral is found from the area under the thermoluminescence peak beyond T, as shown in figure 3.11.

The principal difficulty with this method is that it requires complete isolation of the peaks. A method to achieve this, which is useful with some glow-curves, is discussed by Taylor & Lilley (1978) who describe how to separate the individual peaks by starting at the higher temperature end of the curve. It requires a clean descending portion of the last peak in the glow-curve and, therefore, is restricted to those specimens which do not exhibit continuous thermoluminescence all the way up to temperatures where black-body emission is dominant. Subtraction of the black-body signal may, of course, be attempted, but extra errors are inevitably introduced.

The glow peaks from LiF:Mg between room temperature and 220 °C provide a suitable example (figure 3.12). As described by Taylor & Lilley, the procedure begins by heating the specimen to the maximum intensity of peak 5, and quenching back to room temperature. Reheating of the specimen produces a 'cleaned' peak 5 – i.e., without the contributions

Figure 3.11. Schematic plot of $I(T)$ versus T for an isolated glow peak. The shaded area is $n = (1/\beta)\int_T^\infty I(T)\mathrm{d}T$.

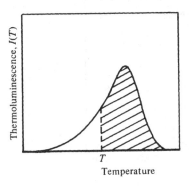

Thermoluminescence, $I(T)$

T

Temperature

from peaks 2 to 4 (peak 1 is not shown) – although of reduced intensity (figure 3.12a; the intensity of peak 5 is shown multiplied by a factor of three).

The shape of peak 4 is determined by heating the re-irradiated sample to 175 °C in order to remove peaks 2 and 3. Reheating now produces a combination of peaks 4 and 5. Using the previously determined shape of peak 5, this peak may be scaled up to fit the decreasing portion of the combined peak. Subtraction of peak 5 from the combined glow reveals the shape of peak 4 (figure 3.12b; peak 4 is shown multiplied by a factor

Figure 3.12. Resolution of the glow-curve from LiF : Mg TLD-100. (a) Resolved peak 5; (b) resolved peak 4; (c) fully resolved glow-curve. Solid circles = calculated intensity values for peaks 4 and 5 using experimentally determined values of E and s. Heating rate = 0.4 °C s^{-1}. (After Taylor & Lilley, 1978.)

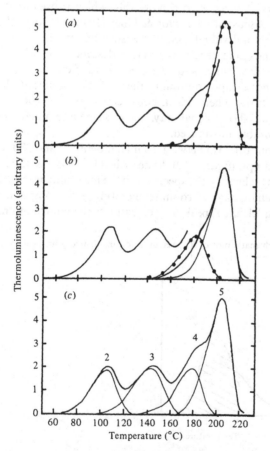

of two for clarity). The process is then repeated to account for each of the remaining peaks in the glow-curve. The final resolved glow-curve for LiF:Mg TLD-100 is shown in figure 3.12c.

The method of resolution is especially useful with just two or three peaks but becomes more difficult the more peaks there are in the glow-curve. Any errors introduced during the resolution of the first peak will be carried over to the separation of the second, and so on. Furthermore, it becomes extremely difficult (if not impossible) when the peaks are characterized by non-first-order kinetics, when a shift in peak position is expected as the trap is partially emptied (see figure 3.2). (The observed shift in the position of a peak for different levels of trapped charge is a good indication that the thermoluminescence is non-first-order.)

3.3.2 *Peak shape methods*

In contrast to the partial or whole peak analyses, methods based on the shape of the peak utilize just two or three points from the glow-curve. Usually, these are the maximum of the peak T_m and either, or both, the low-and high-temperature half-heights at T_1 and T_2 (see figure 3.3). However, because the shape of the peak is strongly affected by kinetic order (cf. figure 3.1) then the methods are dependent on the order of kinetics, unlike the techniques of the previous section.

Since the first of this class of method, suggested by Grossweiner (1953) for first-order kinetics, a great many variations and corrected formulae have emerged, for both the first- and second-order cases. In each of the methods the approximate formulae, equations (1.6) and (1.7), are reduced to simple expressions relating E to T_m and T_1 and/or T_2. To do this the integral appearing in the formulae is reduced to an asymptotic expansion and then simplified by dropping terms after the first or second. In addition to Grossweiner (1953), Dussel & Bube (1967), Lushchik (1956), Keating (1961), Halperin & Braner (1960), Land (1969) and Balarin (1979b) have all produced simple formulae to calculate E, some accounting for first- and second-order kinetics (Lushchik, Balarin), others for the temperature dependence of the frequency factor (e.g., Keating), and others have used inflection points on the thermolumines-cence peak, rather than half-heights (e.g., Land), but the principles remain the same.

However, it is Chen (1969a,b) who has analysed these methods to their fullest extent and, rather than list all of the published formulae, only Chen's equations will be quoted here. Chen (1969a) gives three equations each for first- and second-order peaks, relating the trap depth to either the full width of the peak at its half-height ($\omega = T_2 - T_1$), its low-temperature half-width ($\tau = T_m - T_1$), or its high-temperature half-width ($\delta = T_2 - T_m$). A general formula for E was given namely:

$$E = c_\gamma (k T_m^2/\gamma) - b_\gamma (2 k T_m) \tag{3.26}$$

where γ is ω, τ or δ. The constants c_γ and b_γ for the three methods, and for first- and second-order kinetics, are given in table 3.1. Chen (1969b) developed his analysis further to encompass general-order kinetics by making use of the Halperin & Braner geometrical factor, defined earlier as $\mu_g = \delta/\omega$. Chen numerically solved the general-order expression for thermoluminescence, equation (3.16), and calculated values of μ_g for

Table 3.1. *Values for the constants c_γ and b_γ in equation (3.26). If the temperature dependence of the frequency factor of the type $s \propto T^a$ is suspected, then a term $a/2$ must be added to the constant b_γ (Chen & Kirsch, 1981)*

	First-order			Second-order		
	ω	τ	δ	ω	τ	δ
c_γ	2.52	1.51	0.976	3.54	1.81	1.71
b_γ	1.0	1.58	0	1.0	2.0	0

Figure 3.13. Calculated geometrical factor, μ_g as a function of kinetic order, b. The solid line is the average value, whilst the dotted lines correspond to the largest possible variation (for various values of E and s). (After Chen, 1969b.)

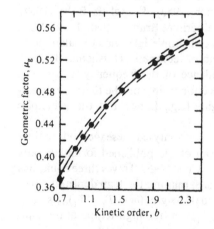

different values of b. The resultant variation of μ_g with b is given in figure 3.13. In order to find the values of c_γ and b_γ for $b \neq 1$ or 2, Chen used the known values for μ_g for first- and second-order kinetics (namely, 0.42 and 0.52, respectively) in order to arrive at modified values of the constants used in equation (3.26). Chen's final general-order coefficients are given in table 3.2.

If, instead of the general-order expression of equation (3.16), the mixed-order equation (3.22) is preferred, equation (3.26) can still be used to calculate E because the geometrical factor μ_g is seen to vary from 0.42 to 0.52 as α varies from 0 to 1.0 (Chen & Kirsh, 1981).

A test of the accuracy of these methods of calculating E was provided by Shenker & Chen (1972) and Kivits & Hagebeuk (1977). Both sets of authors conclude that the methods yield trap depths to within 5% of the actual value. The tests were carried out on theoretical thermoluminescence curves of known E. Their application to an experimental glow-curve, however, may be much more hazardous due to the fact that the calculation relies heavily on the accurate determination of just two (or three) experimental points. Any slight inaccuracies (due, for example, to inefficient thermal cleaning of the lower temperature 'satellite' peaks) will inevitably produce erroneous values for the trap depth. In particular, the methods which rely on the measurement of the high-temperature half-width are open to error because of the greater difficulty experienced in obtaining a clean descending part of a glow peak. Furthermore, there is an additional difficulty which sometimes occurs. Thermal quenching has been shown to alter the glow peak shape to the extent that E values calculated by these methods are often erroneous and unreliable (Sunta & Bapat, 1982). Thus application of these methods must be made with caution.

Table 3.2. *Values for the constants c_γ and b_γ in equation (3.26) for the general-order case. Clearly the values reduce to the first order values given in table 3.1 when $\mu_g = 0.42$, and to the second-order values when $\mu_g = 0.52$. A term $a/2$ is to be added to b_γ if one needs to account for the temperature dependence of the frequency factor*

γ	c_γ	b_γ
τ	$1.51 + 3(\mu_g - 0.42)$	$1.58 + 4.2(\mu_g - 0.42)$
δ	$0.976 + 7.3(\mu_g - 0.42)$	0
ω	$2.52 + 10.2(\mu_g - 0.42)$	1.0

3.3.3   *Peak position methods*

The calculation of trap depths based on the positions of the thermoluminescence peaks (i.e., temperatures of the maxima) are of two basic types: (i) those which give a direct relationship between E and T_m, of the sort $E = \text{constant} \times T_m$; (ii) those which calculate E from the change in T_m caused by changing the rate at which the sample is heated ('heating rate' methods).

The first type can only be considered to be very approximate because they ignore the effect on T_m of the frequency factor s. Randall & Wilkins (1945a) assumed that at $T = T_m$, the probability of charge release from a trap is unity, i.e.,

$$s \exp(- E/kT_m) = 1,$$
or
$$E = kT_m \ln(s). \tag{3.27}$$

If the same value of s can be assumed for each peak in the glow-curve, then E is directly proportional to T_m, with the constant of proportionality being different from sample to sample. An earlier method of this same type (Urbach, 1930) assumed that the constant of proportionality was the same no matter what the sample was and used:

$$E = T_m/500 = 23kT_m. \tag{3.28}$$

However, this kind of calculation can be grossly inaccurate because of the non-constant value of s between glow peaks. Furthermore, the effect of the heating rate, β, on the value of T_m, for a given E and s, is completely ignored. The importance of β can be demonstrated for the first-order case by differentiating equation (1.6) with respect to temperature.

Figure 3.14. Variation in T_m with E for various values of s/β ($°C^{-1}$) for first-order kinetics (Selected data from Christodoulides & Ettinger, 1971.)

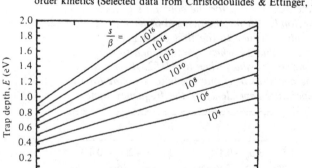

Equating the derivative to zero at $T = T_m$ gives

$$\beta E/kT_m^2 = s \exp(-E/kT_m). \tag{3.29}$$

Christodoulides & Ettinger (1971) computed first-order glow-curves for various values of E, s and β and plotted the variation of T_m with E for different values of the ratio (s/β) (figure 3.14). Clearly, for a given peak, β has a powerful influence on the position of the peak.

This relationship between E, s, β and T_m has spawned a number of methods for calculating E by noting the change in T_m for a certain change in β. Using equation (3.29), Hoogenstraaten (1958) suggested employing several different heating rates and noting the value of T_m each time. From a plot of $\ln(T_m^2/\beta)$ against $1/T_m$ a straight line of slope E/k and intercept $\ln(E/sk)$ is obtained, from which both E and s can be calculated.

Chen & Winer (1970) extended these concepts to the general-order case and derived

$$\beta E/kT_m^2 = s[1 + (b-1)\nabla_m] \exp(-E/kT_m) \tag{3.30}$$

where $\nabla_m = 2kT_m/E$. The term $[1 + (b-1)\nabla_m]$ can be considered to be approximately constant and so a plot of $\ln(T_m^2/\beta)$ versus $1/T_m$ will still yield a good value for E. In fact, Chen & Winer (1970) also showed that the heating function, $T = $ function (t), does not have to be linear, i.e., β does not have to be a constant. In this case the value of β at the maximum is all that is required.

It can also be shown, from differentiation of the general-order equation (3.16), that the maximum intensity of a thermoluminescence peak (I_m) is related to β by

$$I_m^{b-1}(T_m^2/\beta)^b = (E/bks)^b \exp(E/kT_m), \tag{3.31}$$

which suggests that a plot of $\ln(I_m^{b-1}(T_m^2/\beta)^b)$ against $1/T_m$ will give a straight line of slope E/k. However, a serious disadvantage here is that the value of b needs to be known beforehand, thus equation (3.30) is more useful than equation (3.31).

Chang & Thioulouse (1982) developed an expression for the glow-curve based on a thermally assisted tunnelling model and from this expression arrived at equation (3.29) to describe the variation in the peak position with heating rate. Finally Chen & Winer (1970) discussed the effect of the temperature dependence of s ($s \propto T^a$) on the heating rate method to show that the energy calculated from a plot of $\ln(\beta/T_m^2)$ against $1/T_m$ is $(E + akT_m)$, rather than E.

The major advantage of the heating rate methods is that they only require data to be taken at a peak maximum (T_m, I_m) which, in the case of a large peak surrounded by smaller satellites, can be reasonably accurately determined from the glow-curve. Furthermore, the calculation of E is not affected by problems due to thermal quenching, as with the initial rise method. However, when it is the activation energies of the smaller peaks that need to be determined, it becomes a difficult technique to use. For first-order curves, thermal cleaning might separate the peaks enough to determine T_m (but not I_m which has been affected by the cleaning process). With second-order kinetics even T_m is unreliable because the peaks will move due to the reduction in trapped charge during cleaning. Similar arguments apply when two (or more) peaks are so closely overlapping that they appear as one composite peak.

A recent variation on the heating rate methods may prove to be extremely useful in these complex cases. Sweet & Urquhart (1980, 1981) start with equation (3.5) and consider the temperature dependences of R, and s to be small, so that the equation can be rewritten,

$$I(t) = f(n)s\exp(-E/kT),$$

or

$$I(T) = (f(n)s/\beta)\exp(-E/kT). \tag{3.32}$$

Here, $f(n)$ is a function of n, which, in turn, is given by the area under the thermoluminescence curve thus, $n = (1/\beta)\int_T^\infty I(T)\mathrm{d}T$ (cf. figure 3.11). By varying the heating rate and selecting those temperatures at which the values of n, and therefore $f(n)$, are equal, Sweet & Urquhart determine that

$$I(T_x) = \text{constant} \times \exp(-E/kT_x) \tag{3.33}$$

where T_x is the temperature at which the areas from T_x to ∞ are equal for each value of β, and x represents the fraction of the total area. Thus, by changing β, T_x is seen to vary and a plot of $\ln(I(T_x))$ against $(1/T_x)$ can be made in order to determine E. Sweet & Urquhart (1981) computed glow-curves with known values of E and found that this method gave an accurate value of E right across the thermoluminescence peak – i.e., for all x.

The biggest advantage of this method is when there are two (or more) overlapping peaks producing a composite glow-curve with only one, perhaps ill-defined, maximum. Instead of needing to determine areas under individual peaks, the area under the *whole* glow-curve is calculated. For small fractions of this area, the activation energy for the lower temperature peak is determined. For large fractions, the energy for the higher temperature peak is determined.

This was demonstrated by Sweet & Urquhart (1981) for two computer-generated, overlapping peaks as shown in figure 3.15a. The results of their analysis, for different fractions (x) of the total glow area, are shown in figure 3.15b. For more peaks a 'staircase' shaped curve of E versus x should result. This method is independent of the order of kinetics and has high resolving power – in the quoted case it is able to separate two peaks which are no more than 4 or 5 degrees apart. As with the fractional glow technique, the main experimental requirements of the heating rate methods are that the heating rate is sufficiently high (to give a strong signal), that there is good thermal contact between the sample and the heating strip and that there are no thermal gradients across the specimen. Reliable temperature control is essential.

Many authors when describing their thermoluminescence results quote glow peak temperatures and heating rates. This is necessary in order to compare results from different groups on the same material. For a specimen that exhibits several peaks, it is of little use merely referring to 'the peak at X °C' without giving the heating rate used.

It is interesting to compare the results obtained by different authors.

Figure 3.15. (a) Computer-generated curves with $E_1 = 18.0$ meV, $s_1 = 3 \times 10^5 \text{s}^{-1}$, relative concentration $= 64$; $E_2 = 22.5$ meV, $s_2 = 3 \times 10^5 \text{s}^{-1}$, relative concentration $= 100$. $R_1 = R_2 = 90$. β values (1) 0.072 °Cs^{-1}, (2) 0.054°Cs^{-1}, (3) 0.046°Cs^{-1}, (4) 0.025°Cs^{-1}, (5) 0.016°Cs^{-1}. (b) Calculated trap depths expressed as a fraction of the total area. The E values used in the computation are shown by arrows. (After Sweet & Urquhart, 1981.)

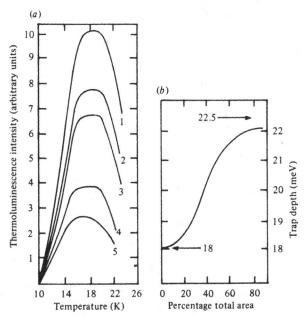

This is exemplified in figure 3.16 where a conventional 'heating rate plot' (i.e., $\ln(T_m^2/\beta)$ versus $1/T_m$) is displayed using data collected from the published literature for LiF:Mg. For this material, the most readily available data concern thermoluminescence peak 5 (cf. figure 1.4). The plot reveals a startling spread in the data. According to the theory of first-order glow peaks all the data should be on, or near, the same straight line. Clearly, this is not so. It is evident that although different authors use the same heating rate, the peak does not appear in the same position.

Figure 3.16. 'Heating rate plots' of T_m^2/β versus $1/T_m$ for LiF:Mg peak 5 using data obtained from the literature, as follows: (1) Taylor & Lilley, 1978; (2) Sagastibelza & Alvarez Rivas, 1981; (3) Lilley & McKeever, 1983; (4) McKeever, 1980a,b; (5) Kathuria & Sunta, 1979; (6) Klick *et al.*, 1967; (7) Jain, 1981a; (8) Jain, 1981b; (9) Townsend *et al.*, 1983b; (10) Kos & Nink, 1977; (11) Buckman & Payne, 1976; (12) Kathuria & Sunta, 1981; (13) Podgoršak *et al.*, 1971; (14) Ziniker, Rusin & Stoebe, 1973; (15) Moran & Podgoršak, 1971; (16) Grant, Stow & Correl, 1968; (17) Miller & Bube, 1970; (18) Fairchild *et al.*, 1978b.

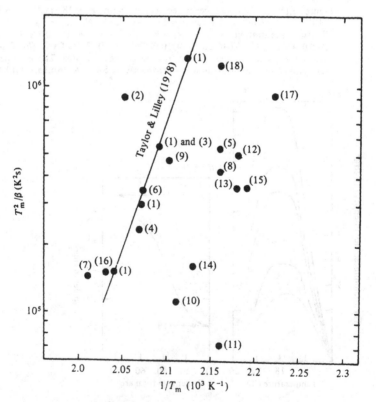

The most obvious explanation for this is poor temperature control –
i.e., inaccurate estimates of temperature and heating rate. This undoubtedly will explain some of the spread of data shown in the figure, but
it cannot be invoked as an explanation for all the observed spread. For
example, some authors (e.g., Taylor & Lilley, 1978) obtain a large degree
of self-consistency such that their data (points numbered 1 in figure 3.16)
all fall on a straight line. From this line they obtain $E = 2.06$ eV which,
within experimental uncertainty, is close to the value obtained by them
using other methods. However, other authors who also use the same
value for the activation energy (e.g., Moran & Podgoršak, 1971 (number
15); Fairchild *et al.*, 1978*b* (number 18)) produce data which do *not* lie on
the Taylor & Lilley line. This is not easy to explain by simply invoking
temperature inaccuracies because one would not expect them to
calculate the same E value in this case. Alternative explanations must be
sought.

Townsend, Clarke & Levy (1967) performed a heating rate analysis on
LiF:Mg and observed that repeated application of the method produced
a peak shift such that, although a straight line on a $\ln(T_m^2/\beta)$ versus $1/T_m$
plot was still obtained, it was not the same straight line as on the first
measurement. It appears that thermal treatments or repeated irradiations cause some readjustment of the peak position. Additionally,
many of the LiF:Mg samples referred to in figure 3.16 are from different
sources. This may be a relevant factor.

It is emerging that the thermoluminescence from LiF:Mg is a complex
function of the impurity content and of the sample's thermal history.
Some peaks are known to be related to large defect complexes which
change their concentration with thermal and radiation treatments (see
chapter 5). It has already been discussed in chapter 2 how such
complexes might affect the activation energy. Furthermore, in some
materials, changing the concentration of an impurity has been observed
to alter the trap depth (e.g., Lee & McGill, 1975). Thus, it is perhaps not
surprising to observe that different groups, ostensibly studying the same
material, produce a wide range of disparate data, considering that they
are using samples from different origins and may have treated the
specimens in similar, but not identical, ways. These factors ought to be
borne in mind when monitoring peak positions in thermoluminescence
measurements.

3.3.4 *Curve fitting*
Instead of employing simple formulae to derive values for the
parameters E, s and b (or α), methods of computerized curve fitting have
been used on several occasions with apparent success. The procedure is

to establish the approximate positions of the most prominent peaks in the glow-curve and to estimate initial values of E, s and b by using one of the analytical methods discussed so far. A theoretical curve is then computed using the general-order equation (3.16) for $b \neq 1$, or the first-order equation (1.6) for $b = 1$. The computed curve is then compared with the actual experimental curve and a root-mean-square (RMS) deviation between the two is calculated. The procedure continues by sequentially changing the E, s and b values until a minimum value of the RMS deviation is obtained. The method was described for isolated first- or second-order peaks by Mohan & Chen (1970) and for general-order curves by Shenker & Chen (1971). The method requires a numerical procedure for solving the integral $\int_{T_0}^{T} \exp(-E/kT) \times dT$ and several suitable ones are suggested by Chen (1969c), Balarin (1977) and Jenkins (1978).

The usual problem of how to deal with complex overlapping peaks still presents itself. Kirsh, Kristianpoller & Chen (1977) deal with this by analysing only the prominent peaks in the glow-curve. This can be seen in figure 3.17 where the thermoluminescence curves from γ-ray- and ultra-violet-irradiated MgO samples are displayed (*a*). Only four of the prominent peaks are analysed (*b*). This is done by selecting the data in each of the four regions and fitting the computed curves to them. Kirsh & Kristianpoller (1977) examined the complex glow-curve from SrF_2 doped with Tb. They thermally cleaned each peak and fitted the parameters E, s and b to the cleaned peaks. Clearly, the curve fitting can be used only on the increasing side of the peak and a small amount of the decreasing part, depending upon how well the peaks are separated.

In case any errors may be introduced by less than thorough cleaning, some authors prefer to use the complete glow-curve without separation procedures. Thus, Fairchild *et al.* (1978*b*) did not attempt peak resolution of the glow-curve from LiF TLD-100 but instead guessed the number and position of the peaks present in the glow-curve and estimated initial values of E, s and b for each of them. The quantity n_0 (cf. equations (1.6) and (3.16)) is taken to be proportional to the area under each peak and an initial value is guessed for this too. The whole glow-curve is then computed using the guessed values, it is compared with the experimental curve and an RMS deviation is determined. By sequentially changing the values for these parameters, for each peak in turn, a final computed glow-curve is arrived at. The method appears to work well in that the values obtained for the parameters closely agree with those obtained by other methods (Taylor & Lilley, 1978). The same method is used by Levy (1979) to analyse the thermoluminescence from quartz and various geological specimens.

McKeever (1980*a,b*) uses a very similar computerized curve-fitting procedure, but prefers to use the $T_m - T_{STOP}$ method to obtain an initial estimate of the number and positions of the component peaks. The combined $T_m - T_{STOP}$ and curve-fitting procedures are seen to work very well when analysing complex second-order glow-curves from meteorite

Figure 3.17. (*a*) Thermoluminescence in MgO observed after (A) γ irradiation; (B) Vacuum ultra-violet illumination of a non-irradiated sample; (C) Vacuum ultra-violet illumination after γ irradiation at room temperature and heating to above 550 K. (*b*) Measured glow peaks (dots) at 108, 135, 334 and 516 K, compared with the computed peaks, normalized to the maximum intensities of each peak. (After Kirsh, Kristianpoller & Chen, 1977.)

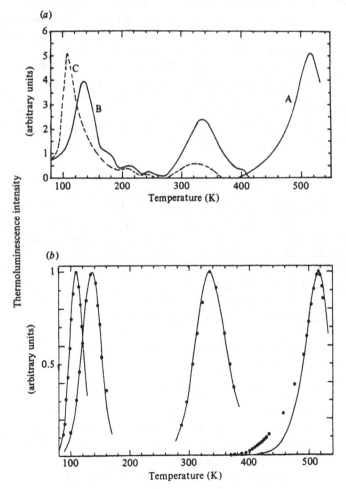

samples (McKeever, 1980*a,c*; see also figure 8.4). Although experimentally and computationally tedious, the combined T_m–T_{STOP}/curve-fitting method is recommended because it avoids all difficulties associated with peak separation procedures.

An important point which should not be overlooked is made by Townsend (1968) and later by Fairchild *et al.* (1978*a*). When using curve-fitting or indeed when using any of the methods of analysis so far discussed, what should be analysed are curves of dn/dT versus T. Instead, what is usually measured is light detector output (most commonly photomultiplier output) versus T. Only if there is always a constant relationship between dn/dT and detector output can the analyses give completely reliable data. Unfortunately, the sensitivity of most detectors varies markedly with wavelength; thus, if the emission changes wavelength during the occurrence of a glow peak the detector output will change accordingly and the signal obtained would not be suitable for kinetic analysis. For LiF (TLD-100) this does not appear to be a problem (as can be seen from the papers of Fairchild *et al.*, 1978*b*, Taylor & Lilley, 1978, and McKeever, 1980*b*), but the possible effects should be considered (especially as recent data for this material show that there is some difference in the spectral emissions between peaks 2 and 3, and peaks 4 and 5 (e.g., figure 2.14)).

3.3.5   *Isothermal analysis*

In the methods of analysis discussed so far E is calculated from the form, or shape, or position of the glow peak which itself is produced during a non-isothermal rise in sample temperature. Quite a different method of analysis has been devised in which the sample temperature is kept constant and the luminescence intensity is followed as a function of time. This method of analysis, therefore, deals with phosphorescence rather than thermoluminescence, but by raising the temperature at which the phosphorescence decay is being monitored, the deep traps which take part in thermoluminescence can be sampled and their trap depths determined.

For monomolecular luminescence at constant temperature equation (3.8) may be used to give

$$I(t) = I_0 \exp\left(-s\exp\left\{-E/kT\right\}t\right) \tag{3.34}$$

where I_0 is the initial intensity at $t = t_0$ (i.e., $I_0 = n_0 s \exp\left(-E/kT\right)$). This equation shows that at a constant temperature T, the luminescence will decay exponentially with time and a plot of $\ln\left(I/I_0\right)$ against t will give a straight line of slope $m = s\exp\left(-E/kT\right)$. If the decay is monitored at several different temperatures a plot of $\ln(m)$ versus $1/T$ will give a straight line of slope E/k from which E can be determined. The intercept will give $\ln(s)$. This procedure was first described by Randall & Wilkins (1945b). The main assumption in this analysis is that the decay is first-order. If the order of kinetics takes on any value other than one a non-exponential decay is expected. This can be demonstrated by using the general-order expression, equation (3.15). Rearranging this equation gives

$$dn/n^b = -s'\exp(-E/kT)dt$$

which may be integrated to give

$$(n/n_0)^{1-b} = 1 + n_0^{b-1}(b-1)s'\exp(-E/kT)t. \qquad (3.35)$$

Using $I_0 = n_0^b s'\exp(-E/kT)$, equation (3.35) becomes

$$(I/I_0)^{(1-b)/b} = 1 + n_0^{b-1}(b-1)s'\exp(-E/kT)t. \qquad (3.36)$$

Equation (3.36) clearly shows that I does not decay exponentially with t; instead, a plot of $(I/I_0)^{(1-b)/b}$ versus t will give a slope $m = n_0^{b-1}$ $(b-1)s'\exp(-E/kT)$. This demonstrates an important characteristic of non-first-order luminescence decay, namely, that the rate of decay is dependent upon n_0, the initial trapped charge population. Furthermore, a plot of $\ln(m)$ versus $1/T$ will again give the activation energy E. It is clear that a straight line on the $(I/I_0)^{(1-b)/b}$ versus t plot will only be obtained when the correct value of b is inserted (May & Partridge, 1964; Takeuchi, Inable & Nanto, 1975), thus the kinetic order of the decay can be estimated. When second-order kinetics are considered and the value $b = 2$ is inserted, equation (3.36) reduces to the bimolecular expression for luminescence decay, given in equation (1.5), with $\alpha = s'\exp(-E/kT)$. A similar analysis by Moharil & Kathuria (1983) has demonstrated that the value for R in equation (3.4) can also be found from an isothermal decay curve.

In many circumstances the actual decay of luminescence is seen to follow neither the monomolecular nor the bimolecular laws. Instead, a more general t^{-x} law is often found. The special case, which appears to be quite common, of $x = 1$ deserves some discussion. Randall & Wilkins (1945b) considered the possibility that the traps are uniformly distributed in energy. If there are N_E traps between E and $E + dE$, then the luminescence decay is given by

$$I = \int_0^\infty N_E s\exp(-E/kT)\exp(-st\exp\{-E/kT\})dE$$

which, upon integration, yields

$$I = n_0 kT/t, \qquad (3.37)$$

100   *Thermoluminescence analysis*

assuming that $\exp(-st) \ll 1$. Thus, a uniform distribution of energies will yield the t^{-1} law directly. Medlin (1961) and Visocekas and colleagues (Visocekas *et al.*, 1976; Visocekas & Geoffroy, 1977) appear to have found evidence of a temperature-dependent t^{-1} decay for luminescence from calcite and interpret it in terms of a distribution of traps. In general, a t^{-x} behaviour can be explained by the presence of trap distributions with the specific value of x depending upon the exact nature of the distribution.

However, in several recent articles, a temperature-*independent* (or nearly so) t^{-1} decay law has been observed which is inconsistent with any of the above models involving thermal detrapping (e.g., Visocekas *et al.*, 1976; Hama *et al.*, 1980; Avouris *et al.*, 1981; Avouris & Morgan, 1981; and the references therein). The model proposed to account for this law is that of quantum mechanical tunnelling from one site to another nearby (cf. figure 2.6). Sometimes the tunnelling only takes place after the thermal excitation of the donor electrons from a deep level to a shallower one ('thermally assisted' tunnelling). Delbecq, Toyozawa & Yuster (1974) use the thermally assisted tunnelling model of Mikhailov (1971) to demonstrate that a t^{-1} law naturally emerges and apply the law to explain the athermal decay of luminescence in KCl:Ag and KCl:Tl (see section 4.4).

Calcite is a good example of a material which displays tunnelling, and figure 3.18 illustrates that part of the glow-curve of calcite that decays athermally at liquid nitrogen temperature (Visocekas, 1979). The phenomenon has also been clearly observed in zinc silicate (Avouris & Morgan, 1981; Avouris *et al.*, 1981), spinel (Lorincz *et al.*, 1982) and

Figure 3.18. Thermoluminescence of calcite: solid line – after 3 min decay at liquid nitrogen temperature; dashed line – after 2 hours 20 min at liquid nitrogen temperature. (After Visocekas, 1979.)

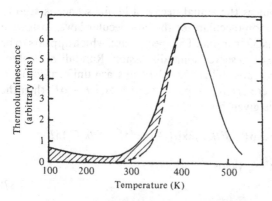

appears to be widespread in many amorphous materials, both organic and inorganic (e.g., Kieffer, Meyer & Rigaut, 1971, 1974; Doheny & Albrecht, 1977; Hama *et al.*, 1980; Charlesby, 1981).

There are two ways of conducting an isothermal decay experiment. One method (possibly the most popular) is to irradiate the specimen and then to maintain it at a steady temperature for a certain period, after which the glow-curve is recorded and the size of the thermoluminescence peaks noted. By repeating the experiment on a freshly irradiated sample the variation of the glow peak height (or area) with post-irradiation storage time can be obtained. The whole experiment may be repeated at different temperatures. The alternative method is to hold the irradiated specimen in the temperature range in which a thermoluminescence peak is observed and to monitor the phosphorescence decay in real time. In both cases accurate temperature control is extremely important.

One potential problem with both methods of analysis which is not often discussed or considered in the published literature is the phenomenon of defect clustering reactions and impurity precipitation. For example, in alkali halide crystals the energy associated with the incorporation of divalent metallic impurities (M^{2+}) into the lattice may be reduced if the impurities and their charge-compensating alkali ion vacancies associate firstly into pairs, and then into higher-order complexes. The kinetics of these defect reactions have been the subject of much discussion over the last 20 years and are considered in more detail in chapter 5; nevertheless, they appear to be third-order, implying a reaction sequence of the type

$$M^{+2} + V^- \rightleftharpoons M^{2+}V^-_{PAIR}, \tag{3.38a}$$

$$3[M^{2+}V^-]_{PAIR} \rightleftharpoons [M^{2+}V^-]^3_{TRIMER}. \tag{3.38b}$$

Furthermore, precipitation reactions are known to result in the formation of a second phase in these materials. This type of clustering and precipitation reaction has the effect of converting one type of defect into another type. In recent years there has been much controversy over the exact role played by these defects in thermoluminescence of the alkali halides (discussed in chapter 5) but what is clear is that: (i) holding the specimens at high temperatures for long times either pushes the reactions in (3.38) to the left – i.e., dissociation (also dissolution of precipitates) or to the right – i.e., association (also precipitation), depending upon the temperature chosen; and (ii) both effects have a drastic influence on the size of the thermoluminescence peaks, presumably because these defects are acting as charge traps and recombination centres. It is not yet proven that clustering reactions take place with those defects which have actually trapped the charge, but evidence to support this proposal is emerging (Taylor & Lilley, 1982*a,b,c*). The net effect of this is that the thermoluminescence peaks will be changing with storage time due to both detrapping of charge and to the defect reactions. For example, Zimmerman *et al.* (1966) annealed LiF:Mg samples at temperatures of between 124 °C and 143 °C for times in excess of 10 hours after irradiation. This is exactly the heat treatment necessary to induce a significant amount of precipitation of Mg in this material (Bradbury & Lilley, 1977*a,b*). Similar heat treatments were used by Blak & Watanabe (1974) and Sagastibelza & Alvarez Rivas (1981) in their analyses of the LiF:Mg glow-curve. The result is that the peaks associated with Mg in solution (e.g., peak 5, cf. figure 1.4) decay more rapidly than would be expected from thermal detrapping alone. Thus, the activation energies calculated will be too small. However, if the same experiment is performed by holding the specimens at higher temperatures but for much shorter times (e.g., Taylor & Lilley, 1978) the degree of precipitation is much less with the result that the isothermal decay is due only to thermal detrapping, at least in the early stages. Thus, Zimmerman *et al.* (1966) obtained an activation energy of 1.19 eV whilst Taylor & Lilley (1978) obtained 2.06 eV – a significant difference. Several independent analyses of a different type indicate that the result of Taylor & Lilley is the correct one (e.g., Fairchild *et al.*, 1978*b*, McKeever, 1980*b*).

Other effects of defect aggregation can be seen in the isothermal data of Ausin & Alvarez Rivas (1974) who actually observed an increase in the luminescence at some temperatures in NaCl, before the more usual decay is seen. Inabe & Takeuchi (1977) examined this in more detail in KCl:Sr and explained it by invoking the dissociation of impurity-vacancy complexes (e.g., trimers) into smaller units (pairs). An increase in luminescence cannot be explained by the normal laws of isothermal decay and, as discussed by Inabe & Takeuchi, misleading results will accrue unless proper account is taken of all of the underlying processes.

Analogous concepts apply to all of the alkali halides but such effects cannot be ruled out from other materials as well. For example, defect clustering has also been observed in several other ionic crystals (e.g., the alkaline earth fluorides: Ong & Jacobs, 1979; Jacobs & Ong, 1980; Puma *et al.*, 1980) and the possible effect of this on the thermoluminescence of these materials has been discussed (Bangert *et al.*, 1982*b*).

We have already seen how an isothermal decay experiment may not necessarily produce an exponential decay. In a real glow-curve there may be a reason for this other than non-first-order kinetics, namely, overlap of two or more thermoluminescence peaks. If the temperature is such that two peaks will decay together, but at different rates, then fast-

and slow-decaying components will be seen. In favourable circumstances it may be possible to separate these graphically, but with several closely overlapping peaks, this clearly becomes very difficult. In fact, the occurrence of peak overlap between peaks 4 and 5 in LiF:Mg recently led Kathuria & Sunta (1979) to suggest that the thermoluminescence from this material did not in fact follow first-order kinetics, but rather has a varying order, stabilizing at 1.6. The concept of varying-order kinetics must be treated seriously. The kinetic order is, as we have seen, dependent upon the relative values of the rates of detrapping and recombination. When a trap is full one might expect that retrapping is negligible, but as the trap empties, the rate of retrapping must increase. If it becomes comparable with the recombination rate, then the reaction order will increase as the trap empties. However, the data of Kathuria & Sunta (1979) were interpreted in a much simpler fashion by Lilley & Taylor (1981) who accounted for peak overlap and precipitation to show that the complex decay curve exhibited by the thermoluminescence from LiF is consistent with the generally accepted first-order kinetics for the glow peaks from this material. The data produced by Lilley & Taylor (1981) to support their case were confirmed by Lilley & McKeever (1983), cf. figure 3.19.

When two or more first-order glow peaks overlap, the overall decay curve for the luminescence can be interpreted as the addition of two or more straight lines on a semi-log plot. However, for this to be so, the decay rates must be different for the peaks. The isothermal decay rate for first-order thermoluminescence peaks can be conveniently described by a 'half-life', i.e., the time taken for the peak to decay to one-half of its original intensity (analogous to the well-known 'half-life' of radioactive decay). The concept of a half-life cannot be applied to non-first-order kinetics owing to the lack of an exponential decay in this case. Hence, for two first-order thermoluminescence peaks we would expect that the half-life for the higher temperature peak would be greater than the half-life for the low-temperature peak – i.e., we expect the high-temperature peak to be thermally more stable. Unfortunately this need not always be so. The half-life is defined as

$$\tau_{1/2} = \ln(2)/\exp(E/kT)s \tag{3.39}$$

where all the terms have their usual meaning. From this equation it becomes clear that $\tau_{1/2}$ depends upon s, which is constant, and $\exp(-E/kT)$, which is temperature dependent. The following example serves as a good illustration of the importance of the relative values of s and $\exp(-E/kT)$. Consider two traps with $E_1 = 0.80\,\text{eV}$ and $s_1 = 10^{12}\,\text{s}^{-1}$, and $E_2 = 0.979\,\text{eV}$ and $s_2 = 10^{15}\,\text{s}^{-1}$ respectively. At a heating rate of $10 \, \mathrm{K \, s^{-1}}$, one thermoluminescence peak would occur at 339 K in a glow-curve, whilst the other would be at 331 K. However, at a post-irradiation storage temperature of 300 K, both peaks have the same half-life ($\sim 19 \, \mathrm{s}$). Thus, only one straight line would be obtained on a semi-log plot and the peaks could not be resolved. However, if the storage temperature was increased to 350 K, the high-temperature peak (peak 1) would have a half-life which is approximately 2.5 times longer than peak 2. Conversely, if the storage temperature was decreased to 250 K, the high-temperature peak would actually decay faster (~ 4.0 times faster) than the low-temperature peak. Thus, it must not be assumed that the fastest decaying component belongs to the lowest temperature peak; nor must it be assumed that a single straight line on a

Figure 3.19. Plot of the log of thermoluminescence intensity versus time for LiF TLD-100 at 165 °C following annealing at 140 °C for 30 min. (A) Combined decay of peaks 4 and 5 (cf. figure 1.4); (B) decay of peak 5; (C) precipitation effects increasing decay of peak 5; (D) decay of peak 4, obtained by subtracting the extrapolation of (B) from (A). (After Lilley & McKeever, 1983.)

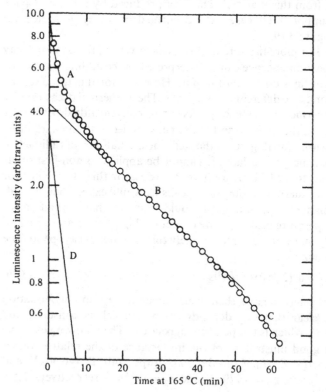

semi-log plot indicates that only one thermoluminescence peak is present.

Even stranger effects can be obtained with non-first-order glow peaks because not only do E and s govern the decay rate, but the initial level of trap filling, n_0, is also important, as was mentioned in connection with equation (3.36).

It is clear from the foregoing that the correct interpretation of isothermal decay curves is not straightforward, and experimentalists must be aware of all the possible sources of error. For this reason, isothermal analysis *alone* should be discouraged and only when supporting data from the other studies is at hand can any firm conclusions be reached from the observed results.

3.3.6 *Energy distributions*

Most analyses of thermoluminescence results proceed on the assumption that the traps which are associated with a particular species of defect all have the same trap depth. Such an assumption may be valid for crystalline materials for which it may be argued that the surroundings of a given type of trap are uniquely defined, and thus each trap has the same, discrete activation energy. However, this picture may not easily be extended to vitreous or amorphous materials in which there will exist random differences among the configurations between nearest neighbours (e.g., differences in bond angles and nearest neighbour distances) with the result that the activation energies tend to be 'smeared out' (e.g., Kikuchi, 1958). The band gap of an amorphous material is known to contain a finite density of available states due to the lack of long-range order and Mott & Davies (1979) discuss the existence of a band (or bands) of localized states near the Fermi level. Given the non-crystalline nature of the lattice in such materials, several different kinds of energy distribution have been discussed for these localized states, including uniform, exponential and Gaussian (for examples and further discussions see Lampert & Mark, 1970; Roberts *et al.*, 1980; Kao & Hwang, 1981).

Polymeric compounds are examples of materials in which the traps are thought to be distributed in energy. In these materials thermoluminescence is believed to be associated with the onset of different forms of molecular motion (Partridge, 1972). Such motions are not expected to be uniquely defined by a single activation energy and an energy distribution is commonly assumed.

Evidence for energy distributions comes from a variety of experimental techniques, the most common being *Space Charge Limited Currents* thermally stimulated conductivity and, not surprisingly, thermoluminescence. Many analyses begin by simply assuming a form to the distribution and examining the results obtained in the light of the assumed function. For example, space charge limited currents from semiconducting amorphous selenium have been interpreted in terms of a uniform trap distribution and a Gaussian distribution (Grenet *et al.*, 1973); Owen & Charlesby (1974) use an exponential distribution to explain space charge limited current results from a wide variety of insulating solids; thermally stimulated conductivity has been used to highlight both discrete (Garofano, Corazzari & Casalini, 1977) and distributed traps (Helfrich, Riehl & Thoma, 1964) in anthracene. The proposed Gaussian distribution near the Fermi level in amorphous selenium (Grenet *et al.*, 1973) is thought to overlap with an exponential distribution near the valence band and the suggested distribution functions for this material are illustrated in figure 3.20. Many other examples have been reported, some of which can be found in the review article by Roberts *et al.* (1980) and in the book by Kao & Hwang (1981).

Recombination processes between distributed states have been dealt with in general by Fowler (1956) and Randall & Wilkins (1945*b*) have discussed the analysis of luminescence, including thermoluminescence.

Figure 3.20. (A) Gaussian, and (B) exponential trap distributions near the Fermi level in amorphous selenium. Gaussian function: $N(E) = 7 \times 10^{14} \exp(-[(E - 0.88)/950k]^2)$. Exponential function: $N(E) = 5.23 \times 10^{18} \exp(-E/870k)$. E is the energy depth above the valence band. (After Grenet *et al.*, 1973.)

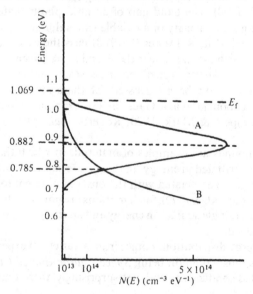

The problem of how to determine the form of the energy distribution from a thermoluminescence glow-curve has been addressed by several authors and generally the methods arrived at are extensions of the techniques discussed thus far for the determination of discrete activation energies. Some examples are the initial rise technique (e.g., Ikeda & Matsuda, 1979), the fractional glow technique (e.g., Rudlof *et al.*, 1978), curve-fitting procedures (e.g., Pender & Fleming, 1977*a,b*) and isothermal analysis (e.g., Visocekas *et al.*, 1976). Analysis of isothermal decay curves has also been advocated as a means of determining the kinetic order in the presence of a distribution of trap energies (Hagekyriakou & Fleming, 1982*b*).

Examples of thermoluminescence peaks arising from a distribution of trap depths are shown in figure 3.21 for amorphous inorganic (*a*) and organic (*b*) materials. For a generalized distribution of trapped charges $n(E)$ the simplest way to regard the glow-curves in figure 3.21 is as the addition of luminescence intensities due to many independent groups of traps, each with its own value of E. Thus, the first-order equation (1.6),

Figure 3.21. Thermoluminescence from (*a*) $ZnIn_2S_4$ (after Bossachi *et al.*, 1973) and (*b*) polystyrene (after Pender & Fleming, 1977*b*). The $ZnIn_2S_4$ sample has been irradiated at different temperatures ranging from 78 K (1) to 190 K (7), whilst the polystyrene curves have undergone several thermal cleaning cycles (experimental data, solid circles; fitted glow-curve, solid curve; and quenching cycles, broken curves).

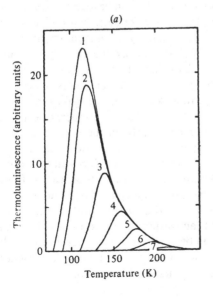
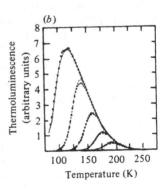

for example, becomes

$$I = \int_{E_A}^{E_B} n(E)s\exp(-E/kT)\exp\left[-(s/\beta)\int_{T_0}^{T}\exp(-E/kT)\mathrm{d}T\right]\mathrm{d}E.$$

For a uniform distribution between E_A and E_B,

$$n(E) = n_t[(E - E_A)/(E_B - E_A)] \tag{3.40}$$

where n_t is the constant concentration of trapped charge between E_A and E_B. For an exponential distribution, from $E_A = 0$ to E_B

$$n(E) = n_e \exp[(E - E_B)/kT_c] \tag{3.41}$$

where n_e is a constant and T_c is a characteristic temperature for the distribution. For a Gaussian distribution,

$$n(E) = n_m \exp(-[(E - E_m)/d]^2) \tag{3.42}$$

where n_m is the maximum concentration at energy E_m and d is a constant. Typical examples of equations (3.41) and (3.42) are given in figure 3.20.

One effect of a distribution $n(E)$ of trapped charge is that an isothermal plot of either $\ln(I)$ versus t, for first-order kinetics, or I versus t^{-2}, for second-order kinetics, will not be linear. However, in the previous section several other causes of non-linear isothermal plots were mentioned and possibly the best evidence for a distribution comes from a $T_m - T_{STOP}$ curve. As discussed in section 3.2.1 a continuous distribution of traps will produce a straight line in a $T_m - T_{STOP}$ plot of slope $\simeq 1$ (cf. figure 3.7). The peak maximum is seen to shift continuously to high temperatures as T_{STOP} is increased. Examples can be found in Avouris & Morgan (1981) for thermoluminescence in $ZnSiO_4:Mn$ and in figure 3.21b in which the continuous shift of the glow maximum to higher temperature is clearly seen in polystyrene as the thermal cleaning temperature is increased (Pender & Fleming, 1977b). Two variations on this theme are: (i) to irradiate the specimen at different temperatures (the temperatures being close to the glow maximum). Once again a shift in T_m to higher values is observed as the irradiation temperature (T_I) is increased. An example is shown in figure 3.21a for thermoluminescence in $ZnIn_2S_4$ (Bosacchi et al., 1973). (ii) An isothermal decay experiment for a constant time but at different temperatures induces an increase in T_m with increasing storage temperature (T_S). An example can be found in Visocekas et al. (1976) for thermoluminescence from calcite. In all cases, a straight line T_m versus T_{STOP}, T_I or T_S plot will result.

In principle, the spread of activation energies involved in the production of the glow-curve can be determined by increasing T_{STOP} by small amounts and applying the initial rise technique to each segment of the glow-curve. The obvious development of this is to apply Gobrecht &

Hofmann's fractional glow technique and obtain the actual energy distribution $n(E)$. Rudlof *et al.* (1978) have discussed this fully and have simulated the glow-curves and the analysed distribution on a computer for both a quasi-continuous, uniform (equation (3.40)) and a Gaussian distribution (equation (3.42)). Their results are shown in figure 3.22 in which it can be seen that the fractional glow method over-

Figure 3.22. (*a*) Quasi-continuous linear distribution $n(E)$ (1), resulting glow-curve I(t) and $E(T)$ dependence (2), plus determined $n(E)$ from fractional glow analysis (3). (*b*) Same as above, but with a Gaussian instead of a linear distribution. (After Rudlof *et al.*, 1979.)

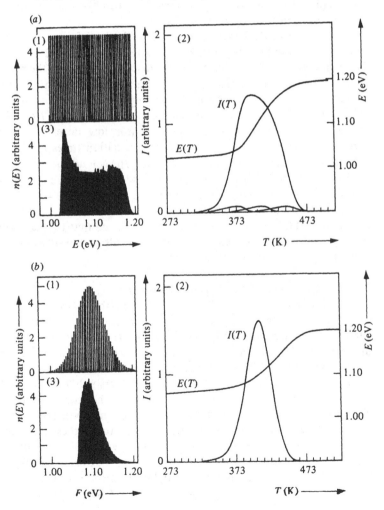

estimates $n(E)$ on the low E side of the uniform distribution and skews the Gaussian distribution to higher E. A similar exercise with E constant but with n distributed uniformly with s (i.e., $n(s)$) indicates that the fractional glow method correctly analyses this case.

Pender & Fleming (1977a) proposed a numerical procedure based on the curve fitting of glow-curves, isothermal decay curves and thermal quenching cycles. Curve fitting of the glow-curve alone does not produce a unique set of s and $n(E)$ parameters; several $s - n(E)$ combinations can be made to give a good fit, but when combined with isothermal decay and thermal quenching curves Pender & Fleming claim good results when applied to first-order glow-curves. They can also account for the possible temperature dependence of s.

The problem of deciding on the order of kinetics for a distributed process was examined by Hagekyriakou & Fleming (1982b). It was shown in the previous section that for a discrete value of E, the rate of decay of thermoluminescence in an isothermal experiment is independent of the initial level of trapped charge n_0 when the kinetics are first-order. Thus, for a given decay time and for first-order kinetics, $I(t)$ will be linearly dependent on absorbed dose D (for low doses) or, more particularly, on dose-rate r, for constant irradiation times t_r ($D = rt_r$). For a kinetic order other than one, however, $I(t)$ will not be any simple function of r, and furthermore this dependence may vary with T_t. Hagekyriakou & Fleming (1982b) extended this principle to a distribution of trap energies and dealt in particular with an exponential distribution. They applied the principle to indicate that the thermoluminescence from polystyrene is first-order, whilst the results for ZnS strongly suggested second-order kinetics.

3.3.7   Calculation of the frequency factor, s

Some of the methods of analysis described so far allow the calculation of the frequency factor s (or s') and some do not. The whole curve analysis (section 3.3.1) enables s' to be determined from the intercept of a $\ln(I/n^b)$ versus $1/T$ plot. Similarly, the heating rate and isothermal decay methods allow an estimate of s from the intercepts of the plots of $\ln(T_m^2/\beta)$ against $1/T_m$ and $\ln(m)$ versus $1/T$, respectively. Curve-fitting techniques involve s (or s') as one of the fitting parameters. The initial rise, fractional glow and peak shape methods, however, do not directly calculate values for the frequency factor. In these cases, the s values are normally calculated by substituting values of E (calculated), T_m (measured) and β (known) into equation (3.29). For non-first-order kinetics, equation (3.30) should be used, but the error introduced in s by using only equation (3.29) is negligibly small (Chen & Winer, 1970).

Furthermore, the temperature dependence of s can easily be accounted for on most occasions.

It is clear from equation (3.29) that any errors in E will be magnified greatly when s is calculated because of the exponential function, $\exp(-E/kT_m)$. This difficulty has led some authors to search for methods for calculating s which are independent of the *calculated* value of E. Moharil (1981) advocates the isothermal decay method for doing this, but experimenters must not be misled into thinking that the wide range of published s values is owed solely to the method chosen for the calculation. The experimental data is related to both E and s by an equation of the form given in equation (3.15) and no matter how the data is manipulated this will always be true (Lilley, 1982; Moharil, 1982). The largest error in s comes not from the chosen method of calculation, but in the accumulation and interpretation (or misinterpretation) of the data. When accurate data are correctly interpreted, consistent values for E and s emerge. The case of LiF is a classic example. Earlier measurements gave wide ranging values for s, but as techniques and knowledge improved, later measurements gave consistent results (Fairchild *et al.*, 1978*b*; Taylor & Lilley, 1978; McKeever, 1980*a,b*).

Normally s values are expected to be of the order of the Debye frequency, namely $10^{12}-10^{14}\,\mathrm{s}^{-1}$ (see section 2.3.1). Occasionally, abnormal s values are obtained which are either very much larger, or very much smaller, than the expected ones. These abnormalities, when they cannot be explained away by experimental inaccuracies, usually call for a reappraisal of the simple concept of thermal stimulation from a trap and several explanations have been forthcoming, depending on the material studied. The most widely known example of an abnormal s value is that for so-called peak 5 in LiF:Mg (figures 1.4 and 3.12) for which reliable estimates give a value between $10^{20}-10^{22}\,\mathrm{s}^{-1}$. So far, two suggestions have been put forward to explain the high s values and although different in detail, in many respects they are quite similar. Fairchild *et al.* (1978*b*) allow not just for the thermal release of, say, trapped electrons (with parameters s_e and E_e) but also for the simultaneous release of trapped holes (with parameters s_h and E_h). In this respect, this model is very similar to the Schön–Klasens model discussed earlier. In these circumstances, Fairchild *et al.* develop a formula for first-order kinetics of the form

$$I = n_{oe}n_{oh}s_es_h\exp(-(E_e+E_h)/kT)$$

$$\times \exp\left[-(1/\beta)\int_0^T (s_e\exp(-E_e/kT)+s_h\exp(-E_h/kT))\mathrm{d}T\right]. \quad (3.43)$$

Fairchild *et al.* (1978*b*) show that this expression describes a glow peak which is of exactly the same form as a 'normal' first-order peak described by equation (1.6) and when analysed as a normal first-order curve the calculated E is approximately $E_e + E_h$, whilst the calculated s is of the order of $s_e s_h$.

Townsend, Taylor & Wintersgill (1978) suggest that the electron trap is in fact a $Mg^{2+}-V_3$-centre complex. Radiative recombination takes place at free V_3-centres, but a non-radiative transition can take place at the $Mg^{2+}-V_3$ complexes by trapping a second electron. As the concentration of non-radiative centres simply equals the concentration of trapped electrons, the thermoluminescence intensity has to be multiplied by an efficiency term containing the expression $s \exp(-E/kT)$. Thus, the equation for thermoluminescene now contains an energy term $2E$ and a pre-exponential factor of s^2. These are the parameters which are actually calculated and thus the high value of the frequency factor is explained.

A third possibility is that peak 5 is caused by the presence of trimer complexes within the LiF lattice (see chapter 5) which are unstable at these temperatures (> 200 °C). Thus, as the temperature is raised, the trimers dissociate into impurity–vacancy pairs plus trapped charge. The trapped charge is not thermally stable at the pairs at temperatures above ~ 160 °C and so it is released to the conduction band. Thus, the activation energy being determined is the sum of the binding energy for a trimer (0.95 eV) and the energy for release of charge from impurity–vacancy pairs (~ 1.1–1.2eV), i.e., the calculated value of E is 2.05–2.15eV, and the corresponding calculated s value is $10^{20}-10^{22}s^{-1}$.

Besides anomalously high values, abnormally low values for s have also been reported. In a detailed investigation of thermoluminescence from alkali halide specimens, Alvarez Rivas and colleagues (Ausin & Alvarez Rivas, 1972a, 1974; Mariani & Alvarez Rivas, 1978; Jiménez de Castro & Alvarez Rivas, 1979) suggest that many of the glow peaks from these compounds result from the recombination of a halogen interstitial atom (i.e., a halide ion which has trapped a hole) and a vacancy which has trapped an electron (e.g., F-centres, M-centres, Z-centres, etc.). During the recombination of the interstitial and vacancy, electron–hole recombination takes place and luminescence emission occurs. It is clear that during this mechanism neither the electron nor the hole are raised to their respective delocalized bands. Furthermore, the first-order kinetics found for the thermoluminescence emissions suggest a localized, or correlated recombination between the vacancy and interstitial centres. In this way, this mechanism is similar to the centre-to-centre recombination models discussed earlier. For this model, Chen & Kirsh (1981) developed the equation

$$I(t) = (ns\gamma/(\gamma + s))\exp(-E/kT) \tag{3.23}$$

where γ is the recombination probability. The first-order nature of the glow peaks is consistent with this equation. The important point is that when $s \gg \gamma$, equation (3.24) becomes

$$I(t) = n\gamma\exp(-E/kT) \tag{3.44}$$

and thus analysis of the glow peak yields γ, not s. This, therefore, may be the explanation for the unusually low frequency factors calculated for these peaks (see Alvarez Rivas, 1980). For example, Mariani & Alvarez Rivas (1978) quote s values in KI from $4 \times 10^8\,\text{s}^{-1}$ down to $2\,\text{s}^{-1}$.

It should be emphasized that if a thermoluminescences peak is caused by a mechanism which is not described by an expression of the form given in equation (3.15) then an analysis based on this equation will yield meaningless values for both E and s. It may be difficult to decide whether or not an equation of this form is applicable, especially if the calculated value for E is 'reasonable'. In one or two instances, however, the inappropriate application of this type of equation has been highlighted by impossibly high s values. A class of materials in which this is especially true is polymers. Alfimov & Nikol'skii (1963) analysed thermoluminescence peaks in polybutadiene and, using equation (3.29), determined values for s of 10^{42}–$10^{45}\,\text{s}^{-1}$. Similarly, the same school (Nikol'skii & Buben, 1960; Tochin & Nikol'skii, 1969) calculated s for some glow peaks from polyethylene to be $10^{27}\,\text{s}^{-1}$ to $10^{61}\,\text{s}^{-1}$. Clearly, these values do not correspond to any simple physical process and call into question the validity of using equation (3.15) and its daughters for the analysis. In fact, the thermoluminescence peaks examined by Nikol'skii and colleagues correspond to a structural transition known as the glass–rubber transition. For all thermally activated processes below the glass transition temperature, T_g, the usual Arrhenius equation (1.1) can be used to describe the time constant for the process (Hedvig, 1972). At T_g, however, the activation energy of the process is temperature dependent, according to

$$E(T) = kAB/(1 - (T_g - B)/T) \tag{3.45}$$

where A and B are constants. This leads to the Williams–Landel–Ferry (WLF) equation for the reaction constant, namely,

$$\tau(T) = \tau_0\exp\{[-A(T_g - T)]/[T - T_g + B]\}. \tag{3.46}$$

Although the value of T_g varies enormously from system to system, A and B are found to be $17\,°C$ and $51\,°C$ respectively for many glass transitions (Williams, Landel & Ferry, 1955; Hedvig, 1972). The experiments of Alfimov & Nikol'skii (1963) utilize the variation of T_m with the heating rate to calculate the activation energy. Partridge (1972) re-analyses their data using equation (3.46) instead of equation (1.1) and arrives at values of B/A of 1.7 to 2.2, compared with the expected number of ~ 3. The value of τ_0 cannot be obtained without a knowledge of T_g, but assuming $T_g \simeq T_m$ gives moderate values for τ_0^{-1} compared with the enormous quantities quoted earlier by Nikol'skii and colleagues.

Despite the fact that below T_g the thermoluminescence peaks in polymeric materials can be analysed using equation (1.1) (and therefore equations like (3.15)), analysts must still be careful of the interpretation of s. Many of the thermoluminescence peaks are due to the 'shaking off' of electrons from molecular chains. In this case, s is interpreted as the frequency of this motion (α_0) multiplied by the probability that each movement will result in charge release P. Thus, $s = \alpha_0 P$ (Partridge, 1965). P is expected to be temperature dependent (e.g., Pender & Fleming,

Table 3.3. *Methods of peak separation*

Method	Reference	Comments
Peak cleaning	Nicholas & Woods (1964)	Quick and simple. Not easy to ensure that satellite peaks completely removed. With non-first-order kinetics, will change position of peaks
High-temperature cleaning	Taylor & Lilley (1978)	First-order kinetics only. Needs clean descending portion of last peak in glow-curve. Relatively tedious
$E - T_{STOP}$	Nahum & Halperin (1963)	Use initial rise method to calculate E. Needs well-separated E values. Needs increasing E with T. Independent of kinetic order
$T_m - T_{STOP}$	McKeever (1980a,c)	Independent of kinetic order. Good resolution ($\sim 5\,°C$). Accurate T_m not obtained. Many experiments needed. Can be combined with $E - T_{STOP}$ Variations: $T_m - T_i$ (irradiation temperature): $T_m - T_s$ (storage temperature). Can detect trap distribution

1977*a,b*). The molecular chain frequency at temperatures less than T_g is expected to be only $\sim 10^{-1}\,\text{s}^{-1}$ (Partridge, 1972), thus very low *s* values can be expected for these glow peaks. Other thermoluminescence peaks associated with impurities within the polymers can be expected to have 'normal' pre-exponential factors of values near the Debye frequency.

3.3.8 *Summary*

Considering the wide range of analytical methods that are available for calculating the trap depths and frequency factors from thermoluminescence glow-curves, it is useful to summarize the salient features of the main models dealt with in the foregoing sections. The various methods are summarized in the following tables. Table 3.3 deals with techniques for peak resolution. Separation of glow peaks is a necessity in most glow-curves and in complex cases may not be straightforward. If in doubt more than one technique should be employed. Table 3.4 summarizes the main methods for calculating the trap depths and lists their limitations and advantages. As with the peak separation methods, it is advisable to perform several methods of analysis. Clearly, *all* of the methods discussed in these tables require reliable temperature control. In order to hold in check any overconfidence on this on the part of the experimenter, the reader is referred once again to figure 3.16! Also, ideally the analyses should be carried out on dn/dT versus *T*, not just detector output versus *T*.

Finally, it ought to be remembered that arriving at values for the three parameters, *E*, *s* and *b* (or α), which accurately describe the glow-curve behaviour is only a first step. Interpreting the values is quite another problem.

3.4 **Trap filling**

The study of the build-up of thermoluminescence intensity as the imparted dose to an insulator or semiconductor is increased (the so-called 'growth curve') has become a useful tool in investigating the accumulation of trapped charge in phosphors. The same trapping and recombination parameters (capture cross-section, density of available sites, etc.) that control the trap emptying process also control the trap filling process. Thus, a study of the kinetics of trap filling can provide data relating to the charge storage properties of these materials. Unfortunately, the trap filling regime has not benefited from such extensive studies as have been given to the trap emptying regime, but nevertheless much useful information has emerged. As with the study of trap emptying, analysis of the trap filling process begins by examining the rate equations which govern the flow of charge between bands and

Table 3.4. *Methods for calculating the trap depth, E*

Method	Equation (+ graph)	Reference	Comments
Initial rise	$I = \text{constant} \times \exp(-E/kT)$. Plot: $\ln(I)$ versus $1/T$	Garlick & Gibson (1948)	Quick and simple. Independent of kinetic order. Affected by satellite peaks and by thermal quenching. Care must be taken in choice of 'initial' part of curve. Recommended for first estimate. Variation: fractional glow
Fractional glow	As above	Gobrecht & Hofmann (1966)	Excellent, but needs large experimental effort. Obtains full energy level distribution
Whole glow peak	$I = (s'/\beta)n^b \exp(-E/kT)$. Plot: $\ln(I/n^b)$ versus $1/T$	Halperin et al. (1960)	Requires complete peak resolution. Independent of kinetic order. s' calculated from intercept; b determined from value which gives a straight line
Peak shape	$E = c_\gamma(kT_m^2/\gamma) - b_\gamma(2kT_m)$. ($\gamma = \delta, \tau$ or ω)	Chen (1969a,b)	Very quick and simple. Reliance on only two or three data points, therefore highly dependent on obtaining clean peaks. Affected by quenching. c_γ and b_γ dependent on kinetic order
Peak position–heating rate methods	$E/kT_m^2 = s \exp(-E/kT_m)$	Hoogenstraaten (1958)	First-order kinetics ⎫ Only data at peak needed.
	$E/kT_m^2 = s[1 + (b-1)V_m] \times \exp(-E/kT_m)$. Plot: $\ln(T_m^2/\beta)$ versus $1/T_m$.	Chen & Winer (1970)	General-order kinetics ⎬ Unaffected by thermal quenching. Prominent

Method	Equation	Reference	Comments
	$I = \text{constant} \times \exp(-E/kT_x)$. Plot: $\ln(I)$ versus $1/T_x$	Sweet & Urquhart (1980, 1981)	s calculated from intercept, or directly from equation. If E calculated by any other means, this equation can be used to determine s. Recommended. High resolving power. Relatively simple. Independent of kinetic order
Curve fitting	Full equations for TL† (first-, second- or general order)	e.g, Mohan & Chen (1970) (and others)	Computationally tedious, but recommended for more reliable calculation. $T_m - T_{STOP}$ procedure recommended as a preliminary step. s and b determined along with E
Isothermal decay	$I = I_0 \exp(-s\exp[E/kT]t)$. Plot: (i) I/I_0 versus t; (ii) slope of (i) versus $1/T$	Randall & Wilkins (1945a,b)	Many difficulties. Difficult to resolve several processes. Defect diffusion can cause problems. Can be modified for non-first-order kinetics. Can detect trap distribution or tunnelling. s calculated from intercept; b adjusted to give straight line plot
Williams–Landel–Ferry	$E = kAB/(1-(T_g - B)/T)$	Williams, Landel & Ferry (1955)	For structural transformations in polymers

† TL = thermoluminescence.

centres during irradiation. To do this it is first of all necessary to establish which energy level model is being examined. We begin with the rudimentary model of one trap and one recombination centre.

3.4.1   *The simple model*

The differential equations which describe the traffic of charge between energy levels during trap filling have already been given in equation (2.16) for the simple one trap–one centre model (figure 2.11). For charge neutrality, equation (2.17) is also considered. I reiterate here that direct band-to-band recombination is ignored and that the thermal release of trapped carriers during irradiation is considered negligible – i.e., $E \ll kT$ at (constant) $T = T_1$. These equations have been examined in a series of papers by Maxia and colleagues who attempted to solve the equations for n (Aramu *et al.*, 1975*a*; Aramu, Maxia & Spano, 1975*b*; Maxia, 1977, 1979, 1980). However, as discussed by these authors, the differential equations are analytically intractable as they stand and certain simplifying assumptions have to be introduced into the analysis. Thus, Maxia (1979, 1980) introduces the same non-equilibrium thermodynamic concept as was introduced during the analysis of the trap emptying process, namely, that of a minimum rate of entropy production (see section 3.2.1). Using this principle, Maxia determines that

$$\mathrm{d}n_\mathrm{c}/\mathrm{d}t = \mathrm{d}n_\mathrm{v}/\mathrm{d}t = f/2, \tag{3.47a}$$

and

$$\mathrm{d}n/\mathrm{d}t = \mathrm{d}n_\mathrm{h}/\mathrm{d}t, \tag{3.47b}$$

where f is the electron–hole pair generation rate (per unit volume per second). Using these expressions in equations (2.16) yields

$$n_\mathrm{c} = f/2(A(N - n) + A_\mathrm{r}n). \tag{3.48}$$

Substituting for n_c in equation (2.16*b*) and integrating with respect to time gives finally,

$$(1 - A_\mathrm{r}/A)n - (A_\mathrm{r}N/A)\ln(1 - n/N) = ft/2. \tag{3.49}$$

It is to be noted that for the special case of $A = A_\mathrm{r}$, equation (3.49) reduces to

$$n = N[1 - \exp(-ft/2N)]. \tag{3.50}$$

Equation (3.49) (or equation (3.50)) represents the simplified expression for the build-up of trapped charge n with total generation, $R = ft$. (N.B. The total imparted dose D is related to R by $D = \phi R$, where ϕ is the efficiency of electron–hole generation ($\mathrm{cm}^{-3}\,\mathrm{rad}^{-1}$). Similarly, dose rate $r = \phi f$.)

The equalities given in equations (3.47) were justified by Maxia on the basis of non-equilibrium thermodynamics. Aramu *et al.* (1975a) used quite a different set of approximations and in fact preferred to assume quasi-equilibrium throughout the excitation period, i.e.,

$$dn_c/dt \ll dn/dt, \text{ and } dn_v/dt \ll dn_h/dt. \tag{3.51}$$

Using these inequalities, Aramu *et al.* derived

$$(1 - A_r/A)n - (A_r N/A)\ln(1 - n/N) = ft \tag{3.52}$$

which differs from equation (3.49) by the factor of 1/2 on the right-hand side. Similarly, when $A_r = A$,

$$n = N[1 - \exp(-ft/N)]. \tag{3.53}$$

It is important to assess the limitations of the approximations introduced by Maxia and by Aramu *et al.* Clearly equation (3.47a) cannot be maintained throughout the irradiation period because it allows for the continuous accumulation of charge in the delocalized bands during excitation. The question must be asked, what happens to this accumulated delocalized charge once the irradiation ceases? Furthermore, will dn_c/dt necessarily equal dn_v/dt throughout the irradiation? At first sight, therefore, the quasi-equilibrium assumption may appear to offer a better

Figure 3.23. Growth of $n(t + t^*)$ and of $n(t)$ with total generation R for the one trap–one centre case. The importance of the additional time t^*, at the end of the irradiation, for relaxation of free charge carriers is indicated by the fact that $n(t + t^*) > n(t)$. In addition, $n(t)$ is seen to exhibit supralinear growth, while $n(t + t^*)$ does not. The parameters chosen in this figure are: $N = 10^{15} \text{cm}^{-3}$; $N_h = 3 \times 10^{14} \text{cm}^{-3}$; $A = 10^{-17} \text{cm}^3 \text{s}^{-1}$; $A_r = 10^{-13} \text{cm}^3 \text{s}^{-1}$; $A_h = 10^{-15} \text{cm}^3 \text{s}^{-1}$; $f = 10^{14} \text{cm}^{-3} \text{s}^{-1}$. (After Chen *et al.*, 1981b.)

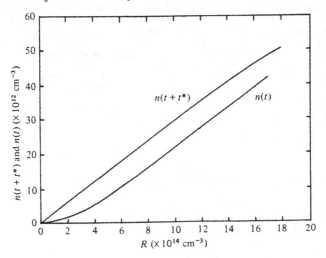

alternative because this approximation does not involve charge accumulation in the bands. The best way to test the applicability of the various simplifications is to produce an exact numerical solution of equations (2.16a–d) similar to that produced by Kemmey *et al.* (1967) and by Kelly *et al.* (1971), and by others for the trap filling case. This analysis has been performed by McKeever *et al.* (1980) and by Chen *et al.* (1981b) who solve equations (2.16a–d) numerically using a Runge–Kutta sixth-order predictor–corrector method. An important concept introduced into the analysis used by these authors is the relaxation of any free charge into the traps and recombination centres in the period following irradiation. This relaxation adds charge to the traps but also allows for recombination of free electrons and trapped holes. Any electrons 'lost' in this way cannot then take part in the subsequent thermoluminescence. To simulate the relaxation period, Chen *et al.* continue the calculation for a time t^* beyond the end of the irradiation. For this part of the calculation, f is set to zero and $n(t + t^*)$ is calculated. Clearly, at $t^* = 0$, $n(t + t^*) = n(t)$. A value of t^* of 50 s was used. The calculated 'growth-curves' of $n(t)$ and $n(t + t^*)$ against $R = ft$ are shown in figure 3.23. For the chosen parameter values, $n(t + t^*)$ is seen to be greater than $n(t)$ which in fact means that free charge has accumulated in the conduction band during irradiation. This in turn means that the inequalities (3.51) are not good assumptions in this case. However, equations (3.47a, b) are also inappropriate because they do not allow for the relaxation of the accumulated charge at the end of the irradiation. Only in the circumstances when

Figure 3.24. The dependence of $n(t + t^*)$ on f for different R. The range of f over which $n(t + t^*)$ is non-constant increases as R increases. As R increases the behaviour changes from one of decreasing $n(t + t^*)$ to one of increasing $n(t + t^*)$ with increasing f. $N_h = 3 \times 10^{14}$ cm^{-3}; $N = 10^{15}$ cm^{-3}; $A = 10^{-17}$ cm^3 s^{-1}; $A_r = 10^{-13}$ cm^3 s^{-1} and $A_h = 10^{-15}$ cm^3 s^{-1}. (After Chen *et al.*, 1981b.)

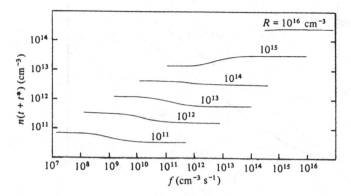

$n(t + t^*) = n(t)$, which could be obtained for some parameters, can the condition $dn_c/dt \ll dn/dt$ be said to be met.

The numerical analysis also enables the variation of $n(t)$ and $n(t + t^*)$ to be followed as functions of f, for constant R. A typical set of results is shown in figure 3.24 where a significant 'dose rate' dependence can be observed. This is more clearly seen in figure 3.25 where growth curves obtained for three different dose rates are shown. Chen *et al.* show that the dose rate dependence is very sensitive to values of N_h which is consistent with their suggestion that this dependence arises from competition for free electrons between traps and recombination centres during irradiation. This highlights another inadequacy of the simplified expressions, equations (3.49) and (3.52), in that these relationships between n and R do not predict a dose rate dependence. Over certain ranges of the parameter values, the simplified expressions do represent good approximations for the variation of $n(t + t^*)$ with R; for example, at low f Chen *et al.* show that equation (3.52) accurately predicts the behaviour. However, in general the approximations are restrictive in that they are valid for only a part of the range of feasible parameters. Fitting experimentally produced growth curves to equations (3.49) and (3.50), or (3.52) and (3.53) is, therefore, of limited value (e.g., Aramu *et al.*, 1975a).

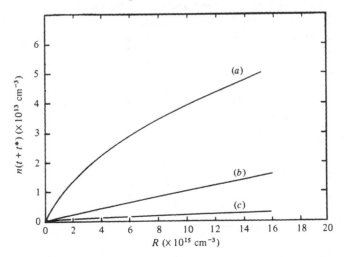

Figure 3.25. Growth of $n(t + t^*)$ for three values of f, namely (a) 10^{13} cm^{-3} s^{-1}; (b) 10^{15} cm^{-3} s^{-1}; (c) 10^{17} cm^{-3} s^{-1}. For the set of parameters chosen, the efficiency of trap filling decreases as the generation rate increases. The chosen parameters are; $N = 10^{15}$ cm^{-3}; $N_h = 10^{17}$ cm^{-3}; $A = 10^{-17}$ cm^3 s^{-1}; $A_r = 10^{-13}$ cm^3 s^{-1}; $A_h = 10^{-15}$ cm^3 s^{-1}. (After Chen *et al.*, 1981b).

3.4.2 *Additions to the simple model*

The model discussed so far is of the simplest kind, but it none the less serves to demonstrate that the exact numerical solution of the governing rate equations is more revealing than the different approximate analyses. It now becomes useful to extend the model in the hope of gleaning further insights into the behaviour of charge storage levels during irradiation. In an attempt to explain the supralinear (or superlinear) growth of thermoluminescence with imparted dose in some luminescent materials, Suntharalingam & Cameron (1969) discussed the possibility of an extra trapping level in the energy band model. These authors thus considered the band diagram of figure 2.17, although the stipulation that the extra traps are thermally disconnected is not necessary in this discussion.

The model of Suntharalingam & Cameron (known as the 'competing trap' model) requires that both traps A (concentration N) and B (concentration M) begin to be filled during the irradiation. At a certain level of absorbed dose, D_1, it is postulated that the level of charge trapped in B saturates (i.e., this trap becomes full) and thus more electrons are available for trapping in A. Thus, if the thermoluminescence intensity due to charge in A is monitored against dose absorbed, it will be seen to increase linearly up to D_1, whereupon it will begin to increase faster at a new linear rate. The transition from the slow to the fast rate would produce supralinear growth. A schematic example of a supralinear growth curve is shown in figure 3.26. Saturation cor-

Figure 3.26. Schematic representation of a thermoluminescence growth curve showing the linear, supralinear and sublinear regions. *N.B.* Absorbed dose $D = \phi R$, where ϕ is the efficiency of electron–hole generation $(cm^{-3}rad^{-1})$.

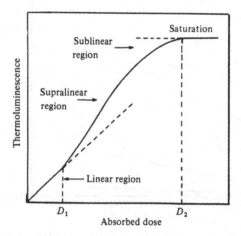

responds to all traps full, i.e., $n = N$, at dose D_2. Actual examples for thermoluminescence in several materials are shown in figure 3.27.

Bowman & Chen (1979) have attempted an analytical solution of the rate equations which govern the charge flow in this case by using the assumption of quasi-equilibrium (i.e., inequalities (3.51)). The equation derived by Bowman & Chen to describe the build-up of charge in the shallower trap is

$$M[(A_r - A_m)/A_m][(1 - n/N)^{A_m/A} - 1] = [(A - A_r)/A]n$$
$$- (NA_r/A)[M/N + 1] \ln[1 - n/N] = ft \qquad (3.54)$$

which is equivalent to equation (3.52) in the one trap–one centre case and can be made the same as this equation by putting $M = 0$. A_m is the trapping coefficient of the extra trap for free electrons. Once again, it can be seen that this equation does not account for a dose rate dependence, nor does it include the additional charge accumulated during relaxation time, t^*.

Using the above equation Bowman & Chen arrived at a set of inequalities which they contend must be obeyed in order to obtain a supralinear growth curve with the competing trap model. These are:

$$A_m > A_r, \qquad (3.55a)$$
$$MA_m^2/NA > A_r, \qquad (3.55b)$$

and

$$A_m > A. \qquad (3.55c)$$

Chen *et al.* (1981*b*) tested these inequalities, and the quasi-equilibrium assumption, by solving the necessary differential equations numerically. Qualitatively, some agreement was found in that if one or more of the

Figure 3.27. Examples of supralinear growth in (*a*) LiF (after Zimmerman & Cameron, 1968); (*b*) CaSO$_4$:Dy (Lakshmanan *et al.*, 1981*a*). The dashed line gives the line of linearity in each case.

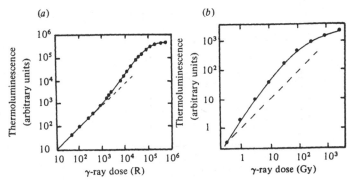

above inequalities were invalid, then the growth of $n(t + t^*)$ with R did not exhibit any supralinear regions. If all three inequalities are appropriate, supralinearity always results. However, the agreement between the numerical and approximative solutions is not quantitative, as can be seen in figure 3.28 in which a selection of growth curves according to equation (3.54) are compared with those arrived at from the numerical analysis. In each of these cases, one or more of the inequalities (3.55) is invalid and so $n(t + t^*)$ versus R does not exhibit supralinearity. In all cases though the quantitative agreement between $n(t + t^*)$ and the solution of equation (3.54) is poor.

Figure 3.28. Examples of growth curves of (1) $n(t + t^*)$ and (2) $n(t)$ with R for which one or more of the inequalities (3.55a–c) are not obeyed. $n(t + t^*)$ does not exhibit supralinearity, but $n(t)$ does in all cases. Furthermore (1) does not agree quantitatively with either (3) or (2). The parameters used are: $N = 10^{15}\,\text{cm}^{-3}$: $M = 10^{15}\,\text{cm}^{-3}$; $N_h = 10^{15}\,\text{cm}^{-3}$; $A_h = 10^{-13}\,\text{cm}^3\,\text{s}^{-1}$. In (a) $A_m = 10^{-16}\,\text{cm}^3\,\text{s}^{-1}$; $A_r = 10^{-16}\,\text{cm}^3\,\text{s}^{-1}$; $A = 10^{-17}\,\text{cm}^3\,\text{s}^{-1}$. In (b) $A_m = 10^{-16}\,\text{cm}^3\,\text{s}^{-1+}$ $A_r = 10^{-16}\,\text{cm}^3\,\text{s}^{-1}$; $A = 10^{-15}\,\text{cm}^3\,\text{s}^{-1}$. In (c) $A_m = 10^{-17}\,\text{cm}^3\,\text{s}^{-1}$; $A_r = 10^{-15}\,\text{cm}^3\,\text{s}^{-1}$; $A = 10^{-16}\,\text{cm}^3\,\text{s}^{-1}$. (3) shows the solution to equation (3.54) in each case. (After Chen et al., 1981b.)

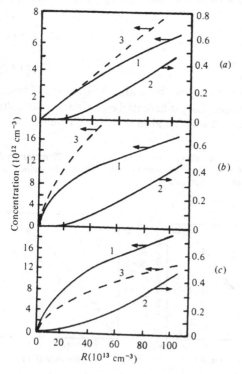

The dose rate dependence predicted from the numerical analysis of the simple model broadly agrees with the experimental findings of Groom *et al.* (1978) who observed dose-rate-dependent effects in the thermoluminescence of natural quartz crystals. More detailed work on quartz by Valladas & Ferreira (1980) has shown that while one wavelength component of the thermoluminescence emission decreases with increasing dose rate, another increases with increasing dose rate. This type of behaviour cannot be explained by the models discussed so far. As a result, Chen *et al.* (1981*b*) also looked at the case of one trap and two recombination centres. The numerical solution of the equations for this case revealed that the concentration of holes trapped at one centre could decrease with increasing dose rate, whilst the concentration of holes at the other centre would increase over the same variation of dose rate. Thus, radically different behaviour can be obtained for different thermoluminescent components.

The clear message that emerges from the exact treatment of the trap filling process is that those analyses which assume a growth curve of the form in, say, equations (3.50) or (3.53) (a popularly assumed growth curve shape) without first establishing the applicability of the model from which these equations are derived, are probably subject to considerable error. Figure 3.27, for example, clearly demonstrates that a wide variety of growth curve shapes can be obtained and to assume a simplified description of the curve can be misleading (e.g., none of the simplified analyses predict a dose rate dependence). So far, numerical analysis of the growth of thermoluminescence with imparted dose has only been reported for the simple energy level models discussed above, but there is no reason to assume that making the model more complex in order to make it more realistic will remove any of the principal findings discussed here.

Possibly the most surprising of these findings is the possibility that the intensity of thermoluminescence for a given absorbed dose can, in some circumstances, be dependent upon the dose rate. Many materials appear to show dose-rate-independent thermoluminescence over a wide range of dose rates; for example, LiF has been examined in detail for this type of behaviour, but none has been found (Karzmark, White & Fowler, 1964; Tochilin & Goldstein, 1966) except when very high doses were used (Goldstein, 1972; Gorbics, Affix & Kerris, 1973). Quartz, however, appears to exhibit important dose rate effects (Groom *et al.*, 1978; Valledas & Ferreira, 1980). It may also be that some earlier thermoluminescence results may be explicable by considering possible dose rate dependencies (for example, the decrease in thermoluminescence yield from semiconducting diamond at lower dose rates as reported by

Halperin & Chen (1966) and the apparent non-additive effects of radiation in LiF when irradiated at different dose rates described by Wintersgill & Townsend (1978a)). Furthermore, Nagpal (1979) reports exposure rate dependence of thermoluminescence from several types of phosphor irradiated with ultra-violet radiation, although the physical mechanism may be different to that dealt with in this chapter.

4 Additional factors governing thermoluminescence

In this chapter several other effects which are often observed in thermoluminescence experiments are discussed. Correct interpretation of these effects is important for a more complete understanding of the thermoluminescence data. To begin with some explanations of the phenomenon of supralinearity are examined as alternatives to the competing trap model detailed in the previous chapter.

4.1 Further discussions of supralinearity

4.1.1 Multi-stage reaction models

The supralinear growth curves described in figures 3.26 and 3.27 are characterized by a linear region, followed by a region of supralinear growth. In some cases, the opposite effect is seen in which the supralinear region precedes the linear part. Such behaviour is generally less common than that shown in these figures, but has been observed in several classes of material. An often-quoted example is the thermoluminescence from quartz extracts from pottery, for example figure 4.1 in which the low-dose supralinear region presents difficulties when using thermoluminescence as a method of age determination (see chapter 7). In general, supralinear growth can manifest itself in a growth curve in which the thermoluminescence intensity, I, increases as a function of R^l, where l is not necessarily one, nor is it necessarily an integer, and R is the absorbed

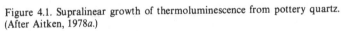

Figure 4.1. Supralinear growth of thermoluminescence from pottery quartz. (After Aitken, 1978a.)

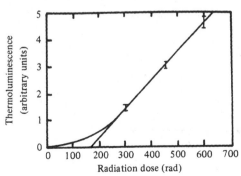

dose. Halperin & Chen (1966) find that $l = 3$ for growth curves from semiconducting diamond and conclude that the growth proceeds via a three-stage transition, as depicted in figure 4.2. A two-stage transition, as argued by Halperin & Chen, would give $I \propto R^2$. The absorption of a radiation photon is said to raise a valence electron to level A_1; for a dose R, this induces a population in A_1 of $n_1 \propto R$. A second photon raises the electrons from A_1 to A_2, where $n_2 \propto R^2$. Finally, a third photon raises the electrons from A_2 to the conduction band from where they become trapped at T; thus $n_T \propto R^3$.

The multi-stage filling model of Halperin & Chen is similar to a class of mechanism known as the multi-hit, or multi-stage reaction models (Larsson & Katz, 1976; Takeuchi *et al.* 1978). These models take their name from the requirement that a trap needs to undergo a two-stage (or more) reaction before it can actively take part in thermoluminescence. For example, Katz & colleagues (Larsson & Katz, 1976; Katz, 1978; Waligórski & Katz, 1980) use track theory to develop an equation relating thermoluminescence intensity to defect concentration for a trap which is only produced after trapping first one, and then a second electron (a two-hit trap). In this way Katz is able to describe the supralinear growth of certain thermoluminescence peaks in LiF. In a very similar theory, Takeuchi and colleagues (Takeuchi *et al.*, 1978; Inabe & Takeuchi, 1980) describe a two-step reaction sequence of the type

$$A + irradiation \rightarrow B \tag{4.1a}$$

$$B + irradiation \rightarrow C \tag{4.1b}$$

Figure 4.2. The multi-stage transition model of Halperin & Chen (1966) to explain supralinear growth of thermoluminescence in semiconducting diamond.

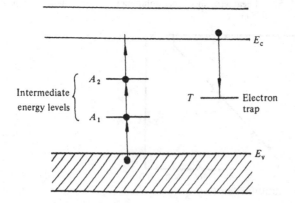

where A, B and C are the defects responsible for the initial, intermediate and final steps, respectively. Centres A and B are allowed to exist before the irradiation, but centre C is a product of the irradiation only. Using this model, Takeuchi established a growth equation which is identical in form to that developed by Katz. In general, this equation is:

$$I = (1 - \delta)(1 - (1 + \gamma)\exp(-\gamma)) + \delta(1 - \exp(-\gamma)) \qquad (4.2)$$

where δ is the normalized concentration of centre A in equation (4.1a) (thus, the concentration of B = $1 - \delta$). Katz gives $\gamma = D/E_0$, where D is the absorbed dose and E_0 is the 'characteristic dose of γ-rays at which there is an average of one hit'; Takeuchi gives $\gamma = \alpha D$, where α is the 'transition probability' for the first reaction (i.e., reaction 4.1a). Clearly E_0^{-1} and α are identical.

Two separate models for thermoluminescence in LiF have been proposed, both of which are consistent with these multi-stage reaction models for supralinearity. Nink & Kos (1976; Kos & Nink, 1979, 1980) in a series of papers have proposed a two-stage Z-centre conversion model in which Z_0-centres (cf. defects A, equation (4.1a)) capture an electron to become Z_3-centres (cf. defect B). The Z_3-centres each then trap an additional electron to become Z_2-centres (cf. defect C). The Z_2-centres are proposed by Nink & Kos to be responsible for thermoluminescence peak 5 in LiF TLD-100 (see figure 1.4) and thus the two-hit model can explain the supralinearity of this peak. It is less easy to explain the supralinearity of the next peak (peak 6), which is said to be due to Z_3-centres, because these defects require only a single reaction to be formed.

Sagastibelza & Alvarez Rivas (1981) explain supralinearity in LiF in terms of their interstitial atom/F-centre recombination model. These authors envisage an impurity centre trapping firstly one interstitial (released by irradiation), and then a second and possibly more, to form interstitial aggregates. The low-temperature thermoluminescence peaks correspond to one-interstitial traps, whereas the higher temperature peaks correspond to additional interstitials. This multi-trapping of interstitials explains the observed decrease in the intensities of the low-temperature peaks as the dose increases, and also explains the supralinear growth of the high-temperature peaks with increasing dose.

A note ought to be made of the fact that Katz's multi-hit, track-interaction theory stems from a consideration of the growth of radiation-induced cell mutation in biological systems. Recently, Burch & Chesters (1979) have questioned the validity of track interaction at low-dose, low linear energy transfer radiation in explaining the observed supralinear growth of cell mutations with dose. They argue that for low linear energy transfer, the track density is such that they are unlikely to interact and thus multi-hit events are improbable. Thus, although Katz's mechanism may explain the changes in supralinearity with increasing linear energy transfer (section 6.1.2) it cannot explain supralinearity at low linear energy transfer. On this basis, Lakshmanan & Bhatt (1981) have criticized the multi-hit model for supralinearity in thermoluminescence. However, the validity of applying arguments which relate the growth of cell mutations to thermoluminescence is questionable. If the trapped species in a phosphor is an interstitial atom (as appears to be the case with alkali halides, including LiF – see chapter 5) then track overlap is not necessary for multi-hit events. The observations of di-interstitial centres in X-irradiated materials is clear support of this (e.g., van Puymbroek & Schoemaker, 1981). Thus, multi-stage trapping may still be a cause of supralinearity in these systems.

4.1.2   *More on competition models*

Multi-stage models of the above type were rejected by Rodine & Land (1971) as explanation for the supralinear growth of thermoluminescence in ThO_2. Instead, these authors preferred to invoke a mechanism of competition, similar to that described in the previous section, but during the trap *emptying* stage rather than during trap filling. This model allows for either retrapping in deeper traps, or for recombination at a luminescence centre (to produce thermoluminescence) and results in an R^2 dependence for the emission intensity. Kristianpoller, Chen & Israeli (1974) treated this model in greater detail by examining the appropriate differential equations and solving them via approximations (i.e., $dn_c/dt \ll dn_h/dt$). This analysis confirmed that competition during heating can yield an R^l dependence with $l = 2$ at low doses. These authors also illustrated that the same growth curve shape will be obtained if n (number of trapped electrons) is taken to be proportional to the glow peak height or to the, more accurate, glow peak area. A very small difference is found for second-order kinetics, but not for first-order.

Stoebe & Watanabe (1975) invoke similar competition mechanisms to explain the supralinear growth of thermoluminescence in LiF doped with Mg (i.e., TLD-100), as do Lakshmanan *et al.* (1981*a*) and Chandra, Lakshmanan & Bhatt (1982) who use competing trap models (both during irradiation and during heating) to explain non-linear effects in a variety of phosphors. In particular, they suggest that the radiation reduces the trapping efficiency of the competing trap by a mechanism of radiation damage and thereby causes either more charge to be trapped at the trapping level of interest (during irradiation) or causes increased recombination at the luminescence centre of interest (during heating).

An observation which is relevant to deciding between competition during trap emptying or during trap filling relates to the growth of optical absorption bands during irradiation. In LiF there have been many observations which relate certain absorption bands to certain thermoluminescence peaks (e.g., Jackson & Harris, 1969, 1970; Bapat & Kathuria, 1981; Sagastibelza & Alvarez Rivas, 1981); however, these absorption peaks are not observed to grow supralinearly (e.g., Crittenden, Townsend & Townshend, 1974a). This has led some authors to the conclusion that the competition can only be taking place during thermoluminescence readout, i.e., during trap emptying (e.g., Horowitz, 1981). However, it must be pointed out that this is not necessarily so. This conclusion assumes that the centres being observed during optical absorption act as the traps for the thermoluminescence process. If they in fact act as the recombination centres (e.g., as required for F-centres in the model for thermoluminescence advocated by Alvarez Rivas and colleagues) then supralinearity can still take place during the trap filling stage without affecting the optical absorption.

When discussing competition during heating the competing centre need not be a trap, but can be a non-radiative recombination centre (a 'killer' centre). As argued by Chen & Kirsh (1981) such centres are necessary to explain the track-interaction model for supralinearity suggested by Attix and colleagues (e.g., Attix, 1974; see also the similar spatial correlation model of Fain & Monnin, 1977). In these models, at low doses the traps and recombination centres are sited far apart, on average, and only local recombination can take place. At high doses, however, the average separation between traps and centres decreases such that more luminescence centres become available which in turn gives rise to an increase in the luminescence efficiency, η, according to equation (2.9). (In this equation P_r is proportional to the number of luminescence centres and P_{nr} is proportional to the number of 'killer' centres.)

4.1.3 Trap creation models

One obvious candidate for the cause of supralinear growth is that of trap creation during irradiation. This model was suggested originally by Cameron and colleagues for supralinearity in LiF (e.g., Cameron *et al.*, 1968) and in this material it requires that the freshly created traps are the same as those originally present, an unlikely event. A later suggestion was that luminescence centres are being created by the irradiation, but again this requires that the new centres are the same as the original ones because no change in the emission spectrum is noted from the LiF phosphor (Zimmerman & Cameron, 1968). However, although trap or luminescence centre creation is an unlikely explanation for the supralinear growth in LiF, it may be applicable to other materials (e.g., quartz; Durrani *et al.*, 1977*a*) and must be generally considered as a possible cause of supralinearity in thermoluminescence phosphors.

It is clear from the above discussions on supralinearity that this phenomenon has several plausible explanations, each of which can produce typical supralinear growth curves. Not all of them can equally predict all of the experimental observations, e.g., the incident energy dependence of the supralinear properties (Horowitz, 1981); however, each of the mechanisms discussed have, on various occasions, been forwarded to explain supralinearity in LiF and with the present level of knowledge it is impossible to decide between them and to determine which mechanism is operative in any one sample. In fact, several mechanisms may be making a contribution to the effect – for example, in LiF, competition during irradiation (and heating) could be working alongside the multi-stage reaction model involving interstitial trapping. It is probable that a definitive statement on the cause of supralinearity will never be made for any material.

4.2 Sensitization

The sensitivity of a thermoluminescence phosphor is measured by the intensity of the luminescence emission per unit of absorbed dose of radiation per unit weight. Thus, if two phosphors of equal weight are irradiated with exactly the same dose then the specimen which gives the more intense thermoluminescence emission on heating is said to be the more sensitive. An important effect which is commonly observed in many thermoluminescent materials is the increase in the sensitivity of the phosphor (more particularly, of a given thermoluminescence peak) following the absorption of radiation. This process, known as sensitization, is important in the field of radiation dosimetry using thermoluminescence (see chapter 6) where the enhancement of sensitivity is a desirable feature. The mechanism of sensitization is closely related to the phenomenon of supralinearity and often the models which have been invoked to explain one can be used to explain certain features of the other. It is the purpose of this section to discuss in broad terms how the sensitization mechanisms work, but it is not worthwhile discussing in great detail any particular mechanism because, firstly, the models depend heavily upon the particular defect proposed to account for the thermoluminescence peak (proposals which themselves are often still being fiercely contended) and, secondly, as with supralinearity, it is almost impossible at this stage to decide between the various models using the available data. The possibility of several mechanisms operating at once should not be overlooked.

4.2.1 Competing trap models

The competing trap model has been extensively discussed as a means of providing increased sensitivity following irradiation. An often-quoted sensitization procedure for LiF is to irradiate the specimen to a dose of approximately 10^3 Gy and to follow this by an anneal of 300 °C for one hour (Cameron *et al.*, 1968). This has the effect of filling the deep traps (i.e., > 300 °C) which may act as competitors to the lower temperature traps (i.e., < 300 °C), most specifically peak 5. Thus, on subsequent irradiation, peak 5 can be filled without the competitive interference of the deeper traps. This procedure increases the sensitivity severalfold. Peak 7 is also said to be sensitized in a similar fashion and here peak 10 is said to be related to the competing trap (Chandra *et al.*, 1981).

Clearly, if the initial radiation causes damage to the competitor, then on subsequent irradiation there will be fewer competitors available, resulting in more charge being trapped at the centre of interest, per unit dose, giving rise to an increase in sensitivity (Lakshmanan *et al.*, 1982*a*). The radiation may cause a decrease in the concentration of 'killer centres', an increase in the concentration of luminescence centres or an increase in the concentration of traps. All of these phenomena could give rise to increased sensitivity and all have been proposed, in one form or another, by different research groups to explain sensitization in a variety of materials.

4.2.2 Centre conversion models

Many different kinds of radiation-induced defect centre conversion mechanisms have been proposed to explain sensitization effects. It is fair to say that all of these mechanisms have been developed with a view to describing the sensitization observed in LiF and, therefore, are rather specialized. It is also fair to say that each relies heavily on the model for thermoluminescence in this material which, as will be seen in the next chapter, has not yet been confirmed and which has been the source of an enormous literature over the years. For example, Nink & Kos (1976) use the two-step Z-centre conversion model to explain the sensitization of peak 5 in LiF. Here Z_3-centres are converted to Z_2-centres (cf. peak 5) by the trapping of electrons thus leading to a higher concentration of Z_2-centres than would normally be present in the unsensitized material. Chandra *et al.* (1980, 1981) suggest that a similar kind of trap conversion mechanism may be taking place between peak 10 and peak 7 defects, leading to the sensitization of peak 7. In a series of papers, Jain (1980, 1981a,b) proposed that the radiation causes the break-up of a complex centre and converts it to both trapping centres and luminescence centres. The increase of luminescence centres causes an increase in the luminescence sensitivity of all the thermoluminescence peaks. Sagastibelza & Alvarez Rivas (1981) suggest a complex trap conversion mechanism for sensitization in LiF. In their model, peak 2 defects convert to peak 5 defects under irradiation. Heavier irradiation results in the conversion of peak 5 defects to the defects responsible for higher temperature peaks by increased interstitial trapping. A similar interconversion relationship would also be appropriate if the peak 2 defects were impurity–vacancy pairs and if peak 5 defects were trimer clusters of these pairs. This is so because the radiation has been observed to induce clustering of pairs into trimers (Muccillo & Rolfe, 1974; Kao & Perlman, 1979). However, without a detailed knowledge of the defects involved in the production of thermoluminescence it is extremely difficult to decide between the various mechanisms and the occurrence of more than one mechanism cannot be dismissed.

4.2.3   Trap creation models (radiation and thermal)

It has already been noted in section 4.1.3 that radiation can create defects by radiation damage. If the same type of defect also existed before irradiation, then obviously the thermoluminescence sensitivity may be increased by this treatment.

It is worthwhile noting that radiation can also decrease the sensitivity. In fact, almost all common phosphors exhibit a decrease in response at very high doses. Examples are shown in figure 4.3 for some glow peaks in natural quartz and in LiF. Durrani et al. (1977a) suggested a complex mechanism to explain both increases in sensitivity at low doses and a decrease at high doses in crystalline quartz. In general, however, it is clear that any of the arguments discussed above can be tailored to produce sensivity decreases.

Changes in sensitivity (via defect creation and/or destruction) can also be induced by thermal treatments, and again LiF has been the subject of most experimental investigations to study these effects. An example of the changes which can be induced in the LiF glow-curve as a result of various annealing treatments is shown in figure 4.4. As will be discussed later, a 400 °C anneal for 1 hour dissipates the Mg impurities in LiF into impurity–cation vacancy pairs. This heat treatment alone gives rise to the characteristic glow-curve given in figure 4.4 (curve A). If this heat treatment is followed by an 80 °C anneal for 24 hours, the impurity–vacancy pairs cluster to form trimer complexes (McKeever & Lilley, 1982) and the glow-curve correspondingly changes shape to that given in figure 4.4 (curve B) – i.e., peaks 4 and 5 grow at the expense of peaks 2 and 3. A high-temperature anneal at 190 °C for 10 hours (figure 4.4, curve C) induces precipitation of the Mg, causing a drastic reduction in the size of peaks 4 and 5 (Bradbury & Lilley, 1977a,b). Clearly, the annealing treatments are inducing defect clustering reactions which are in turn increasing or decreasing the sensitivities of the individual peaks. A combination of radiation and thermal treatments is known to be necessary to induce sensitivity changes in the 110 °C thermolumines-cence peak in quartz (Zimmerman, 1971a,b; Fleming, 1973). These properties have been put to good effect in thermoluminescence dating and the detailed model for the sensitivity change will be discussed in chapter 7.

In view of the above discussion, it is not surprising that supralinearity

Figure 4.3. (a) Changes in thermoluminescence response (A) and sensitivity (B) following proton irradiation in crystalline quartz. The thermoluminescence intensity is the area under the glow-curve between 240 °C and 480 °C. (After Durrani *et al.*, 1977a.) (b) The height of the 200 °C glow peak in LiF as a function of γ dose. (After Charalambous & Petridou, 1976.)

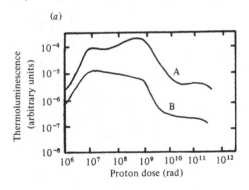

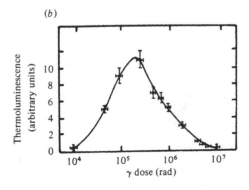

and sensitization properties depend critically upon the thermal treatments both before (e.g., pre-irradiation annealing) and during readout. Clustering reactions among defects can be induced during thermoluminescence production, thereby giving rise to a strong dependence of supralinearity and sensitization properties on heating rate (Taylor & Lilley, 1982a,b,c). This may relate to the often-quoted observation that supralinearity increases with the temperature of the glow peak (e.g., Attix, 1974; Antinucci *et al.*, 1977; Chandra *et al.*, 1981). The different temperature dependencies noted by these authors may then be due to the

Figure 4.4. Glow-curves for LiF TLD-100 after (A) a pre-irradiation anneal at 400 °C for 1 hour followed by a fast quench to room temperature; (B) a pre-irradiation anneal at 400 °C for 1 hour followed by an anneal at 80 °C for 24 hours; (C) a pre-irradiation anneal at 400 °C for 1 hour followed by an anneal at 190 °C for 10 hours.

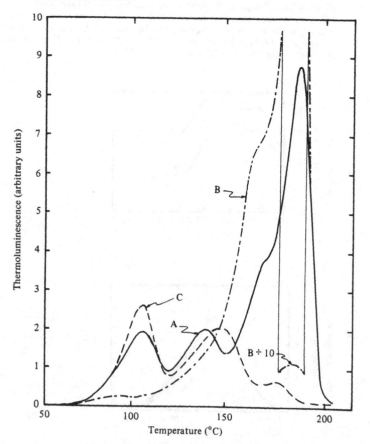

different annealing treatments and heating rates used. Alternatively the assertion that the radiation itself induces defect clustering into higher complexes may also be important. If the lower temperature peaks (i.e., peaks 2 and 3) in LiF are related to impurity–vacancy dipoles and the higher temperature peaks (i.e., 4 and 5) to higher-order complexes of these dipoles (see chapter 5) then one might expect the higher temperature peaks to show a greater degree of supralinearity than those at lower temperatures.

4.3   Optical effects

4.3.1   *Optical stimulation*

The observation of thermoluminescence requires the thermal stimulation of trapped charge carriers, either into the delocalized energy bands or into higher excited states. In a similar fashion, the charge carrier stimulation can be achieved via the absorption of optical energy, rather than of thermal energy. The optical ionization energy (E_0) required is often greater than the equivalent thermal ionization energy (E_T) as can be understood from an examination of the configurational coordinate diagram (figure 2.8). Optical transitions take place vertically, but thermal transitions require multi-phonon interaction and take place over several vibrational states. If a scan over a range of optical wavelengths is carried out the optimum wavelength at which photo-depopulation (optical bleaching) takes place can be determined. If the released carriers are photo-excited into the delocalized bands then a photon-stimulated current can be observed, and by scanning over a range of optical frequencies, the ionization energy, E_0, can be determined (e.g., Thomas & Fiegl, 1970; Brodribb, O'Colmain & Hughes, 1975). The exact interpretation of the depopulation spectra and the evaluation of the trap energies is complicated by a variety of factors, including the scanning rate, the optical band width and the width of the conduction band in the material, but nevertheless the technique has been applied successfully to amorphous and crystalline, organic and inorganic insulators and semiconductors.

The ability to bleach trapped charge optically has an important influence on thermoluminescence measurements. By illuminating an irradiated specimen with light of a specific wavelength (in order to depopulate a particular defect centre) it is possible to observe the subsequent changes which occur in the glow-curve and from these changes arrive at information relating to the sign of the trapped charge involved in its production, and to the relationship between the thermoluminescence traps and the optically activated centres. One of the earliest measurements of this sort was carried out by Stoddard (1960) who illuminated X-irradiated NaCl and observed the subsequent alterations in the glow-curve shape. Stoddard's experimental method, which is now widely followed, was to irradiate the specimen at room temperature and then to cool the sample to liquid nitrogen temperature, at which the specimen is illuminated with monochromatic light of selected wavelength. The sample is then heated and the thermoluminescence glow-curve is recorded in the normal manner. Stoddard found that the illumination diminished the size of some glow peaks which were stable at

Figure 4.5. Stimulation spectra: (*a*) radiation/illumination/thermal sequence necessary for the production of a stimulation spectrum, an example of which is given in (*b*) for two glow peaks (I and II) in X-irradiated KCl. (After Bosacchi *et al.*, 1965 and Fieschi & Scaramelli, 1968.)

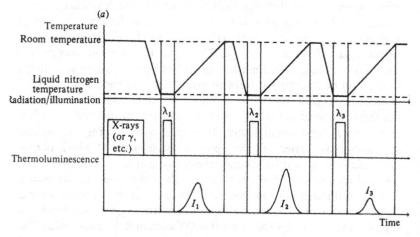

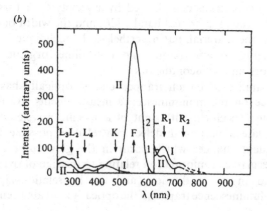

the temperature of X-irradiation (i.e., temperatures greater than room temperature), and induced the presence of some glow peaks which were unstable at the irradiation temperature (i.e., at temperatures less than room temperature). Additionally, the greatest effects were observed to take place following illumination by ultra-violet light corresponding to the F-centre absorption wavelength (F-light; in NaCl this is ∼ 470 nm). Similar observations were then made by several other workers, studying a variety of specimens. Braner & Israeli (1963) studied the effects of illumination on a selection of alkali halides. Here, illumination with monochromatic F- or V-centre light in KBr, KCl, KI and NaCl at liquid nitrogen temperature after X-irradiation at room temperature was observed to excite different glow peaks in the region between liquid nitrogen temperature and room temperature. F-illumination was observed to produce one set of thermoluminescence peaks, whilst V-light produced a different set. Together, these peaks added up to the normal glow-curve obtained after X-irradiation at liquid nitrogen temperature. At that time, similar observations were found in diamond by Nahum & Halperin (1963), in LiF by Tournon & Bergé (1964) and in quartz by Schlesinger (1965).

Most information can be obtained from a stimulation spectrum. Here the illumination/readout sequence is repeated several times, with a different wavelength each time. Care must be taken to ensure that the same number of photons are taking part during each illumination of fixed wavelength. A plot is then taken of the intensity of the induced glow versus illumination wavelength (Bosacchi, Fieschi & Scaramelli, 1965; Fieschi & Scaramelli, 1966). Figure 4.5a shows the radiation/illumination/thermal sequence necessary for the production of a stimulation spectrum, shown in figure 4.5b. From detailed spectra of the sort shown in this figure, Bosacchi *et al.* (1965) were able to show that only illumination with light corresponding to electron-centre absorption (e.g., F-centres, R-centres, etc.) was able to induce thermoluminescence between liquid nitrogen temperature and room temperature. The apparent stimulation using V-centre (trapped holes) light observed by Braner & Israeli (1963) may have been due to stray light during the illumination sequence. This illustrates an important use of optical stimulation. By comparing the wavelength at which optical bleaching occurs with the conventional *optical absorption* spectrum, the nature of the trapped charge may be evaluated. Thus, if it can be determined that only, say, electron centres are being bleached (e.g., by illuminating with F-light to release electrons from F-centres) then any glow peaks which are induced must be due to the presence of electron traps. Thus, the sign of the charge carrier can be determined. In this way, Townsend *et al.* (1967) and Podgoršak *et al.* (1971) were able to identify several electron traps in the LiF glow-curve by using F-centre stimulation at liquid nitrogen temperature following irradiation at room temperature. The released electrons are thought to redistribute themselves among the other traps.

Another useful point can be made with regard to the comparison between stimulation spectra and optical absorption spectra. It is often found that a particular centre does not have a strong absorption coefficient and thus is very weak, and therefore difficult to detect, in an optical absorption measurement. However, its photo-ionization efficiency may be high and thus the centres will plainly show up in a thermoluminescence stimulation spectrum (e.g., excited F-centres in alkali halides (Fieschi & Scaramelli, 1966, 1968)).

Two considerations should be noted with regard to charge carrier identification using optical bleaching. Firstly, the failure to induce a glow peak following colour-centre illumination does not mean that the corresponding traps or recombination centres are not present. The capture cross-section of the defect may be much smaller than the others, giving only a small glow peak compared with the remainder. It may only show after prolonged radiation and/or illumination. Alternatively, it may be that the colour-centres in question do not photo-ionize in the temperature range studied (e.g., V-centres in KCl, Fieschi & Scaramelli, 1968). Secondly, the trap in question must be empty before illumination to be certain that the trap, not the recombination centre, is being filled. For example, results are often reported in which the ultra-violet illumination takes place at the same temperature as the X-irradiation. Following this treatment, some glow peaks are seen to increase their size (compared to the unilluminated specimen) and often this is interpreted as increased trapping at the traps following optical bleaching of a known centre (e.g., Vaz, Kemmey & Levy, 1968). However, this may not be true. If the illumination is with, say, F-light, then the electron released may be retrapped at a defect which serves as a recombination centre for the trap in question (a *hole* trap). Increased thermoluminescence will then result due to the increased luminescence efficiency, η, given in equation 2.9. Only if the trap has been initially thermally emptied can it be said that the *trap* is being refilled. This is often overlooked in experiments of this sort. Thus, a combination of photostimulated thermoluminescence from previously empty traps, and from previously partially empty traps, should enable the experimenter to distinguish between trap filling and recombination centre filling, in addition to determining the sign of the trapped carriers. Additional limitations, as discussed by Townsend *et al.* (1967), are that the absorption bands for hole centres may overlap those of electron centres, making selective photo-depopulation difficult, and that it is impossible to repopulate traps which have been destroyed during the heating.

4.3.2 Phototransfer

Often the illumination is observed not only to induce glow peaks (or increase their size) but it is also seen to diminish others (Vaz *et al.*, 1968; Mayhugh, Christy & Johnson, 1970; Miller & Bube, 1970). This can be interpreted as the release of charge from the traps, or from the recombination centres. In many instances, it is found that the illumination decreases the size of several glow peaks at once, in which case it is almost certainly the recombination centre which is being depleted. Emission spectra measurements to confirm that each glow peak is using the same recombination centre (i.e., that they emit at the same wavelength) would form useful additional information. This increase in some glow peaks at the expense of a decrease in others gives the appearance of a transfer of charge between centres, and in this context it does not matter whether the defects involved are the traps or the recombination centres. The phenomenon is called '*phototransfer*' and following an extensive study in some phosphors (e.g., LiF and quartz) it is now proving to be a useful phenomenon in both dosimetry (Horowitz, 1981; Oberhofer & Scharmann, 1981) and in dating (Aitken, 1978*a*). In LiF, thermoluminescence from the main dosimetry peak (peak 5) can be re-excited, following initial readout, by illuminating with ultra-violet light of several different wavelengths (an example of phototransfer in LiF is given in figure 4.6). The re-excitation is due to the phototransfer of charge from more stable, higher temperature peaks and the size of the re-excited glow peak, for a given illumination time, is said to be proportional to the original absorbed dose. Thus, phototransfer gives LiF the facility to be 're-read'. Its use as an ultra-violet dosemeter has also been noted (Buckman & Payne, 1976). Since the first suggested use of phototransfer to re-read thermoluminescence, by Gower, Hendee & Ibbott (1969), a great effort has been made to determine the precise nature of the phototransfer mechanism (e.g., Sunta & Watanabe, 1976; Kos & Nink, 1977; Mason, McKinlay & Saunders 1977; Jain, 1980, 1981*b*; Chandra *et al.*, 1981; Sagastibelza & Alvarez Rivas, 1981). However, despite an elucidation of the optimum parameters required for phototransfer, the precise mechanism depends heavily upon which defect model is being advocated. In this sense, as with the sensitization and supralinearity phenomenon, it is impossible to give a generalized description of the process in this material.

One fact which has emerged from the work of several authors is that the phototransfer efficiency is temperature dependent. Buckman & Payne (1976), Sunta & Watanabe (1976) and several other workers report that the efficiency increases as the illumination temperature increases, up to a point at which the reservoir trap becomes thermally unstable, when a decrease in efficiency is observed. This temperature dependence has been interpreted as meaning that the transfer actually involves thermal, as well as optical, stimulation. In fact, Sunta & Watanabe (1976) employ the Arrhenius equation to calculate a thermal activation energy of 0.12 eV in LiF. They thus suggest that the optical energy stimulates the trapped charge (electrons in this case) into an excited state, 0.12 eV below the conduction band, from where thermal stimulation into the band is possible.

The observation of phototransfer in quartz by Schlesinger (1965) led to the development of the phenomenon as a useful technique in thermoluminescence dating. Here, phototransfer is proving useful as a means of overcoming thermal quenching (section 4.5.1), anomalous fading (section 4.4), interference by high-temperature, black-body radiation, and non-radiation-induced thermoluminescence ('spurious' signals; discussed in chapters 6 and 7) (Bailiff, 1976; Bailiff *et al.*, 1977*a*; Bowman, 1979; Sasidharan, Sunta & Nambi, 1979).

Besides the heavily studied LiF (and alkali halides in general, e.g., Jiménez de Castro & Alvarez Rivas, 1979, 1981) and the already mentioned studies on quartz, phototransfer and optical bleaching effects have been examined in a variety of materials, including calcite (Vaz *et al.*,

Figure 4.6. LiF glow-curve following irradiation with X-rays (dashed line) at liquid nitrogen temperature (scale *a*), compared to the phototransferred thermoluminescence (full line, scale *b*) obtained by illuminating the crystal at liquid nitrogen temperature with F-band light for 20 min, after X-irradiation at room temperature and heating to 200 °C. (After Podgoršak *et al.*, 1971.)

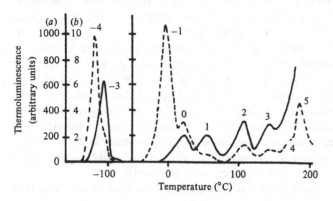

1968), $CaSO_4$ (Caldas & Mayhugh, 1976), CaF_2 (Sunta, 1970, 1979), MgO (Kirsh *et al.*, 1977; Las & Stoebe, 1981), BeO (Hobzová, 1974) and Al_2O_3 (Cooke, Payne & Santi, 1981). Phototransfer has also been observed to take place in a variety of natural minerals (Bailiff *et al.*, 1977a). In MgO doped with transition metal ions, Las & Stoebe (1981) suggest that the transfer takes place between the metal ion and V-centres in a reaction of the type

$$\alpha M^{3+} \cdots + \delta V^{2-} \rightleftharpoons \alpha' M^{2+} \ldots + \delta' V^- + \varepsilon' V_m + \ldots \qquad (4.3)$$

where M is a transition metal ion, V^{2-} represents vacancies, V^- is the intrinsic V-centre and V_m is a V^--centre associated with a transition metal ion; α, δ, α', δ' and ε' are all constants. Several metal ions can take part in the reaction. Similar charge transfer mechanisms have been suggested for rare-earth-doped CaF_2 (Sunta, 1970, 1979) and $CaSO_4$ (Nambi *et al.*, 1974).

Infra-red stimulation is an additional important effect often seen in thermoluminescent phosphors and has been particularly observed in wide band gap semiconductors (Garlick & Mason, 1949; Bull & Mason, 1951). Riehl (1970) suggests that infra-red stimulation from ambient black-body radiation can induce detrapping in ZnS even at low temperature. Zinc oxides, sulphides and selenides, cadmium sulphides and selenides, and doubly activated, alkaline-earth sulphides and selenides (e.g., SrS:Ce,Sm or SrS:Eu,Sm) have all been shown to exhibit strong infra-red stimulation (Keller, Mapes & Cheroff, 1957; Baur, Freydorf & Koschel, 1974).

4.4 Tunnelling and anomalous fading

One of the principal suggested uses of the phototransfer effect in thermoluminescence is to use it to circumvent problems due to the phenomenon of *anomalous fading*, especially in the dating of archaeological artefacts. In equation (1.1), the expected mean lifetime of a charge carrier in a trap of depth E and escape frequency s at a storage temperature T was given by

$$\tau = s^{-1} \exp(E/kT). \qquad (1.1)$$

From kinetic studies of the thermoluminescence process, it is possible to estimate that for charge carriers released at high glow-curve temperatures (say, $> 200\,°C$) the lifetimes at ambient temperatures calculated according to equation (1.1) range from several thousands to several millions of years. However, for many materials it is often found that thermoluminescence 'fades', i.e., the charge carriers are released at rates which are much faster than those expected from this equation. This 'anomalous' fading has serious consequences in a number of applications, especially in thermoluminescence dating, where it is assumed that the accumulated trapped charge population does not undergo any appreciable decrease (fading) during the irradiation period. For example, attempts to date volcanic lava, of known age, by thermoluminescence resulted in age estimations which were an order of magnitude too low (Wintle, 1973). The thermoluminescence used in this dating process appeared at glow-curve temperatures in excess of 350 °C and so, from a consideration of equation (1.1), the expected lifetime of the carriers in the traps is of the order of 10^5 years, at ambient temperatures. The low calculated ages, however, clearly indicate a loss of charge which cannot be explained by thermal fading. Verification that no optical bleaching is taking place leads to the conclusion that the fading is anomalous. Wintle (1973, 1977) observed the effect in a range of natural minerals, including sanidine, fluorapatite, labradorite, andesine, zircon and bytownite. Quartz and limestone, however, did not appear to exhibit the effect to any appreciable extent.

The phenomenon is not restricted to natural minerals and is in fact exhibited, to lesser or greater degrees, by a whole range of material types and thereby shows quite a wide generality. For example, Hoogenstraaten (1958) observed an apparently thermally independent degree of fading of thermoluminescence from ZnS (doped with Cu and Co) in that even at low storage temperatures (liquid nitrogen temperature), a loss of trapped charge occurred. A correlation between the amount of athermal fading and the intensity of the 'afterglow' was also observed. (Afterglow is the luminescence emitted from a specimen immediately after irradiation. It may be thermally dependent, according to equation (1.1), in which case it is properly termed phosphorescence, as discussed in chapter 1. In the case of ZnS, however, as observed by Hoogenstraaten, the luminescence is observed to be essentially temperature independent and therefore the more general term, afterglow, is appropriate. Usually, the emission spectrum for the afterglow is the same as that for the thermoluminescence, indicating that the same luminescence centres are being activated (Visocekas *et al.*, 1976, and the references therein).) The relationship between the degree of anomalous fading exhibited by a sample and the intensity of the afterglow was also emphasized by Zimmerman (1977a, 1979a).

The effect is observed strongly from CaF_2 doped with Mn (Schulman *et al.*, 1969), to a moderate degree from calcite (Visocekas *et al.*, 1976), from lunar material (Blair *et al.*, 1972; Garlick & Robinson, 1972) and ostensibly similar effects have been observed in $ZnSiO_4$ (doped with Mn; Avouris & Morgan, 1981; Avouris *et al.*, 1981), organic glasses and polymers (Hama *et al.*, 1980; Charlesby, 1981), alkali halides, e.g., KCl:Tl (Delbecq *et al.*, 1976) and molecular solids, e.g., solid methane (Racz, 1981).

It is characteristic of anomalous fading that after an initial rapid decay, the decay rate decreases for longer storage times. Nevertheless the level of anomalous fading is often very small when monitored over short periods, and only becomes noticeable after lengthy storage. Thus, the only effective way of detecting anomalous fading is to perform long-term ageing studies of irradiated specimens in order to accumulate a measurable signal loss. If an amount of fading is detected which is over and above the level expected from thermal detrapping alone, taking into account the trap depth, frequency factor and storage temperature, then anomalous fading should be suspected.

Riehl (1970), in a study of the afterglow process in ZnS:Cu, supposed that, even at a temperature of 6 K, infra-red absorption from the cryostat surroundings may cause bleaching of the shallow traps in this material. However, careful screening of the samples pointed to a tunnelling mechanism, first suggested by Hoogenstraaten (1958). An extensive investigation of the phenomenon of afterglow by Visocekas and colleagues (Visocekas *et al.*, 1976; Visocekas, 1979, 1981) has led to confirmation of the relationship between tunnelling, afterglow and anomalous fading in a wide variety of materials. The conclusions reached by Visocekas stem from detailed experiments on calcite in which it is observed that the intensity (I) of the afterglow at liquid nitrogen temperature immediately following the irradiation follows a hyperbolic law, namely, $I \propto t^{-1}$. It has already been noted in section 3.3.5 that a temperature-independent hyperbolic decay of luminescence is symptomatic of a tunnelling mechanism for the reduction of trapped charge. The t^{-1} law arises from a consideration of Mikhailov's (1971) model for tunnelling, illustrated schematically in figure 4.7. Here, an electron in a

Figure 4.7. Schematic diagram of the potential barrier for the tunnelling model of Mikhailov (1971).

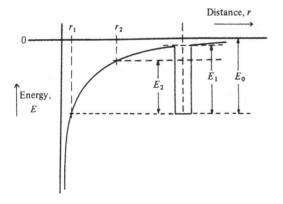

trap of depth E_0, is separated by a distance r from a positive charge centre. Tunnelling through the potential barrier would provide the electron with a means by which recombination with the positive charge could occur, with the resulting emission of a luminescence photon. The tunnelling rate (K) is given by

$$K = v \exp(-\phi r),\qquad(4.4)$$

where r is the electron–hole centre separation, v is a frequency factor, and ϕ is given by

$$\phi = 2(2mE_1)^{\frac{1}{2}}/h.\qquad(4.5)$$

Here, m is the mass of the electron. If the number of trapped electrons (n) decreases according to first-order kinetics (no back reaction) then the luminescence intensity is proportional to the rate of decrease, thus:

$$I = -\,\mathrm{d}n/\mathrm{d}t = Kn = Kn_0 \exp(-Kt)\qquad(4.6)$$

where n_0 is the initial density of trapped charge. Thus, the total luminescence at time t is

$$I = \int_0^t Kn_0 \exp(-Kt)\,\mathrm{d}r.\qquad(4.7)$$

Figure 4.8. Lifetime τ before tunnelling between a trapped charge and a radiative centre as a function of their separation for $E_1 = 3.5\,\mathrm{eV}$ and $E_1 = 0.6\,\mathrm{eV}$. Marked on the graph are the corresponding processes of radiative recombination. Also shown are the concentration of centres which are expected to give centre separation r, assuming a uniform distribution. (After Visocekas, 1979.)

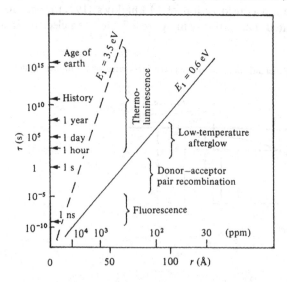

With $dK = -(K/\phi)dr$ from equation (4.4) then

$$I = \int_0^t -[(Kn_0\phi\exp(-Kt))/K]dK$$
$$= -n_0\phi\exp(-Kt)/t. \qquad (4.8)$$

For large r, $K \simeq 0$, so that $\exp(-Kt) \simeq 1$, and therefore,

$$I \simeq -n_0\phi/t, \qquad (4.9)$$

which is the required reciprocal dependence of I on t.

It can be seen from equation (4.4) that the rate constant K is exponentially dependent on distance r, thus I varies with r in a similar fashion. Visocekas (1979) calculates the expected lifetime before recombination between opposite charges, for $E_1 = 3.5\,eV$ and for $E_1 = 0.6\,eV$, and determines its variation with separation r. For $E_1 = 3.5\,eV$, the lifetime varies between $10^{-4}\,s$ for $r = 20$Å, to greater than one million years for $r = 50$ Å (figure 4.8).

Having demonstrated the probable tunnelling origin of the low-temperature afterglow, Visocekas (1979; Visocekas et al., 1976) confirmed the relationship between the luminescence emitted in afterglow and the thermoluminescence lost in anomalous fading (cf. figure 3.18) by demonstrating the conservation of the total light sum. Thus, the light lost in thermoluminescence (shown by the shaded area in figure 3.18) is

Figure 4.9. Remnant thermoluminescence against storage time at low temperatures following irradiation. Open circles, calcite (Visocekas, 1979); open squares, fluorapatite (Wintle, 1973); solid circles, apatite (Zimmerman, 1979a); crosses and triangles, labradorite (Wintle, 1977); solid squares, fluorapatite (Bailiff, 1976). (After Visocekas, 1981.)

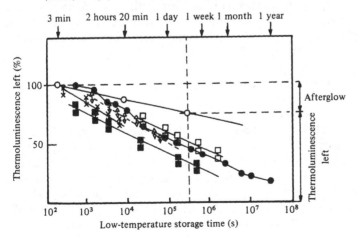

equal to the light emitted during afterglow. Furthermore, the emission spectra are the same. Finally, Visocekas (1979, 1981) confirmed the tunnelling nature of anomalous fading by showing that it too followed a t^{-1} law. Using data from several published papers, it is possible to show that the glow remaining decays as a function of the logarithm of the storage time. In figure 4.9, the remnant thermoluminescence signal is plotted as a function of storage time for several materials, listed in the figure caption. Clearly, a t^{-1} law is demonstrated for each of them.

Visocekas *et al.* (1976) considered the possibility that the electron has to be thermally excited to a higher energy state before tunnelling can take place. From figure 4.7, if the trapped electron is thermally excited by absorbing energy E_2, it still does not have enough energy to escape from the trap completely and can do so only via tunnelling. (This explains how two traps can have the same thermal activation energy, E_2, but different lifetimes (Riehl, 1970). One carrier may have been freed by the excitation, but the other still has to undergo tunnelling. Application of the normal Arrhenius equation to the tunnelling case will yield an abnormal frequency factor, as previously discussed.) This type of tunnelling mechanism is often referred to as 'thermally assisted' tunnelling and may in part relate to the observation that anomalous fading is often less for the deep traps than it is for the shallow traps (Bailiff, 1976). The probability for this alternative tunnelling mechanism is less dependent on r than the mechanism so far discussed. Experimental evidence for this is available (e.g., Racz, 1981). Additionally, the probability of release now includes a Boltzmann term, $\exp(-E_2/kT)$, thus the probability for thermally assisted tunnelling may be smaller than that for tunnelling at low temperatures, but become larger at high temperatures. Tunnelling in calcite appears to be thermally assisted, with $E_2 = 0.62\,\text{eV}$ (Visocekas *et al.*, 1976). This may account for the apparent lack of evidence for significant anomalous fading at low temperatures in this material by other workers (e.g., Wintle, 1977).

4.5   Quenching effects

4.5.1   *Thermal quenching*

It has already been discussed (for example, in section 3.3.1) how thermal quenching can affect the calculation of the trap depth from a thermoluminescence glow-curve. It is worthwhile noting that in addition to overcoming problems due to anomalous fading, the phototransfer technique can also be employed to combat thermal quenching. This was first demonstrated by Bull & Mason (1951) who examined a glow peak at 400 K in ZnS:Mn,Cu which was very difficult to observe during a

Figure 4.10. A: Variation in the phototransferred signal (at 290 K) with decay temperature for ZnS:Mn,Cu. B: Derivative of A. (After Bull & Mason, 1951.)

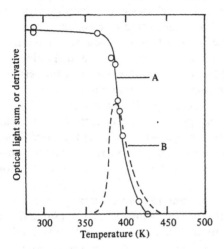

Figure 4.11. Thermoluminescence and radioluminescence in $CaSO_4$:Eu. (After Sunta & Bapat, 1982.)

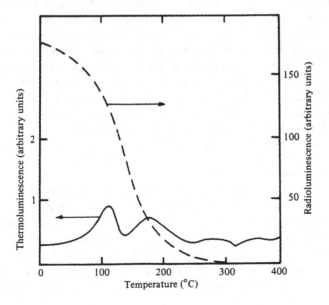

conventional thermoluminescence experiment due to a decrease in luminescence efficiency at these temperatures. However, by phototransferring the charge trapped at the level under consideration (using infrared stimulation) into the lower temperature traps, the properties of the 'hidden' trap could be observed. Bull & Mason illuminated the specimen after irradiation, but between irradiation and illumination the specimen was stored at different temperatures for a given period. The variation in the phototransferred thermoluminescence was plotted against storage temperature and the results are displayed in figure 4.10. At storage temperatures much greater than 400 K, the phototransferred signal is decreased owing to thermal detrapping of the carriers from the source trap. The derivative of this signal gives the approximate position and shape of the 'glow peak' which is not observable in the normal glow-curve.

Similar effects have been examined more recently by Sunta and colleagues. For example, thermal quenching in $CaSO_4$ (Sunta & Bapat, 1982) means that thermoluminescence beyond $\sim 200-250\,°C$ is difficult to analyse. This can be understood by looking at the variation in the luminescence emitted during irradiation (called radioluminescence by Sunta & Bapat) with irradiation temperature, and comparing this with a thermoluminescence glow-curve. An example is shown in figure 4.11 for $CaSO_4$:Eu. Clearly, analysis of the high-temperature glow will present problems due to the decreased luminescence efficiency as measured by the radioluminescence intensity. Similar observations were made for LiF:Mg (TLD-100; Kathuria & Sunta, 1981) and for meteorites (McKeever *et al.*, 1983a). In LiF, glow peak 12 (at $\sim 385\,°C$) is difficult to observe because of thermal quenching. By using phototransfer to transfer charge from peak 12 to the lower temperature peaks, Kathuria & Sunta estimate the amount of thermal quenching taking place in the temperature range of peak 12. They calculate that at 385 °C the luminescence efficiency is ~ 85 times less than the efficiency in the temperature range up to $\sim 300\,°C$. Earlier measurements have indicated that a small degree of quenching may even be taking place at these lower temperatures (Gorbics, Nash & Attix, 1969).

4.5.2   Concentration quenching

Quenching of luminescence can also be achieved by means other than thermal. Nambi *et al.* (1974) and Schmidt, Linemann & Giessing (1974) studied the effect of Dy and Tm doping on the thermoluminescence output from $CaSO_4$. Dy and Tm are known to be important activators of luminescence in this system, such that the thermoluminescence intensity increases (initially) as the dopant concentration increases. However, at larger concentrations (i.e., greater than $\sim 0.1\%$) the thermoluminescence output decreases gradually as the dopant levels are raised further. This effect is known as *concentration quenching* and has been extensively treated by Johnson & Williams (1950) and Ewles & Lee (1953). The expression given by Ewles & Lee (1953) to describe the effect is

$$\eta = K/(1 + \alpha c^{-1} \exp(nc)) \tag{4.10}$$

where K and α are constants and c is the dopant concentration. The coefficient n is equal to $(1/c_{max})$, where c_{max} is the dopant concentration corresponding to η_{max}, as illustrated in figure 4.12. As usual, η is the luminescence efficiency. Medlin (1968) describes a similar quenching effect on increasing the Mn concentration in calcite, with η_{max} being achieved at a dopant level of $\sim 0.3\%$. Medlin says that this behaviour is characteristic of isolated activator centres. When one activator ion is located within a certain radius of another, the luminescence is quenched. From the calcite results, Medlin estimates that the quenching takes place in this material when one Mn^{2+} ion is within 2.5 lattice units of another.

Figure 4.12 illustrates the concentration quenching effect on lumines-

Figure 4.12. Concentration quenching for rare earth emissions in $CaWO_4$ (Dy: 573 nm; Er I: 552 nm at room temperature; Er II: 552 nm at 77K; Eu: 614 nm; Tb I: 545 nm; Tb II: 436 nm). (After van Utiert, 1960; Nambi, 1979.)

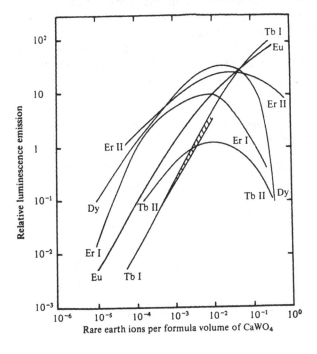

cence from $CaWO_4$, doped with various rare earths (Nambi, 1979; van Utiert, 1960). For Eu^{3+}, the 554 nm and 614 nm emissions are quenched after association of three or more Eu ions. The 510 nm emission (not shown) quenches after pairing of Eu ions. Er^{3+} emissions do not quench sequentially, as with Eu^{3+}, but instead all emissions quench uniformly, at the same time. For Tb^{3+}, the quenching of the emission from the higher of the two energy states (Tb II) permits energy transfer to the lower state (Tb I) thus enhancing its emission. In fact, all of the quenching mechanisms associated with concentration effects appear to be associated with exchange interactions of this sort (van Utiert, 1960).

4.5.3   *Impurity quenching*

A similar, but separate effect, known as *impurity quenching*, is thought to occur via the action of 'killer' centres. Upon introducing certain elemental impurities into a material, especially heavy metals such as Cu, Fe, Co, Ni and Cr, the intensity of the luminescence emission is seen to reduce drastically. An example is shown in figure 4.13 for thermoluminescence from $CaSO_4$: Dy where the metallic impurities are acting as 'killers' of the emission, presumably by introducing competitive, non-radiative recombination paths. Medlin (1968) lists several metals which act as 'killers' in a variety of natural minerals and notes that the 'killing' efficiency is related to the valency. For example, iron is most effective in quenching when it is in the trivalent state; the divalent state is generally less efficient. However, this rule is by no means always true, making a simple interpretation difficult. Na and Mg are observed by Hutcheon *et al.* (1978) to reduce the luminescence efficiency in lunar feldspar.

Figure 4.13. Impurity quenching: 'killing' effect of Cu, Cr, Co and Fe. (After Schmidt *et al.*, 1974.)

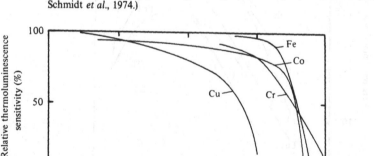

5 Defects and thermoluminescence

5.1 General introduction

Among the list of uses for thermoluminescence outlined in chapter 1, the study of the defect structure of materials using this technique is perhaps the most difficult, in that interpretation of the results is not straightforward. In fact, on its own, thermoluminescence is only of limited value in arriving at a true characterization of the defect structure of the solid under study. Greatest benefit can be attained when the technique is used in conjunction with other experimental methods (dielectric loss, electrical conductivity, ITC/TSPC, optical absorption, electron spin resonance, etc.) in which case the information gained can be extremely useful. It should be realized, however, that these different experimental techniques may be sensitive to different types or collections of defects, so that conclusions derived from one method should be applied cautiously to the results obtained from another.

Some areas requiring special caution ought to be noted. The techniques mentioned above are normally used to give information about specific defects. If correlations can be found between various defect concentrations and certain thermoluminescence peaks, then such results are normally used to argue in favour of a particular defect being responsible for a particular peak. However, it is important to remember that thermoluminescence requires *two* principal defect types – a trap and a luminescent centre. In the techniques listed above it is not known which one, if either, of these defects is being monitored. Furthermore, the intensity of a thermoluminescence peak is proportional to the smaller of the two concentrations (i.e., the concentration of trapped charge and the concentration of recombination centres). Thus, if, say, the recombination centre concentration is the smaller and if, in addition, it remains constant with, say, thermal treatment, then the glow peak intensity will remain constant, despite the fact that the concentration of traps responsible for that peak may be varying. One might even arrive at the situation where the trap concentration is increasing, yet the glow peak associated with it is decreasing.

Additionally, it must be remembered that in a sample where many different types of trap are operative, competition effects are bound to occur. It is not really a case of 'are they occurring?', but more one of 'how strong are they?'. The complex defect structure of the alkali halides is a perfect example, such that if defect A competes with defect B, then increasing the concentration of A may reduce the thermoluminescence due to B, despite the fact that the concentration of B is unaffected.

It is unfortunate that many of the methodologies used to assign atomistic processes to given thermoluminescence peaks appear to overlook complications of this sort, and usually broad relationships between defects and peaks are all that can truthfully be made. Many examples can be found in the literature where speculative assignments have been made without the necessary supporting data from other techniques and, regrettably, this has given the impression that thermoluminescence is able to do more than it can. In fact, it can only ever be regarded as a useful supporting technique and not as an analytical tool to be used independently of others. The discussion which follows is based on two papers which are included in a recently published collection of articles on thermoluminescence materials (edited by McKeever, 1984). It is intended to highlight how different kinds of thermally stimulated defect interaction can give rise to thermoluminescence and possibly the best examples can be found among the alkali halides, and quartz and silica. The alkali halides in particular display reasonably well under-stood thermoluminescence properties, owing, undoubtedly, to the enormous amount of scrutiny to which these materials have been subjected.

5.2   The alkali halides

5.2.1   *Structure and defects*

These materials, of the type M^+X^-, where M = Group I alkali metal and X = Group VII halogen, are possibly the best known of all ionic compounds. Research into alkali halides was initiated in the early 1920s in the hope that it would aid our understanding of the behaviour of solids as a whole. Because of their own rather extreme properties, however, they have tended to be set aside from materials of more direct commercial interest, such as semiconductors or low-absorption glasses. Nevertheless, their ease of crystallization and purification have made them the subject of much detailed investigation and from these studies a wealth of valuable information on solid-state behaviour has been gathered. As a consequence, they are often used as prototype materials for the study of other, less well characterized ionic solids (e.g., Barr & Lidiard, 1970).

The structure of the majority of the alkali halides follows that of NaCl, which is composed of two interpenetrating fcc sublattices for both cations (M^+) and anions (X^-). Each M^+ ion is surrounded by six nearest-neighbour X^- ions, and each X^- ion is similarly surrounded by six nearest-neighbour M^+ ions (figure 5.1).

Deviations in the stoichiometry of the perfect crystal lattice constitute structural defects. At thermodynamic equilibrium crystals contain vacant lattice sites and interstitial atoms, the concentration of which, at any given temperature, is governed by a minimization of the crystal's free energy. In an alkali halide crystal of the type M^+X^-, Schottky defects consisting of equal numbers of cation and anion vacancies, are the most numerous type.

Vacancies may also be present in alkali halides for reasons other than thermal. It has been established since the work of Pick (Pick, 1939) that aliovalent impurities enter an alkali halide lattice substitutionally and require the presence of host ion vacancies ('extrinsic') of the appropriate sign in order to maintain charge neutrality. Divalent metallic cations (e.g., Ca^{2+}, Cd^{2+}, Mn^{2+}, Mg^{2+}, Pb^{2+}, etc.) thus require the existence of an equal number of host cation vacancies (see figure 5.2a). At high temperatures Schottky defects dominate (*intrinsic region*) but at lower temperatures the extrinsic vacancies are the most numerous (*extrinsic region*).

At still lower temperatures, association of the divalent impurity and the host vacancy occurs. From a consideration of the lattice strain energy and of the coulombic attraction between the additional positive charge on the impurity and the effective negative charge on the cation vacancy, it may be shown that the total energy of the system is reduced when the impurity and the vacancy associate to form an impurity–vacancy pair (or *dipole* – figure 5.2b).

At even lower temperatures, further clustering among the impurity–

Figure 5.1. Unit cell of NaCl consisting of two interpenetrating fcc lattices, one for Na^+ and one for Cl^- ions.

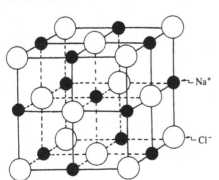

vacancy dipoles occurs. Dryden & colleagues (e.g., Cook & Dryden, 1962, 1981; Dryden, 1963) have made detailed measurements of the concentration of impurity–vacancy pairs in a variety of alkali halide systems. Dryden monitored the decay of a non-equilibrium excess of dipoles at various ageing temperatures and observed that their number decreased with time at a rate governed by third-order kinetics. This was interpreted to mean that three impurity–vacancy pairs were clustering together to form *trimer* complexes.

Much controversy exists over the suggestion of trimer formation. Despite a wide variety of measurements which appear to substantiate Dryden's interpretation (see discussion and references in McKeever & Lilley, 1982) the probability of trimer formation is thought to be low owing to the small likelihood of a three-body encounter compared with the

Figure 5.2. (*a*) Divalent cationic impurity in alkali halide lattice with host cation vacancy for charge neutrality. (*b*) Association of impurity and vacancy to form an impurity–vacancy pair (or dipole).

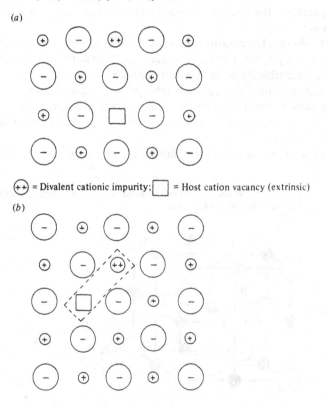

(*a*)

(++) = Divalent cationic impurity; ☐ = Host cation vacancy (extrinsic)

(*b*)

likelihood of a two-body encounter (to form a *dimer* complex). To overcome this Crawford (1970) suggested a two-step reaction in which dimers were formed as a precursor to trimers. If the binding energy for the dimers is low enough, then the second step of the reaction (i.e., dimer + pair = trimer) can dominate the kinetics to give an overall third-order appearance.

Various attempts have been made to estimate the relative binding enthalpies for the different forms of dimer and trimer and in this regard the recent development of sophisticated computational techniques has proved useful (e.g., Catlow, 1976; Catlow *et al.*, 1980; Corish *et al.*, 1981). The binding energy is a function of the particular configuration adopted by the complex (dimer, trimer, etc.) within the lattice. Crawford (1970) discussed some probable configurations, but greater detail is given in the work of Corish *et al.* (1981) who calculated the binding energies for various cluster configurations in a variety of alkali halide–impurity ion systems. The defects examined by these authors are given in figure 5.3. The calculations, in conjunction with the earlier arguments of Crawford (1970) and of Strutt & Lilley (1976), indicate that there are two possible reaction sequences among the defects, depending upon the size of the foreign ion with respect to the host cation. One reaction path involves a two-step process whereby two pairs cluster to form an intermediate dimer (probably type d in figure 5.3) followed by the addition of a third pair to form a stable trimer (in a $\langle 111 \rangle$ plane; type e of figure 5.3). Detection of dimer formation has been claimed by several authors in a variety of

Figure 5.3. The structures of various cluster configurations in alkali halides. Filled circles = cation impurities; squares = cation vacancies. (After Corish *et al.*, 1981.)

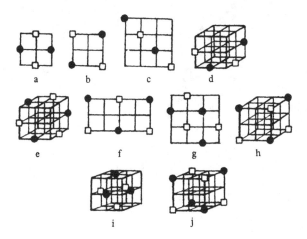

158   *Defects and thermoluminescence*

systems (see Unger & Perlman (1974, 1977) and Cussó and colleagues (Cussó, López & Jaque, 1978)) although these observations have not gone unchallenged (e.g., McKeever & Lilley, 1982).

Thus, there remains considerable uncertainty over the exact reaction steps among the defects. It is not surprising, therefore, that one-to-one correlations between particular thermoluminescence peaks and particular species of impurity defect have not proved easy.

The second possible reaction sequence alluded to above leads to even greater complexity, namely the formation of a precipitate phase. The impurity complexes discussed so far have been in solid solution within the alkali halide lattice. However, under certain conditions of time and temperature, the impurities (I) may precipitate to form a second-phase. This precipitated phase may be the stable $2MX.IX_2$, but the metastable phase $6MX.IX_2$, a vacancy-rich precipitate first discovered by Suzuki (1961) in the NaCl:Cd system, has been shown to be the major phase to precipitate during ageing at moderately low temperatures. Theoreti-

Figure 5.4. Ionic conductivity plot for nominally pure Harshaw NaCl (circles); NaCl doped with 54 ppm Mn (diamonds); 140 ppm (squares); and 1210 ppm (triangles). The 'knees' give the limit on Mn Solubility. (After Bradbury & Lilley, 1977a.)

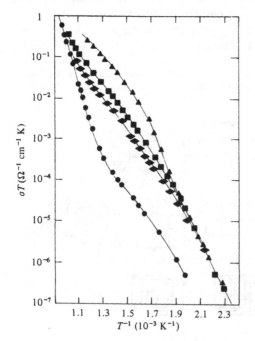

cal calculations of the stability predict that Suzuki phase formation is most likely in systems where the impurity ion is small relative to the host cation. An extensive range of experiments has now revealed the presence of Suzuki phases in a variety of alkali halide systems. The effects of the formation of a Suzuki phase upon the electrical, dielectric and mechanical properties, and particularly important from the point of view of the present discussion, upon thermoluminescence have all been discussed in some detail in the literature, particularly by Lilley & colleagues (e.g., Bradbury, Nwosu & Lilley, 1976; Bradbury & Lilley, 1977a,b).

At elevated temperatures, where the impurity ion and the extrinsic cation vacancy are dissociated, the vacancies will be free to take part in (ionic) conductivity. At these temperatures the conductivity σ will be given by an expression of the form

$$\sigma = X_v \, \text{constant} \, \exp(W/kT) \tag{5.1}$$

where W is the activation enthalpy required for vacancy diffusion and X_v is the vacancy concentration. At these high temperatures, X_v is a constant, equal to the divalent impurity content. At lower temperatures, however, association takes place and so there is a corresponding reduction in the conductivity. In fact, the whole sequence of association–aggregation–precipitation can be followed by ionic conductivity. It is useful to illustrate a typical ionic conductivity plot in order to demonstrate which defects are dominant at which temperatures. Such a plot is shown in figure 5.4 for NaCl doped with Mn^{2+}. At the highest temperatures the intrinsic vacancies dominate and the plots start to merge with that for nominally pure NaCl. At the lowest temperatures precipitation occurs in the impure crystals and the plots coincide. The various changes in slope in between these regions correspond to the association and aggregation regions (Lidiard, 1957; Dreyfus & Nowick, 1962).

5.2.2 *Irradiation effects*

It now becomes instructive to examine the defects produced by the interaction of ionizing radiation with both pure and doped alkali halides. A detailed review of this topic has recently been written by Agulló-López, López & Jaque (1982).

The primary defect production mechanism during irradiation, which is now most widely accepted for ionic crystals, is based on a model involving the non-radiative recombination of an electron with a self-trapped hole (V_k-centre), proposed initially by Pooley (1966) and by Hersh (1966). Irradiation with low-energy photons excites an electron–hole pair (or self-trapped exciton). The hole is localized at two neighbouring halide ions forming an X_2^- molecule along the $\langle 110 \rangle$

160   *Defects and thermoluminescence*

close-packed anion direction. The two ions in this molecule are displaced with respect to their normal positions as a result of the exciton absorption. At low temperatures they return to their normal lattice positions via the radiative recombination of the electron and the hole, resulting in characteristic luminescence emission – the so-called π (low-energy triplet state) and σ (high-energy singlet state) emissions, depending on the orbital state of the excited electron (Pooley & Runciman, 1970; Song, Stoneham & Harker, 1975). At higher temperatures the halide ion separation is such that non-radiative recombination is possible. Under these circumstances kinetic energy is transferred to the lattice and a displacement sequence is initiated along the $\langle 110 \rangle$ direction.

The result of this displacement is that, within a few picoseconds, a ground-state F-centre (anion vacancy with a trapped electron) is formed at the beginning of the displacement collision chain and an H-centre (a halogen atom in an interstitial position) is formed at the end of the chain (figure 5.5). Recent developments of the Pooley–Hersh mechanism allow for appreciable exciton diffusion before relaxation (Townsend, 1976). The relaxation probably takes place near defect sites, so there is a strong possibility that F–H pairs form near impurities (see recent reviews by Williams, 1978, and Townsend, 1982).

It should be noted that the displacement sequence involves the movement of an X_2^- molecule (V_k-centre) in an excited state which requires only a small activation energy of 0.02–0.03 eV, so that substantial separation of the F- and H-centre is possible (Itoh & Saidoh, 1973; Saidoh & Townsend, 1975; Townsend & Agulló-López, 1980). Once separated, the H-centre may undergo further diffusion (with an activation energy of the order of 1.0 eV) to form a stable, primary F–H

Figure 5.5. (a) Self-trapped hole (V_k-centre) and free electron after irradiation. (b) Dissociation of self-trapped exciton. X_2 molecule departs in $\langle 110 \rangle$ direction following non-radiative recombination of e^- and V_k. (c) H-centre (or X_4^{3-} crowdion interstitial) forms some distance from the F-centre.

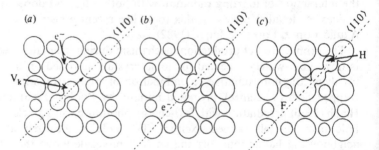

pair, or it may undergo a correlated diffusion back towards the F-centre and recombine, with the emission of luminescence. The F-centre colouring efficiency is thus seen to be temperature dependent and to compete with F–H recombination luminescence (e.g., Hayashi, Ohata & Koshino, 1980). Only π luminescence emission is observed when F- and H-centres recombine, as opposed to both π and σ emission during electron and V_k-centre recombination (e.g., Itoh, Stoneham & Harker, 1977; Tanimura & Okada, 1980).

The presence of both monovalent and divalent impurities is known to enhance the F-centre production during irradiation (e.g., Sonder & Sibley, 1972). This is thought to be due to secondary reactions involving the primary F- and H-centres with the impurities. In fact, F–H production is most likely to occur at defect sites by trapping mobile excitons (Townsend, 1976, 1982). The major features of the F-centre colouring curve (i.e., the three-stage growth, the dose-rate dependence and the impurity dependence) can be explained in terms of trapping of the H-centres at impurities and of heterogeneous nucleation of the interstitial clusters (e.g., Aguilar, Jaque & Agulló-López, 1980; Comins & Carragher, 1980). Thus, monovalent alkali impurities e.g. Li^+ or Na^+ in KCl, KBr, KI, etc.) which have trapped interstitials give rise to a so-called V_1 optical absorption band (Kolopus *et al.*, 1967; Itoh & Saidoh, 1969). Divalent impurities on the other hand give rise to the so-called D bands (e.g., D_3 = two interstitials trapped at a divalent impurity, probably an impurity–vacancy pair; D_2 = trapped interstitials at divalent impurity trimers; Ishii & Rolfe, 1966). Hayes & Nichols (1960) observed an absorption band believed to be due to two interstitials trapped at a divalent impurity (Marat-Mendes & Comins, 1977). A wide variety of interstitial–impurity centres have been suggested (Sonder & Sibley, 1972; Saidoh & Townsend, 1975; Marat-Mendes & Comins, 1975, 1977; van Puymbroeck & Schoemaker, 1981; and many others, summarized in the review by Agulló-López *et al.*, 1982).

The existence of free interstitial point defects, forming the complements to the F-centres, is easily inferred following low-temperature (i.e., liquid helium or liquid nitrogen) irradiation, but they are not observed after room temperature irradiation. The inference is that at high temperatures the interstitials cluster together to form interstitial aggregates. Recent advances in electron microscopy have enabled these interstitial clusters to be studied in detail and the conclusion is that they are in the form of interstitial dislocation loops (Hobbs, Hughes & Pooley, 1972, 1973; Hobbs, 1975). Such extended interstitial defects are thought to give rise to the V_2, V_3, and V_4 optical absorption bands (Ishii & Rolfe, 1966; Itoh, 1972; Hughes, 1978).

Impurities, both monovalent and divalent, can prevent, or slow down, the aggregation process by interstitial trapping. The early stages of irradiation hardening are thought to be due to the stabilization of halogen interstitials by impurities (traps), even at room temperature (Cagon, 1973). The later stages of hardening are believed to be the result of the large interstitial loops presenting a barrier to the dislocation flow (Hobbs, 1975). The impurities are particularly effective at suppressing interstitial aggregation at low temperatures, but divalent impurities are capable of preventing aggregation completely even at room temperature, provided the impurity content is large enough ($> 10^3$ mole ppm; Hobbs, 1975).

For those impurities which act as strong electron traps (e.g., Pb^{2+}, Mn^{2+}, Tl^+) a contrasting effect on the F-centre production rate can take place. At large impurity concentrations (again $\geq 10^3$ mole ppm) these impurties can trap electrons and thus prevent them recombining with the V_k-centres and thereby prevent the initiation of the defect creation sequence. If the electron is released from the impurity, the F–H production sequence can then proceed (Hall, Hughes & Pooley, 1976). In samples of nominally 'pure' or lightly doped alkali halides enhancement of the F-centre colouring rate is normally observed when small amounts of impurity are added.

An additional effect of impurtities is to perturb the intrinsic luminescence from the exciton decay. For example, in NaCl at very low temperatures ($\sim 4\,K$) the normal σ (5.4 eV) and π (3.37 eV) emissions are observed. At temperatures greater than $\sim 60\,K$ however, the σ emission disappears whilst at temperatures of up to 300 K, the character of the π emission is altered. In particular, emission around 2.8 eV overlaps with the intrinsic π luminescence. The interpretation placed on this is that the excited (π state) exciton becomes trapped by monovalent alkali impurities probably arising as a result of exciton diffusion at the higher temperatures (Townsend, 1976; Townsend *et al.* 1976). The 2.8 eV band is termed π_A emission. Similar mechanisms are operative in other alkali halides and these can lead to a suppression of the F-centre formation by monovalent impurity doping (Tanimura & Okada, 1976). Thus impurities influence both the primary process of colouring (by altering the exciton decay mechanism) and the secondary processes (by interstitial trapping).

It is instructive to speculate on what might happen to pre-existing impurity–vacancy dipoles during irradiation. Quite a wide variety of possibilities have been suggested in the literature. The most common observation is that the concentration of impurity–vacancy dipoles (as measured by either *Dielectric Loss* or *Ionic Thermocurrents*) decreases with imparted dose (e.g., Muccillo & Rolfe, 1974; Marat–Mendes &

Comins, 1975). From the foregoing discussion on interstitial trapping it is straightforward to see that such an observation may result from the trapped interstitial destroying, or altering, the dipole moment (or orientation energy) of the impurity–vacancy pair such that its ITC or dielectric loss signal changes. At the same time, the D bands will be observed to grow (Ishii & Rolfe, 1966; Inabe & Takeuchi, 1977). Interstitial trapping at impurity–vacancy dipoles and low-order clusters has recently been discussed by Comins & Carragher (1980) in relation to the first stage of F-centre growth.

Alternative or additional mechanisms have been suggested. For example, in materials doped with impurities with a large capture cross-section for electrons (e.g., Pb^{2+}) a clear possibility is that the charge state of the impurity is reduced by electron trapping (to either Pb^+ or Pb^0) and thereby the dipolar nature of the defect is destroyed. Stott & Crawford (1971) produced evidence for both electron and interstitial trapping at Pb^{2+}-vacancy dipoles in NaCl and KCl. Similarly, Marat-Mendes & Comins (1975), although producing strong evidence for interstitial trapping at Sr^{2+}-vacancy dipoles in KCl, cannot compeletely rule out electron trapping. Kao & Perlman (1979) favour electron trapping for the destruction of Eu^{2+}-vacancy dipoles in KCl, whilst Delbecq *et al.* (1976) use electron paramagnetic resonance spectra to gain evidence of electron trapping at Sn^{2+} in KCl.

Other workers (e.g., Beltrami, Cappelletti & Fieschi, 1964; Nink & Kos, 1980) believe that dipoles may be destroyed via the radiation-induced conversion of the cation vacancy into an F-centre or a Z-centre (i.e., an F-centre perturbed by the presence of an impurity). Additionally, some authors believe that the radiation induces clustering of the dipoles into unstipulated higher-order complexes, presumably due to enhanced impurity diffusion (e.g., Muccillo & Rolfe, 1974). Rubio *et al.* (1982) in particular present strong evidence to support this proposal in NaCl:Eu. In contrast, other authors prefer the view that the radiation in fact breaks up the aggregates to reform dipoles (Watterich & Voszka, 1979; Nakamura *et al.*, 1981). Clearly the situation is quite complex and correct interpretation is not aided by the uncertainty over which impurity complexes are stable in a given material at a given temperature (cf. pairs/dimers/trimers/precipitates). It should be noted that the above alternatives are not mutually exclusive and dipole destruction could be occurring via more than one mechanism.

5.2.3   *Thermoluminescence from* KCl, KBr, KI *and* NaCl, *irradiated at* 4 K

Armed with the above background knowledge of the defect structure of alkali halides it becomes possible to offer interpretations of many of the thermoluminescence properties exhibited by these materials. The main emphasis will be placed on the most recent work and much of the early studies will not be mentioned, unless they offer pertinent information not available elsewhere.

As stated by Alvarez Rivas (1980), the main purpose of studying the nature of a thermoluminescence process is to identify the three basic elements of that process, namely: the trap, the recombination centre and the mobile entity. To this end, the determination of the activation energies, frequency factors and order of kinetics (chapter 3) is a useful step but still leaves the experimenter a long way short of this goal. In the alkali halides, identification of the centres and the charges involved can be simplified by performing the irradiation in different temperature regimes. Thus, irradiation at liquid helium temperatures prevents any thermally activated secondary processes from taking place so that only the primary products of the irradiation (F-, H-, V_k-centres) are involved in the production of thermoluminescence. In addition to these, α-centres (anion vacancy), F'-centres (anion vacancy with two trapped electrons) and I-centres (halide ion interstitials) may also be formed. Electron or hole trapping at impurity sites or other lattice defects should also be considered.

Heating after irradiation should initiate recovery processes such as the recombination between F- and H-centres, between α- and I-centres, or between electrons and V_k-centres. If these processes are radiative, thermoluminescence will result. The F–H recombination is of particular interest. Firstly, because the activation energy for interstitial atom migration along the $\langle 110 \rangle$ direction is expected to be small, any thermoluminescence peaks which result from this recombination will be stimulated at very low temperatures, say 20–50 K. Itoh *et al.* (1977) calculated that the stable F–H-centre pair is in an energy state which is higher than the triplet self-trapped exciton π state, but lower than the singlet σ state, and thus π emission may be observed during thermoluminescence caused by the F–H recombination. However, the efficiency of this process is thought to be extremely low (Agulló-López, 1981). Furthermore, thermoluminescence and optical absorption work on irradiated alkali halides has revealed, firstly, that there is more than one recovery step in the F- and H-centre annealing curve, and secondly, that the thermoluminescence which is excited during some of these annealing steps exhibits *both* π and σ emissions. Other glow peaks exhibit π luminescence only (Tanimura *et al.*, 1974*a*; Purdy & Murray, 1975; López *et al.*, 1980*a*). Typical glow-curves and F- and H-centre annealing steps are shown in figure 5.6. Also shown are the emission spectra associated with the glow-curves and these spectra exhibit both the π and

the σ emissions (see Kabler, 1964, and Pooley & Runcimen, 1970, for a
list of π and σ wavelengths for various alkali halides). It should also be
noted that F–H recombination should not result in thermally stimulated
conductivity, but very often such currents are observed along with
thermoluminescence during F- and H-centre annealing (Cape & Jacobs,
1960; Fuchs & Taylor, 1970a,b). Similar results to those shown in
figure 5.6 have also been obtained for KCl and KI.

The kinetics of these low-temperature glow peaks is found to be pure
first-order by most workers (e.g., Fuchs & Taylor, 1970a,b; López
et al., 1980a; Tanimura & Okada, 1980) although some results suggest a
second-order process (Itoh, Royce & Smoluchowski, 1965; Davidson &
Kristianpoller, 1980. N.B. These latter authors used the peak shape
method to calculate the order, but treated the glow peak in question as a
single peak. If the glow is in fact in the form of a double peak (see below),
the peak shape method will not be applicable.) First-order kinetics are
consistent with correlated recombination between an H-centre and its
associated F-centre along the $\langle 110 \rangle$ direction. A second-order process
would imply motion of the H-centre throughout the crystal, recombining with any F-centre.

The 29 K peak in KBr is a suitable case to illustrate the processes
which might be at work. Fuchs & Taylor (1970a,b) give evidence to
indicate that this peak is composed of two components. The low-
temperature component is calculated by Fuchs & Taylor (1970a) and by
Tanimura et al. (1974a) to have an activation energy of 0.06 eV, which
agrees well with the initial rise value of 0.06 eV obtained by Davidson &

Figure 5.6. Simultaneously measured glow-curves and colour centre (F- and
H-centre) annealing curves for (a) NaCl: glow-curve measured at 230 nm and at
$3.4\,\mathrm{K\,min^{-1}}$; (b) KBr: glow-curve measured at 280 nm and $3.4\,\mathrm{K\,min^{-1}}$. The
emission spectra for the NaCl glow at 16 K and the KBr glow at 29.5 K are
shown in (c). (After Tanimura & Okada, 1980.)

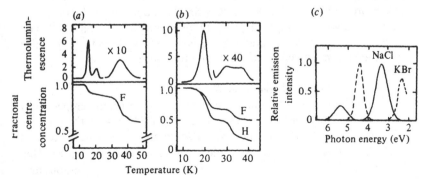

Kristianpoller (1980) and with the value obtained from the annealing data of Itoh *et al.* (1965). This value corresponds to the motion energy for I-centres (Br$^-$ interstitials) in this material. Fuchs & Taylor (1970*a,b*) also observe thermally stimulated conductivity at this temperature. In an effort to explain the presence of both π and σ emissions during thermoluminescence, Aboltin *et al.* (1978) suggested a complex process involving, for example, I + F-centre recombination releasing electrons (to give thermally stimulated conductivity) which in turn radiatively recombine with V$_k$-centres to give π and σ emission. (Most I-centres recombine below 22 K, but $\sim$ 20% are estimated to be still available to take part in thermoluminescence at 29 K). An alternative mechanism might be H–F$'$ recombination, again freeing electrons to recombine at V$_k$-centres. In a very detailed analysis of the 29 K glow from KBr, Tanimura & Okada (1980) were able to separate out the emission at this temperature into two components, one of which gave both π and σ emission, and one which gave π only. The combined π/σ luminescence was explained in the same manner as above, i.e., I + F $\rightarrow$ e$^-$; e$^-$ + V$_k$ $\rightarrow$ π + σ. The π-only component was said to result from F–H recombination. In this way, all of the major observations can be explained. Although differing in detail, most of the low-temperature peaks in these materials may be explained by similar processes.

The main effect of adding impurities to the crystals is to reduce the size of some of the glow peaks and to introduce some others. The characteristic emission from Tl$^+$ has not been observed at these temperatures. However, the characteristic emissions from divalent Ni^{2+} and Mn^{2+} (López *et al.*, 1980*a*) and from Eu^{2+} (Opyrchal & Nierzewski, 1979) have been observed. The reduction in size of the peaks is understood on the basis of diminished F-centre production due to electron trapping by impurities. Thus, those impurities which can easily change their valence state during irradiation (e.g., Mn^{2+} + e$^-$ $\rightarrow$ Mn$^+$) are seen to be more effective in reducing the thermoluminescence than the more stable impurities (e.g., Cs, Sr, Ba; López *et al.*, 1980*a*). This provides a good example of impurity quenching (cf. section 4.5.3). The new peaks may be associated with electron release from the impurities producing luminescence via recombination with the V$_k$-centres.

5.2.4   *Samples irradiated at 80 K*

Irradiation at higher temperature (up to room temperature) allows several secondary processes to take place following the primary F- and H-centre production. At these temperatures the H-centres become highly mobile and are able to move through the lattice. This results in F–H pair dissociation, some H-centre aggregation and interstitial trapping (forming a variety of H-, V- and D-centres). V_k-centre and exciton diffusion also become possible. The net result of these processes is to produce a rather more complex thermoluminescence glow-curve, in terms of the number, position and size of the peaks present, depending upon the irradiation temperature.

Many of the same kind of processes which are operative following low-temperature irradiation can also take place and produce thermoluminescence at these higher temperatures – in particular interstitial–vacancy recombination mechanisms remain important factors. However, many of the higher temperature glow peaks cannot be explained by such mechanisms and 'conventional' electron–hole recombination appears to be taking place.

In KBr crystals irradiated by monochromatic ultra-violet light at 80 K, Kristianpoller & Israeli (1970) observed glow peaks at ~ 103 K, 120 K and 158 K when heated at 25 K min^{-1} (figure 5.7). The 120 K and 158 K peaks could both be stimulated by F-band light (~ 600 nm) which was interpreted by the authors to mean that the peaks are related to the

Figure 5.7. Thermoluminescence in KBr after ultra-violet irradiation at 80 K with light of: A, 187 nm; B, 188.5 nm; C, 192 nm; D, 203 nm. Heating rate = 25 K min^{-1}. (After Kristianpoller & Israeli, 1970.)

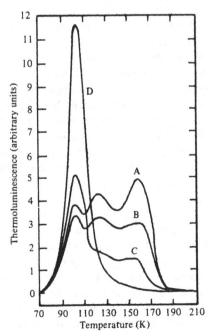

thermal release of electrons (note the limitation of F-light stimulation as discussed in section 4.3.1). The 103 K peak could not be stimulated by ultra-violet irradiation into any of the known F- or V-type centres. The excitation spectrum for the 103 K peak indicated stimulation by the long-wavelength tail of the first exciton band (~ 183 nm) and a large maximum corresponding to the α-band (103 nm). The authors concluded that this thermoluminescence peak is caused by the radiative recombination of a trapped interstitial and a vacancy centre. Although this temperature region coincides with the thermal annealing of the V_1-band in KBr, these authors claim that the interstitial involved is unlikely to be an atom (i.e., a Br atom released from a monovalent alkali impurity) because of the later observations by Kristianpoller & Davidson (1972) of thermally stimulated conductivity at this temperature (~ 110 K). Thus, it was concluded that this peak results from the radiative recombination of charged Frenkel pairs, i.e., I_A–α recombination. (Here, I_A refers to a trapped interstitial ion at an alkali impurity. The equivalent nomenclature for an interstitial atom would be H_A-centre, but in this book the alternative name of V_1-centre is used.)

Tanimura, Okada & Suita (1974b) also observe thermoluminescence in KBr:Na at 102 K and 116 K. These authors relate the 102 K peak to the recombination of V_1-and F-centres and the 116 K peak to recombination between I_A- and α-centres. There is clearly some confusion over the peak positions, and it is possible that the 103 K glow peak observed by Kristianpoller & Israeli (1970) is the same as the 102 K peak observed by Tanimura *et al.* (1974b), whilst the 110 K thermally stimulated conductivity peak of Kristianpoller & Davidson (1972) may be related to the 116 K glow peak of Tanimura *et al.* Thermoluminescence in the higher temperature regions (120–130 K and 158–162 K) has also been related by Tanimura *et al.* to F–V_1-centre recombination. The latter designations must be in some doubt, however, owing to the observation of thermally stimulated conductivity in this temperature range (Kristianpoller & Davidson, 1972). First-order kinetics are indicated for all thermally stimulated conductivity and glow peaks.

More recently, however, Dauletbekova, Akilbekov & Elango (1982) present strong evidence from thermally stimulated and photostimulated luminescence measurements to suggest that although both F–V_1 and I_A – α recombination take place in this temperature range; the I_A – α reaction does not stimulate luminescence. Also, during F – V_1 recombination, weak π emission is noted, but the authors suggest that the dominant emission (2.9 eV) arises from the relaxation of an excited (V_1 + e)-centre, the electron having tunnelled from a nearby F-centre.

Recently obtained glow-curves and thermally stimulated conductivity curves from KCl irradiated at 80 K are shown in figure 5.8. After detailed peak separation analysis using thermal cleaning techniques (see section 3.3.1), Jiménez de Castro & Alvarez Rivas (1979) conclude that, apart from the peak at ∼ 200 K, there are no obvious correlations between the thermoluminescence generated by X-irradiation at 90 K and that generated by F-light irradiation following room temperature irradiation with X-rays, indicating that electron traps are not responsible for the X-induced glow-curve. These authors observed that for each glow peak induced by X-irradiation, there is a corresponding F-centre annealing process, except for the peak at 200 K which clearly has a strong thermally stimulated conductivity signal associated with it. Except for this peak, none of the thermoluminescence peaks have an equivalent thermally

Figure 5.8. (a) Selected glow-curves and (b) selected thermally stimulated conductivity (TSC) curves for KCl from Jiménez de Castro & Alvarez Rivas (1979). Curves A are following X-irradiation at 80 K; curves B are from F-light stimulation at 80 K after X-irradiation at room temperature. Heating rate $= 8.4 \, \text{K min}^{-1}$.

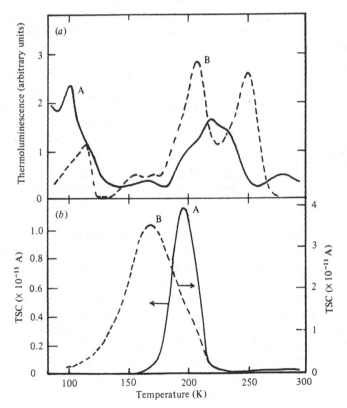

stimulated conductivity signal, which would be expected if the electrons or holes were the mobile species released into the delocalized bands during heating. Itoh (1972) reports annealing of various V_1-centres at 120 K (Na^+) and 240 K (Li^+) in KCl and it is, therefore, tempting to relate the glow peaks at 120 K, and 230 K respectively to the recombination of F-centres with interstitial Cl atoms released from these impurity sites. Similarly, Itoh (1972) reports the annealing of an H_H-centre (interstitial Cl atom trapped at a F^- impurity) at 170 K, which agrees closely with the position of the 165 K glow peak. Indeed, all of the observed thermoluminescence processes (except that at 200 K) appear to be related to various forms of F–interstitial recombination.

All the glow peaks, including the 200 K peak, emit at wavelengths corresponding to the π exciton emission. An explanation of the 200 K peak, offered by Jiménez de Castro & Alvarez Rivas (1979), is that it is caused by annihilation of V_k-centres which are mobile at these temperatures. However, their calculated activation energy of 0.14 eV is far below that required for this process (namely, 0.5 eV). A possible alternative is I_A–α recombination. In this regard, it is interesting to note that Jiménez de Castro & Alvarez Rivas (1979) report thermoluminescence emission at wavelengths corresponding to the excited state of the α-band.

The kinetics of all the peaks appear to be first-order, but of special interest are the remarkably low values for the frequency factors calculated by Jiménez de Castro & Alvarez Rivas (1979), ranging from $7 s^{-1}$ to $5 \times 10^7 s^{-1}$. This is support for correlated recombination between the interstitial and vacancy centres, as previously discussed in detail in section 3.3.7.

Figure 5.9 shows the X-irradiated and F-light-stimulated glow-curves and thermally stimulated conductivity curves for NaCl after irradiation at 80 K. Here, the F-stimulated glow-curves exhibit peaks at 89, 98, 117, 211 and 236 K which are also observed in the X-irradiated samples, implying that these peaks are related to the release of trapped electrons. Each of these peaks can be discerned in the thermally stimulated conductivity spectra, some more clearly than others, but none of them can be correlated to F-centre annealing steps. However, glow peaks at 145, 170, 253 and 285 K do correlate with F-centre annealing processes, suggesting once again a recombination of F–H pairs. Isothermal decay analysis indicates first-order kinetics and low values for the frequency factors (ranging from $35 s^{-1}$ to $2 \times 10^6 s^{-1}$) even for the electron- and hole-related peaks, suggesting correlated recombination of the type shown in figure 2.6. (But note that thermally stimulated conductivity is not expected from this model.)

López *et al.* (1980*b*) have suggested that the thermoluminescence process at 170 K in pure NaCl is due to V_k–e recombination. They suggest that the V_k-centre becomes mobile in this temperature region and that an electron tunnels to the V_k-centre, from an F′-centre, to recombine with the hole and emit light. This suggestion is supported by the observation of both π and σ emission at this temperature and by recent *Thermally Stimulated Exoemission* results (Bräunlich & Dickinson, 1979; Kamada & Tsutsumi, 1981).

Doping alkali halides with impurities introduces new emissions into the thermoluminescence spectra, characteristic of the impurity. Thus, Cu^+ doping in NaCl induces an emission at 350–355 nm which is observed from all of the X-ray and F-light-induced glow peaks (López

Figure 5.9. (*a*) Selected glow-curves and (*b*) selected thermally stimulated conductivity (TSC) curves for NaCl from Jiménez de Castro & Alvarez Rivas (1980). Curves A are following X-irradiation at 80 K; curves B are from F-light stimulation at 80 K after X-irradiation at room temperature. Heating rate = 8.4 K min⁻¹.

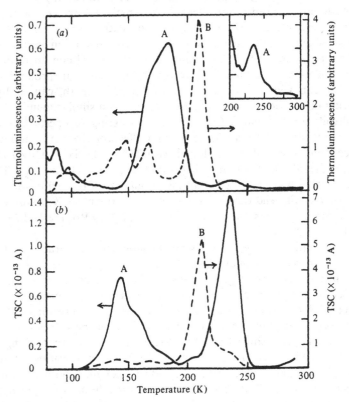

et al., 1980*b*). The normal intrinsic π emission is still detectable, but the weaker σ emission is not (López *et al.*, 1980*b*). Herreros & Jaque (1979) and Kamada & Tsutsumi (1981) report an intense glow peak at ~ 180 K which they relate to the presence of Cu^+.

Doping with Tl^+ gives rise to some interesting thermoluminescence phenomena. For example, in $KCl:Tl^+$ the emission is dominated by the 305 nm emission characteristic of the $3P_1-1S_0$ transition in thallium (Johnson & Williams, 1952). (This emission wavelength is seen at all glow-curve temperatures above room temperature, particularly for the peak at ~ 530 K (Mattern *et al.*, 1970). Beyond this temperature, the 305 nm emission is thermally quenched (cf. figure 2.8)). Below room temperature (for specimens irradiated at 77 K) an emission at 500 nm is seen to play an important part in the glow-curve. From a combination of *electron spin resonance* and glow-curve measurements, Toyotomi & Onaka (1979*a*,*b*) determine that thallium can act as both electron and hole traps by forming centres such as Tl^0, Tl^{2+}, Tl_2^+, Tl_2^{3+}, etc. These centres induce several glow peaks which are not observed in pure KCl. At low dopant concentrations, the characteristic 305 nm emission is observed due to isolated Tl^+, but at higher concentrations, this emission is quenched and the 500 nm emission is observed instead. Furthermore, the glow peaks move to lower temperatures as the Tl concentration increases. This is interpreted by Toyotomi & Onaka as a disturbance of the band gap width, and, therefore, of the trap depth, caused by clustering of Tl^+ defects into larger complexes in a similar manner to that already discussed in section 2.3.3. The clustering also perturbs the 305 nm emission (to 500 nm). Concentration effects on the thermoluminescence were observed in KI doped with Tl^+ by Barland, Duval & Nouailhat (1980). In this latter work, evidence for V_k-Tl^0 and V_k-Tl^+ recombination was presented from measurements of thermally stimulated conductivity and thermoluminescence on specimens X-irradiated at 77 K. Similar results are discussed for $KBr:Tl^+$ by Roth & Halperin (1982).

Divalent cation doping induces similar changes in the character of the thermoluminescence, namely, altering the wavelength of the emission, reducing the intensity of glow peaks which are readily seen in pure samples, and inducing new peaks not observed in pure specimens. Mn^{2+} doping in NaCl provides a useful example. Detailed emission spectra studies reveal that for the 170 K peak observed in pure NaCl the σ emission is largely reduced and that the π emission is either strongly perturbed, from 360 to 410 nm, or that a new, stronger transition is now taking place (López *et al.*, 1980*b*). In addition, an emission at 580 nm, characteristic of Mn^{2+}, is observed. Furthermore, the activation energy associated with this peak is 0.37 eV (López *et al.*, 1981). The interpretation placed upon this peak is that V_k-centres recombine with electrons from Mn^0-centres, rather than F- or F'-centres, as in pure specimens. Also, two peaks, one at 108 K and one at 222 K (heating rate $= 2 K$ min^{-1}) are observed to grow with increased Mn^{2+} doping.

It is interesting to observe the growth and thermal annealing behaviour of optical absorption bands introduced by the addition of impurities. Thus, in $KCl:Sr^{2+}$ X-irradiated at 80 K, Jiménez de Castro & Alvarez Rivas (1981) are able to correlate many of the thermoluminescence peaks with the decay of trapped-interstitial centres, such as interstitials trapped at Sr^{2+}-cation vacancy dipoles (e.g., D-centres and Hayes–Nichols centres). These authors interpret their results in a similar fashion to their interpretation of thermoluminescence from pure KCl (Jiménez de Castro & Alvarez Rivas, 1979), namely, recombination between interstitials (this time released from traps associated with Sr^{2+}) and F- or F'-centres. Strong emission at 450 nm is interpreted as π emission perturbed by Sr^{2+}-vacancy dipoles.

5.2.5 Samples irradiated at room temperature

By far the most popular type of thermoluminescence experiment on alkali halides is to produce a glow-curve following irradiation at room temperature. At this temperature, all of the defects discussed so far are unstable with the result that the thermoluminescence arises from a whole new set of charge recombination processes. The nature of the glow-curve from alkali halides above room temperature has been the subject of intense discussion for many years. Most of the early interpretations have taken the viewpoint that the recombination processes involve conventional electron and hole movement. It has been recognized for sometime that F-centres are intimately involved in the production of thermoluminescence, but their exact role has been the centre of much discussion and conjecture. One school of thought has suggested that the glow peaks result from the thermal release of holes from different kinds of traps, recombining with electrons at F-centres (e.g., Klick *et al.*, 1967). Another body of opinion has discussed the possibility of electron release from F-centres recombining with trapped holes at the luminescent sites (e.g., Jain & Mehendru, 1970; Mehendru, 1970a; Hageseth, 1972). The possibility of a single trap and a range of different recombination sites has also been forwarded (Hill & Schwed, 1955). More recently, however, the suggestion of interstitial–vacancy recombination has been strongly pressed (e.g., Ausin & Alvarez Rivas, 1972a, 1974). (It is possible, of course, that both electron–hole and interstitial–vacancy recombination mechanisms are taking place (e.g., Murti & Sucheta, 1982).)

The involvement of interstitials in the thermoluminescence process is difficult to deny since some glow peaks are not observed from additively coloured alkali halides (Jain & Mehendru, 1970; Ausin & Alvarez Rivas, 1972*a*), but their precise role is not easy to confirm. At room temperature, the interstitials are mobile resulting in the formation of large interstitial clusters. Electron microscopy has revealed that the form of these clusters is that of neutral halogen molecules (X_2), formed from H-centres. The halogen molecules occupy anion–cation vacancy pairs which have given up interstitials to form dislocation loops (Hobbs *et al.*, 1972, 1973). The halogen molecule centres give rise to various optical absorption V-bands. During heating following irradiation the F-centre and V-centre absorption bands disappear at temperatures up to 550 K. Hughes (1978) proposes the following type of reaction for this annealing:

$$2F + X_2 \rightarrow \text{empty vacancy pair.} \qquad (5.2)$$

Thus, reaction (5.2) leaves the crystal with empty vacancy pairs and dislocation loops. At much higher temperatures (600–700 K) the dislocation loops annihilate the vacancy pairs, thus:

$$\text{Vacancy pair} + \text{interstitial pair in loop} \rightarrow \text{perfect lattice.} \qquad (5.3)$$

In this way, full recovery is completed. Vignolo & Alvarez Rivas (1980) have produced evidence from ionic conductivity in KCl irradiated at room temperature to support this mechanism. They note that reaction (5.3) appears to occur at ~ 600 K in KCl but add that a third annealing stage occurs at even higher temperatures (~ 800 K) which they suggest is related to the formation of divacancy clusters.

From the point of view of thermoluminescence it is important to realize that the *total* area under the glow-curve of a sample irradiated at room temperature is directly proportional to the F-centre concentration. Thus, it is not surprising to note that thermoluminescence appears only below ~ 550 K in KCl (Ausin & Alvarez Rivas, 1972*a*, 1974) – i.e., the temperature range in which reactions of the type given in equation (5.2) appear. Thus, it seems that recombination of F-centres and halogen atoms is a plausible source of the thermoluminescence emission. In support of this proposal Alvarez Rivas and colleagues have produced an extensive series of papers which interpret the glow-curve in this fashion for KCl, KBr, KI, NaF, NaCl (Ausin & Alvarez Rivas, 1972*a,b*, 1974; Mariani & Alvarez Rivas, 1978; Delgado & Alvarez Rivas, 1979). An essentially identical model has been suggested by Ang & Mykura (1977).

An important feature of Alvarez Rivas's argument is that for every glow peak, there is a corresponding annealing step in the F-centre concentration, an example of which is shown in figure 5.10. At sufficiently high temperatures (> 400 K in NaCl) F-centres become mobile and form aggregates (M-centres, R-centres, etc.) and colloidal centres (Hughes & Jain, 1979). Thus, it is necessary to modify reaction (5.2) as the cause of thermoluminescence to include recombination between interstitial halogen atoms and F-aggregate centres, or vacancy centres in general. The observation that the colloid optical absorption band possesses greater thermal stability than the F-band presents a possible source of difficulty for this model (Lerma & Agulló-López, 1973; Ausin & Alvarez Rivas. 1974).

A pertinent observation is that the lower temperature peaks, after first increasing with absorbed radiation dose, progressively disappear as the dose increases whilst glow peaks at higher temperatures grow correspondingly bigger. Thus, the precise shape of the curve depends strongly on the dose delivered. The interpretation of this is based on the observation of increased interstitial aggregation at higher doses (e.g., Hobbs *et al.*, 1973). Thus, the different peaks are believed to result from

Figure 5.10. Annealing of F-centres and corresponding glow peak in KCl irradiated at room temperature. Heating rate $= 13.5\,°C\,min^{-1}$. (After Ausin & Alvarez Rivas, 1972*a*.)

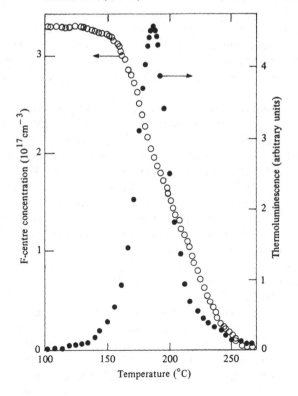

the release of halogen atoms from bigger, more tightly bound aggregates (Ang & Mykura, 1977). The kinetics of most of the peaks appears to be first-order, but some of the higher temperature peaks do not obviously conform to first-order behaviour. An estimate of the activation energies has been obtained from isothermal decay curves, but it is difficult to be confident of the values obtained because of the complexity of the glow-curve shapes. This is particularly true of those glow peaks in KCl which have been induced by plastic deformation (Ausin & Alvarez Rivas, 1972*b*). Tanimura *et al.* (1978) have determined from flow stress measurements in KBr that both I- and H-centres are effective as hardening agents. Thus, the complex kinetics of the deformation-induced glow peaks are interpreted as release of interstitials from several sites associated with the dislocation debris. However, to be certain of the number of individual processes involved in the production of these thermoluminescence peaks (Ausin & Alvarez Rivas suggest at least five), more rigorous analysis of the glow-curve is needed, as recently pointed out by Murti & Sucheta (1982).

The thermoluminescence from doped alkali halides has been extensively studied for several years and many published articles exist in which are discussed a wide range of possible mechanisms for the production of thermoluminescence. It clearly would not be feasible to present a detailed discussion of all of these systems and instead some specific examples will be chosen which illustrate the mechanisms at work.

One of the important observations emerging from the glow-curves of pure materials is that the intrinsic π luminescence, which might occur during F–H-type recombinations, does not occur above room temperature. Instead, all of the glow peaks emit in approximately the same wavelength region depending on the host material (e.g., NaCl – 425 nm; KCl – 440 nm; KBr – 460 nm; KI – 530 nm (Ausin & Alvarez Rivas, 1972*a*; Mariani & Alvarez Rivas, 1978)). These emissions are thought to be π emissions perturbed by trace impurities (Ikeya, 1975; Delgado & Alvarez Rivas, 1981, 1982) but, whatever their origin, they are considered to be characteristic of F–H-type recombination.

In impure materials, the emissions become slightly altered, presumably due to perturbations characteristic of the dopant material. Thus, Mn^{2+} doping in NaCl induces emissions at 400 nm, plus a more intense emission at 595 nm known to arise from internal transitions in the Mn^{2+} ion (López *et al.*, 1977; Delgado & Alvarez Rivas, 1982). Cu^+ doping in the same material produces 365 nm and 445 nm emission, said to relate to F–H recombination and V_k–Cu^- recombination, respectively (Delgado & Alvarez Rivas, 1981). In KCl doped with Ca^{2+},

Sr^{2+} or Ba^{2+}, a single emission band at 445 nm (Ca) or 450 nm (Sr or Ba) is observed (Rascón & Alvarez Rivas, 1978, 1981). In these cases the emission from pure KCl (440 nm) is thought to be perturbed by the impurity–vacancy dipole associated with the impurity. Perturbation effects on luminescence in general by divalent impurities have been treated recently by Egemberdiev, Zazubovich & Nagirnyi (1982a) and Egemberdiev *et al.* (1982b).

From evidence gathered in ITC and electron spin resonance measurements (e.g., Marat-Mendes & Comins, 1975, 1977; van Puymbroeck & Schoemaker, 1981) impurities which do not readily change their valence state during irradiation function primarily as traps for interstitials, although electronic trapping cannot be ruled out. Thus, Rascón & Alvarez Rivas (1978, 1981) show that thermoluminescence from KCl: Ca^{2+}, Sr^{2+} or Ba^{2+} results from the release of trapped Cl interstitial atoms from impurity sites. In particular, the glow peaks are seen to correlate well with annealing steps in both the F-centre and the D_3-centre concentrations. Inabe & Takeuchi (1977, 1978) discuss the effect of impurity aggregation on the glow-curve properties in KCl:Sr^{2+}.

With specimens doped with impurities which easily trap electrons, electron–hole recombination processes also become important. In Mn^{2+}-doped specimens of NaCl each glow peak emits at both 400 nm and 595 nm. The latter emission results from a freed hole being trapped at an Mn^+ site. The hole trapping creates Mn^{2+} in the first excited state, which decays to the ground state with the emission of thermoluminescence (López *et al.*, 1977). The origin of the 400 nm emission is more complex but may result from the F–H-type recombination, as suggested by Delgado & Alvarez Rivas (1982) or may result from the release of trapped V_k-centre recombining with Mn^+ (López *et al.*, 1980c). However, in this latter mechanism, F–H-type recombination is still required to account for the F-centre destruction, but in this model it does not stimulate light emission. Thus, to account for both 400 nm and 595 nm emission at each glow peak, both hole and interstitial release are required.

5.2.6 *Thermoluminescence from LiF*

To complete the description of thermoluminescence from alkali halides, a discussion of the LiF glow-curve is required. This material is treated separately from the other alkali halides because it has been subjected to more detailed investigation, stemming undoubtedly from its importance as a radiation dosemeter (chapter 6). An early review of the topic was given by Stoebe & Watanabe (1975) with a more recent one by Jain (1982). This section is not intended to give a historical development of the arguments relating to the origin of the glow-curve in LiF, but will concentrate on the more recent advances in the understanding of the defects which give this material its characteristic thermoluminescence properties. These properties are often discussed in isolation from those of the other alkali halides, which is unfortunate since much benefit can be derived from such a comparison. Emphasis will be placed upon the thermoluminescence obtained after irradiation at room temperature, in particular upon the glow peaks between room temperature and about 250 °C, since it is this region of the total glow-curve which is important in practical applications.

Since the first use of LiF: Mg, Ti as a dosemeter there have been many studies aimed at understanding the nature of the trapping sites responsible for the complex behaviour of the LiF glow-curve, particularly following different radiation and thermal annealing treatments. To assist in this understanding, dielectric loss, ionic conductivity, optical absorption and electron spin resonance have all been employed. However, when complicating factors regarding glow peak intensities and defect concentrations (mentioned in the introduction to this chapter) are taken into account, and when it is realized that the trap and recombination centre in LiF may be in close association (Sagastibelza & Alvarez Rivas, 1981; Taylor & Lilley, 1982b), it can be seen that we are faced with a situation of immense complexity and it is not surprising to find that a large degree of confusion still surrounds the interpretation of the experimental observations.

Following this warning, one can now attempt to describe a selection of the more helpful recent results on the properties of LiF. Some of the early colour-centre literature is highly relevant. From the earliest studies, it

Figure 5.11. Thermoluminescence and 310 nm-centre annealing in TLD-100, irradiated to 10^3 Gy (heating rate = 16 °C min^{-1}). (After Sagastibelza & Alvarez Rivas, 1981.)

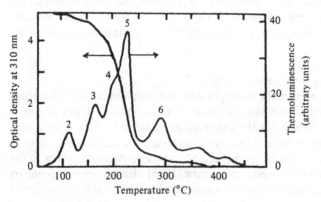

has become apparent that the glow peaks are related to a variety of optical absorption bands. In particular, F-centres are obviously important in the production of thermoluminescence (Christy, Johnson & Wilbarg, 1967; Klick *et al.*, 1967; Mayhugh, 1970; Mayhugh *et al.*, 1970; Mayhugh & Christy, 1972; Crittenden *et al.*, 1974*a*). Additional measurements gave strong evidence of a relationship between glow peaks 2 and 3 and an absorption band at 380 nm (said to be related to Mg) and between peaks 4 and 5 and an absorption band at 310 nm (also related to Mg – see the above cited works, plus Miller & Bube (1970) and in particular Jackson & Harris (1969, 1970)). Furthermore, Mayhugh and colleagues (Mayhugh, 1970; Mayhugh & Christy, 1972) emphasized the disappearance of the V_3-band during thermoluminescence readout. Some of these relationships have been confirmed by recent results (Sagastibelza & Alvarez Rivas, 1981) in which F-centre and 310-nm-centre annealing steps were observed in TLD-100 during glow-curve readout. An example is shown in figure 5.11.

The origin of the 310 and 380 nm absorption bands is still uncertain, but both are related to Mg (e.g., Mort, 1965), and some experimenters have suggested that they might be Z-type centres. Z-centres are normally only formed in alkali halides, when doped with divalent cation impurities, by bleaching the irradiated crystal with F-band light. LiF, however, appears to be an exception if we accept that the Z-centres form easily, along with F-centres, during primary irradiation. The centres are a result of interaction between the divalent impurity and the F-centre. Five different types of Z-centre are known ($Z_1 \dots Z_5$) although their detailed structure and their thermal annealing behaviour are, even now, improperly understood (Hashizume, 1979; Paus & Strohm, 1979, 1980; Strohm & Paus, 1980; Nishimaki & Shimanuki, 1980; Paus & Scheu, 1980). The relationship between Z-centres and thermoluminescence in other alkali halides has often been discussed (e.g., Gartia, 1976; Moharil, Kamavisdar & Deshmukh, 1979; Kamavisdar & Deshmukh, 1981; Reddy, Rao & Babu, 1982; Sridaran *et al.*, 1981).

In LiF, a mechanism to describe the thermoluminescence process has recently received much attention in the literature. The so-called 'Z-centre model' of Nink & Kos (1976) is based upon the identification of an optical absorption band at 225 nm with a Z_3-centre (an Mg^{2+}-F complex). This identification was made by Mort (1966) using a Mollwo–Ivey relationship for Z_3-centres in alkali halides. A glow peak at 250 °C (peak 6) is then identified with the thermal annealing of Z_3-centres (Gartia, 1977; Kos & Nink, 1979) and from the interrelationships between the 310 nm, 380 nm and 225 nm bands, Nink & Kos conclude that the 310 nm band is caused by Z_2-centres (Mg^{2+}-F'). Thus, the Z-centre model views the thermoluminescence at peak 5 as being stimulated by the release of one electron from a Z_2-centre, converting this to a Z_3-centre. Peak 6 results from the release of the last electron. The luminescence emission may arise by an earlier mechanism suggested by Mayhugh (1970; see also Caldas, Mayhugh & Stoebe, 1983), whereby the released electrons recombine with V_3-centres, converting these to V_k-centres. The V_k-centres are unstable at these high temperatures, so the holes are immediately released to recombine at Ti sites with electrons which have tunnelled from nearby F-centres.

Mehta, Merklin & Donnert (1977) extend the Z-centre model to suggest that the 380 nm band is caused by a Z'_2-centre (i.e., a Z_2-centre with an extra trapped electron). Peak 2 is then said to result from the thermal release of this additional electron (converting the centre to a Z_2-centre). An apparent relationship between the intensity of peak 2 and the concentration of Mg^{2+}-vacancy dipoles (e.g., Grant & Cameron, 1966) is explained by Kos & Nink (1980) in terms of a dipole $\rightarrow$ Z-centre conversion mechanism. Figure 5.12 shows the positions of the absorption bands with respect to the F-centre.

However, more recently, strong criticism of the Z-centre model has accumulated. Moharil (1980) points out the inconsistent behaviour of the 225, 310 and 380 nm bands with respect to the Z-centres observed in other alkali halides (e.g., Paus & Strohm, 1979, 1980 Strohm & Paus,

Figure 5.12. Absorption spectrum for irradiated LiF:Mg,Ti. Dashed lines: calculated Gaussian absorption bands. Solid line: sum of dashed lines. Dots: experimental points. (After Nink & Kos, 1980.)

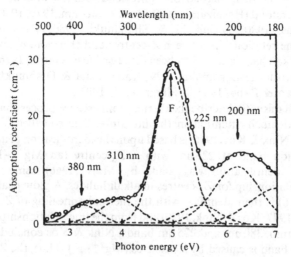

1980). Nepomnyachikh & Radyabov (1980) suggest that the 225 nm band is caused by a Mg atom on an anion site associated with a cation vacancy, whilst Radyabov & Nepomnyachikh (1981) assert that the properties of the 310 nm band are consistent with the behaviour of Mg^+-cation vacancy–anion vacancy centres. Horowitz (1982) points out that, according to the Z-centre model, the growth characteristics of the 380 and 310 nm bands should be identical to those of peaks 2 and 5 respectively, whereas experimental observations reveal that they are quite different (e.g., Horowitz, 1981). Jain (1981a), Bapat & Kathuria (1981) and Lakshmanan, Chandra & Bhatt (1982) relate the 225 nm band to a peak at 400 °C, i.e., not peak 6. Finally, Taylor & Lilley (1982b,c) show that the totally different thermal annealing properties of peaks 2 and 5 are at variance with the Z-centre model modification introduced by Mehta et al. (1977).

When discussing possible origins of the glow peaks in LiF, the observation by Grant & Cameron (1966) mentioned briefly above, namely, that of an apparent relationship between peak 2 and the concentration of Mg^{2+} impurity–vacancy dipoles, is deserving of closer scrutiny. Grant & Cameron monitored the decay of a non-equilibrium excess of impurity–vacancy dipoles at 67 °C following a 400 °C, 1 hour anneal and rapid cool (cf. section 5.1). The measurements were taken using dielectric loss and by comparing the decay curves with similar curves for the intensity of peak 2, Grant & Cameron concluded that impurity–vacancy dipoles were the traps responsible for this glow peak. However, the direct proportionality found by these authors between impurity–vacancy dipoles and peak 2 is only present near 70 °C, and not at higher ageing temperatures (Dryden & Shuter, 1973). Similar annealing studies on peaks 2 to 5 were carried out earlier by Harris & Jackson (1969) but without the dielectric loss measurements. These and other works illustrate that by annealing the LiF crystal in the association region of the ionic conductivity plot (cf. figure 5.4), in order to induce the presence of large numbers of impurity–vacancy dipoles, the intensity of peaks 2 and 3 could be enhanced. Annealing at lower temperatures in order to induce aggregation into clusters reduced peaks 2 and 3 and increased peaks 4 and 5. Treating the crystal so as to induce precipitation reduces all of the peaks except for peak 2, which shows a slight enhancement and then a decrease (Bradbury & Lilley, 1977a,b; Rao, 1974). The effects that the various treatments have on the TLD-100 glow-curve have already been shown in figure 4.4.

These heat treatments have recently been studied in further detail by Taylor & Lilley (1982a,b). Some of the important features of these measurements are summarized in figure 5.13 in which it can be seen that peak 2 and the impurity–vacancy dipole concentration do not follow each other identically and that, when precipitation occurs, peak 2 actually exhibits an initial increase in intensity. This led Taylor & Lilley (1982*b*) to doubt that peak 2 was related to Mg^{2+} at all, and they suggested that it is due to Ti complexes. This suggestion, however, is not favoured by detailed work on LiF:Ti crystals (Mieke & Nink, 1980; Townsend *et al.*, 1983*b*). LiF:Ti exhibits just two glow peaks, both of which exhibit completely different behaviour to peak 2 in TLD-100.

The thermal annealing studies appear to indicate that the peak 5 defects are clusters of whatever defects cause peak 2. Cooling rate studies lend support to this identification (Bradbury *et al.*, 1976; Dhar, DeWerd

Figure 5.13. Selected data from Taylor & Lilley (1982*b*) showing the decay of thermoluminescence peaks 2 and 5, along with the decay of Mg^{2+} impurity–vacancy dipoles (deduced from dielectric loss) and the growth of precipitates (ppt), at temperatures of (*a*) 80 °C and (*b*) 170 °C in TLD-100.

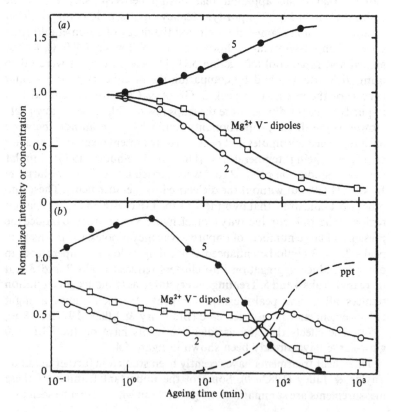

& Stoebe, 1973). Detailed investigations of clustering kinetics support the view that the clusters involved in peak 5 are trimers, thus implying that the defects involved in peak 2 are dipoles. As discussed in section 3.3.7 a dipole–trimer model can provide an explanation for the high value of the trap depth and frequency factor for peak 5. However, we cannot ignore the data of figure 5.13 which clearly throw doubt on this simple assignment. It appears that although a one-to-one correlation between peak 2 and dipoles is not evident, a relationship of some sort must exist.

Some of the features of figure 5.13 may be due to competition between the defects, but a clearer view of the situation is afforded by an examination of the emission spectrum. (However, before doing this it is useful to pause for a moment to examine an interesting question which now presents itself. If annealing at 400 °C for 1 hour, followed by a rapid quench to room temperature, distributes all of the Mg^{2+} in impurity-vacancy dipole form, how can peaks 4 and 5 be present in a specimen which has undergone this pre-irradiation treatment? The answer was provided by Taylor & Lilley (1982c) who extended earlier work by Booth, Johnson & Attix (1972) to suggest that clustering of pairs into trimers actually occurs during glow-curve readout. A heating rate of $\sim 10\,°Cs^{-1}$ is necessary to avoid appreciable clustering and $\sim 0.5\,°Cs^{-1}$ to avoid precipitation. An additional factor may be the possibility that the radiation itself induces clustering into aggregates (Muccillo Rolfe, 1974; Rubio et al., 1982).)

The emission spectrum of nominally pure LiF exhibits a main emission at 415 nm for all the glow peaks up to ~ 525 K when irradiated at room temperature (Sagastibelza & Alvarez Rivas, 1981). A less intense emission (at 320 nm) is seen to be operative at temperatures less than ~ 470 K and greater than 525 K. Low-intensity emission occurs within the red part of the spectrum (600–700 nm; thought by Takeuchi et al. (1982) to be due to internal transitions in the Mg ion on an anion site). For heavily irradiated specimens this increases in intensity, peaking at ~ 663 nm. Sagastibelza & Alvarez Rivas suggest that the complex emission in LiF is dependent upon background impurities. This suggestion is supported by the observation that, at sufficiently high doses, the glow-curve for 'pure' LiF resembles the glow-curve from LiF:Mg, Ti (figures 1.4 and 4.6).

The 415 nm emission is in the same spectral range as that observed for other alkali halides (viz., 425 nm (NaCl); 440 nm (KCl); 460 nm (KBr)) and it was discussed in the last section that this emission may result from interstitial–vacancy recombination. More specifically, it was suggested that it may be the intrinsic π emission (350 nm in LiF) perturbed by trace impurities (Ikeya, 1975; Delgado & Alvarez Rivas, 1981, 1982). This is consistent with a vacancy–interstitial recombination mechanism despite arguments which suggest that the efficiency of the emission will be very low (Agulló-López, 1981).

Perturbation of this emission is suggested in LiF specimens intentionally doped with divalent impurities. The main emission band in TLD-100 (i.e., LiF:Mg,Ti) is variously reported to lie between 400 and 430 nm. This is confirmed by detailed spectral measurements during thermoluminescence readout, resulting in isometric plots (Fairchild *et al.*, 1978*a*; Townsend *et al.*, 1983*b*). A typical isometric plot for TLD-100, along with its associated contour plot, was shown in figure 2.14. Detailed analysis of these plots indicates that the maximum emission shifts to lower wavelengths with increasing peak temperature, in support of previous work (Harris & Jackson, 1970*a*; DeWerd & Stoebe, 1972). Thus, Townsend *et al.* (1983*b*) observe that peak 2 (cf. figure 1.4) emits at 460 nm, peak 3 at 435 nm, peak 5 at 425 nm and peak 6 (at $\sim 270\,°C$) at 425 nm. These authors also note that in LiF:Ti, only one emission is noted for all peaks, at 410 nm. Kathuria (1979) observes that in Mn-, Sm- and Tb-doped crystals, the main emission is also in this same spectral region.

The conclusion arrived at by Townsend *et al.* (1983*b*) is that the emission site is closely associated with the impurities. In LiF:Ti and LiF:Mg,Ti the site is probably a large defect complex associated with Ti. Ti enters the LiF lattice substitutionally for Li^+ in either the Ti^3 or Ti^{4+} state. Charge compensation takes place in several ways. For example, there have been suggestions that oxygen impurities, in the form of O^{2-} ions, replace the nearest-neighbour F^- ions (Crittenden *et al.*, 1974*b*; Hoffmann & Pöss, 1978). Evidence for this is particularly strong from ion implantation studies (Wintersgill, Townsend & Cussó-Perez, 1977; Wintersgill & Townsend, 1978*b*). The oxygen probably arises as a radiation decomposition product of OH^-. More recently, F-centres (or, in Mg-doped crystals, perhaps Z-centres) have been suggested as being in close association with Ti^{3+} (Stoebe, Wolfenstine & Las, 1980). Hydroxyl impurities are known to form $(Ti-OH)_n$ complexes which have often been identified with 200 nm absorption in LiF:Mg,Ti and which are thought to be involved in the thermoluminescence process (Rossiter, Rees-Evans & Ellis, 1970; Watterich, Foldvari & Voszka, 1980; Wachter, Vana & Aiginger, 1980; Wachter, 1982).

Consistent with the idea of vacancy–interstitial recombination in LiF, Mieke & Nink (1980) correlate the thermoluminescence from LiF:Ti with F-centre annealing steps. In view of the fact that Ti may actually be associated with F-centres (Stoebe *et al.*, 1980), it is possible to envisage a defect complex of Ti, possibly with oxygen and/or OH⁻ ions, in close association with F-centres. The thermoluminescence which is emitted by the vacancy–interstitial recombination emits at 410 nm. Without Ti, the emission is at 415 nm. When Mg is added, the interstitials recombine with Z-centres, as well as F-centres (Sagastibelza & Alvarez Rivas, 1981) and the Mg perturbs the emission by an amount which depends upon the form that the Mg is in (dipole, trimer, etc.). For impurities other than Ti or Mg, perturbation of the emission also occurs, but the emission wavelength remains in the same spectral range (Kathuria, 1979).

Thus, the reason that the intensity of peak 2 is not directly related to the impurity–vacancy dipole concentration may be that the peak 2 defect is a complex which includes both the Mg dipoles and Ti. In fact, the trap and the recombination centre can be viewed as different parts of the same large defect complex. This possibility is favoured in recent work by Taylor & Lilley (1982b) and is reinforced by Waligórski & Katz (1980) who, using multi-hit analysis (see chapter 4), found that the trap radii for peaks 5 and 6 are 10 nm and 40 nm respectively. Such large radii imply large defect complexes extending over several lattice sites. Thus, the peak 2 complex will not be expected to have the same thermal annealing behaviour as the free dipoles, as we indeed observed in figure 5.13.

The above discussion of the recombination process assumed that the light emission stemmed from vacancy–interstitial recombination. However, this may not be so and establishing the nature of the mobile species during thermoluminescence in LiF has been a subject of contention and debate for two decades. The model mentioned briefly earlier, namely that of Mayhugh (1970), has become generally accepted, at least until the detailed work of Sagastibelza & Alvarez Rivas (1981) who favour interstitial atom movement rather than electron movement as in the Mayhugh model. The advantage of the Mayhugh model, which was originally developed to describe the thermoluminescence above room temperature, but has recently been extended to describe the glow-curve between 10–300 K (Cooke, 1978; Cooke & Rhodes, 1981), is that it predicts the disappearance of V_3-, F-, 310 nm and 380 nm centres. The Mayhugh model, as modified by Cooke & Rhodes, is illustrated in figure 5.14.

The disadvantages of the model are that it predicts the occurrence of thermally stimulated conductivity peaks corresponding to each glow peak, which have not yet been found (although such currents are difficult to confirm above room temperature because of the high background ionic conductivity); it does not easily explain the observation that the higher temperature peaks are favoured over those at lower temperatures for high doses; it predicts F-centre annealing steps for all glow peaks,

186   *Defects and thermoluminescence*

whereas for some peaks in TLD-100, 310 nm band annealing steps are observed, but not F-band steps (Sagastibelza & Alvarez Rivas, 1981). These latter authors thus proposed an alternative mechanism to electron–hole recombination in order to describe the properties of the glow-curve above room temperature. Consistent with the observations of Alvarez Rivas and colleagues on other alkali halides, an interstitial–vacancy recombination mechanism is favoured by them. Rather than F-centres being involved in the thermoluminescence process for peaks 2 to 5, Sagastibelza & Alvarez Rivas suggest that the 310 nm centres are involved. Further support comes from the clear indication of first-order kinetics for all of the relevant peaks (Taylor & Lilley, 1978; Fairchild *et al.*, 1978*b*; McKeever, 1980*b*) implying a correlated recombination process. However, a problem exists with the explanation of the F-band stimulation experiments reported by many workers, whereby the glow peaks can be regenerated by bleaching with F-band light (254 nm), e.g., figure 4.6. However, a mechanism involving the diffusion of an excited F-centre has been suggested to account for this.

The glow-curve below room temperature has, so far, been interpreted in terms of electron–hole recombinations. A large peak at ~ 138 K (peak 4 in figure 4.6) has been interpreted by most workers as the

Figure 5.14. The Mayhugh/Cooke mechanism for thermoluminescence in TLD-100. Process (1) illustrates the release of trapped electrons (solid circles) and their recombination with trapped holes (open circles) at V_k-centres producing 270 nm emission during thermoluminescence below 160 K. Above 160 K, process (2) illustrates the recombination of electrons with V_3-centres converting these to V_k-centres which release holes to recombine at an activator site with electrons which have tunnelled from F-centres to yield 400 nm emission. The thermal release of shallow-trapped holes (process 3) also stimulates 400 nm emission. Solid arrows, electron transitions; open arrows, hole transitions. (After Cooke & Rhodes, 1981.)

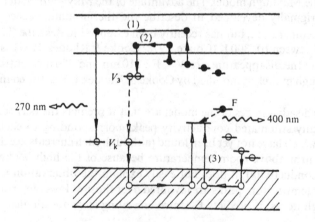

annihilation of V_k-centres by recombination with electrons with light being emitted at 270 nm (Townsend *et al.*, 1967; Podgoršak *et al.*, 1971; Cooke & Rhodes, 1981). A peak at 115 K is attributed by Townsend *et al.* (1967) to the thermal annealing of an H-centre. The rest of the glow peaks below room temperature have been associated with various electron and hole untrapping mechanisms, in accordance with the model shown in figure 5.14.

Finally, mention ought to be made of the effect of deformation on the thermoluminescence from TLD-100. In contrast to KCl (Ausin & Alvarez Rivas, 1972*b*), but in line with observations on other alkali halides (e.g., Mehendru & Radhakrishna, 1969; Mehendru, 1970*b*), plastic deformation does not induce any new peaks into the LiF glow-curve but does reduce the intensity of those already present (Petralia & Gnani, 1972; Srinivasan & DeWerd, 1973). The induced changes may be annealed out by heating to 350 °C (Niewiadomski, 1976*a*). Bradbury & Lilley (1976) argue in favour of the deformation creating competing traps whilst De Werd *et al.* (1976*a*) show that the deformation creates F-centres at the expense of the Mg-related 310 nm and 380 nm centres. The possibility that the deformation induces non-radiative recombination pathways should not be overlooked (Panizza, 1964).

5.3 Quartz and silica

The alkali halides have been at the centre of most detailed investigations into crystalline structure and defect equilibria over the years. However, possibly the first crystal to be subjected to the inquisitive ponderings of a scientific mind was 'rock crystal', or quartz (SiO_2) which was studied by Strabo as early as 64 B.C. (Evans, 1966). From the stand point of thermoluminescence, much less is known about this material's glow-curve properties than the much more intensely studied alkali halides, but this has not prevented it from becoming a thermoluminescence phosphor of great practical use. One wonders how far the technique of thermoluminescence dating (chapter 7) would have progressed without the observation of thermoluminescence emission from quartz grains extracted from ancient pottery (Grögler *et al.*, 1960; Kennedy & Knopff, 1960).

Although it is not necessary to understand fully the defect structure of quartz to be able to use the material in thermoluminescence dating, it is desirable, and even helpful, to appreciate 'what the electrons and holes are actually doing and not be content to only think of them as hopping from one line to another on a simplified band diagram' (Kelly, 1978). A large proportion of the research into the defect structure of both quartz and its amorphous relative, silica, has received its impetus from the need to develop optical and electronic devices which are able to perform reliably in high-background radiation environments. Examples include SiO_2 thin films in metal oxide–silicon devices (e.g., Srour, Curtis & Chiu, 1974; Curtis, Srour & Chiu 1975; Srour *et al.* 1976; van Turnhout & van Rheenen, 1977), silica fibre light guides (Dianov *et al.*, 1980; Sigel *et al.*, 1981; Sigel & Marrone, 1981) and quartz crystal oscillators (King & Sander, 1972, 1975; Paradysz & Smith, 1975; Euler *et al.*, 1978). These papers discuss the devices in the light of their resistance to radiation and include detailed discussions of the SiO_2 defect structure. In the coming section the defects in crystalline quartz and in silica will be discussed and this will be followed by a description of the thermoluminescence behaviour of these materials.

5.3.1   Structure

Quartz and its polymorphs, tridymite and cristobalite, occur naturally in the temperature ranges $< 870\,°C$, $870–1470\,°C$ and $> 1470\,°C$ respectively. Both tridymite and cristobalite are metastable at normal temperatures and both occur as natural minerals. All three minerals have α and β modifications, α-quartz being stable below $573\,°C$ and β-quartz being stable between 573 and $870\,°C$. The $\alpha–\beta$ transition is a displacive phase change, involving small adjustments of atomic positions without any rearrangement of the bonds. As a result, the phase change is very fast and is completely reversible, although crystal twinning occurs as the transition goes from α to β and back to α.

The α form has a rhombohedral structure, sketched in figure 5.15, along with the hexagonal structure of β-quartz. In both forms, the oxygen atoms are tetrahedrally shared between those of silicon. Two of the Si—O bonds in α-quartz make an angle of 66° with the optic axis (perpendicular to the plane of the paper in figure 5.15) and have a bond length of 1.598 Å, whilst the other two Si—O bonds make an angle of 44° and have a length of 1.616 Å. Thus, the oxygens occur as two equivalent pairs.

The difference in electronegativity between silicon and oxygen gives rise to a bond which is $\sim 40\%$ ionic and $\sim 60\%$ covalent (Dienes, 1960; Evans, 1966; Levine, 1973). This covalent nature gives quartz a rather rigid structure and prevents macroscopic diffusion of lattice atoms. Dislocations and intrinsic point defects exist, but have complex structures and are rather poorly defined. Misplaced lattice atoms tend to be located near impurities. The mixed ionic and covalent nature of the bonds in quartz and silica has resulted in the emergence of two basic mechanisms for radiation damage – namely, rupturing of covalent Si—O bonds, and vacancy and interstitial creation via 'knock-on' collisions.

The first requires only ionizing radiation; the second requires energetic nuclear particle radiation.

Most of the defects caused by ionizing radiation appear to occur at pre-existing imperfections – predominantly impurities. Electrons and holes released by the radiation may become trapped and give rise to optical absorption, luminescence and various magnetic and electrical phenomena. Particle radiation can induce large-scale damage by lattice atom displacement, severely altering the material's character. Both quartz and silica are affected. A well-known example is the change in density following neutron irradiation, with the density of quartz decreasing and that of silica increasing, with both densities eventually reaching the same value (Lell, Kriedl, & Hensler, 1966). An extensive literature exists on the topic of permanent damage studies in quartz and silica and the reader is referred to several reviews on the topic (e.g., Lell *et al.*, 1966; Griscom, 1978). It is important to note that the density changes observed can be explained by invoking both ionizing and elastic collision processes (Eernisse, 1974; Eernisse & Norris, 1974) and so too can many other properties of SiO_2, including some luminescence phenomena (e.g., Chandler, Jaque & Townsend, 1979; Jaque & Townsend, 1981). A recent book is useful in describing many of these properties (Pantelides, 1978).

Figure 5.15. Plan of the silicon atoms in the rhombohedral structure of α-quartz. The oxygen atoms are tetrahedrally disposed about those of Si. The positions of the Si atoms in β-quartz are shown as dotted circles.

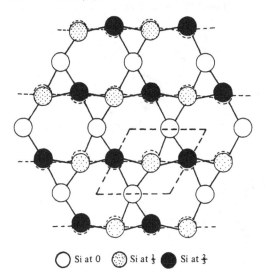

○ Si at 0   ◉ Si at $\frac{1}{3}$ ● Si at $\frac{2}{3}$

5.3.2   *Defects*

Among the numerous defects in pure SiO_2 those associated with oxygen vacancies, called the E'-centres, are possibly the most important in the explanation of the observed physical, optical and magnetic properties. These centres have been found in all forms of SiO_2, including α-quartz (Mitchell & Paige, 1956; Weeks & Nelson, 1960; Silsbee, 1961; Weeks, 1963; Fiegl, Fowler & Yip, 1974; Isoya, Weil & Halliburton, 1981; Bossoli, Jani & Halliburton, 1982; Jani, Bossoli & Halliburton, 1983), amorphous silica (Arnold & Compton, 1959; Ruffa, 1973/74; Sigel, 1973/74; Griscom, Friebele & Sigel, 1974; Dianov *et al.*, 1980) and thin SiO_2 films in metal oxide–silicon devices (Hochstrasser & Antonini, 1972; Jones & Embree, 1976).

The model for the E_1'-centre is that of an oxygen vacancy with an unpaired electron located on one of two non-equivalent Si atoms (Si(I) in figure 5.16) which result from the lattice distortion around the vacancy (Ruffa, 1973/74; Fiegl *et al.*, 1974). The E_2'-centre also consists of an oxygen vacancy, but with an associated proton. In this centre the electron is located on the Si(II) site (Fiegl & Anderson, 1970). A third E'-centre, namely the E_4', has been identified by Isoya *et al.* (1981) to be an oxygen vacancy with a hydride ion bonded to the Si(I) atom. A trapped

Figure 5.16. The model for the E_1'-centre in α-quartz, consisting of an oxygen vacancy (dotted circle) with an unpaired electron located at the Si(I) site. The arrows indicate the asymmetric relaxation of the Si atoms from their normal lattice positions. (From Isoya *et al.*, 1981.)

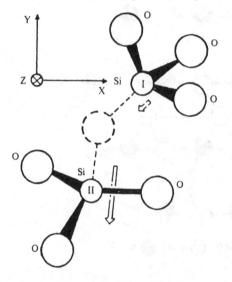

electron is shared between Si(I) and Si(II), spending most of its time on Si(I).

The available optical absorption and electron spin resonance data relate absorption bands at 215 nm (5.8 eV) and at ~ 235 nm (5.3 eV) with the electron spin resonance signals from the E_1'- and E_2'-centres respectively. Figure 5.17 shows the radiation-induced optical absorption spectra for alkali-doped and Al–alkali-doped silica, showing the positions of the E_1'- and E_2'-bands. The E′-bands appear in α-quartz only after previous particle irradiation, but can be observed in as-grown amorphous silica (Levy, 1960; Nelson & Crawford, 1960). An intense absorption band is also observed at 7.6 eV and has been tentatively associated with trapped holes at oxygen interstitials. Thus, the 7.6 eV-centre and the E′-centres are viewed as complements of each other. Hole-trapping Si vacancy centres have also been reported (Nuttal & Weil, 1980).

Apart from vacancy and interstitial centres, broken silicon and oxygen bonds are important defects in SiO_2 networks. Non-bridging oxygen in either the quartz or silica lattices will result in dangling oxygen bonds, which are potential hole traps, and empty Si orbitals which may trap electrons in a similar way to the Si atoms at oxygen vacancies (E′-centres). In this view, the 7.6 eV absorption arises, not from oxygen interstitials, but from the broken oxygen bonds (e.g., Stapelbroek *et al.*, 1979). The broken bonds are expected to be present in as-grown materials, particularly silica, but can easily be created by ionizing radiation and are therefore likely to be commonly observed. In a recent set of articles Greaves (1978) and Lucovsky (1979*a*,*b*, 1980) explain many

Figure 5.17. Optical absorption in silica. Solid line: 0.05 mole % Na; dashed line: 0.05 mole % Na + Al. (From Sigel, 1973/74.)

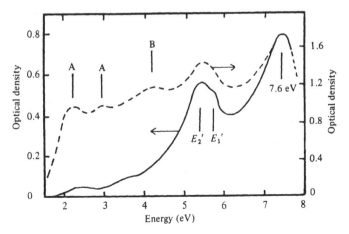

of the electronic properties of vitreous silica on the basis of the defect pair (hole trap and electron trap) produced during bond rupture. Greaves in particular argues that the E'-centres are induced by this mechanism and explains many of the optical absorption bands on the basis of whether or not the defect is charged or neutral.

The introduction of impurities into the SiO_2 lattice gives rise to new optical absorption, dielectric relaxation and electron spin resonance features, whilst at the same time influencing the properties of the intrinsic defects. Monovalent alkali ions (Na^+, Li^+, K^+) can act as network modifiers by inducing non-bridging oxygen without the action of radiation. Such a defect is sketched schematically in figure 5.18a. Absorption in the region of the E'-bands is seen to be dependent on the alkali concentration. Absorption in the region of the B-band is also affected by the presence of alkali ions, its position varying with the type of alkali used (Lell *et al.*, 1966).

Figure 5.18. (*a*) Non-bridging oxygen formed as a result of the incorporation of an alkali ion M^+ into the SiO_2 lattice. (*b*) the Al–alkali centre.

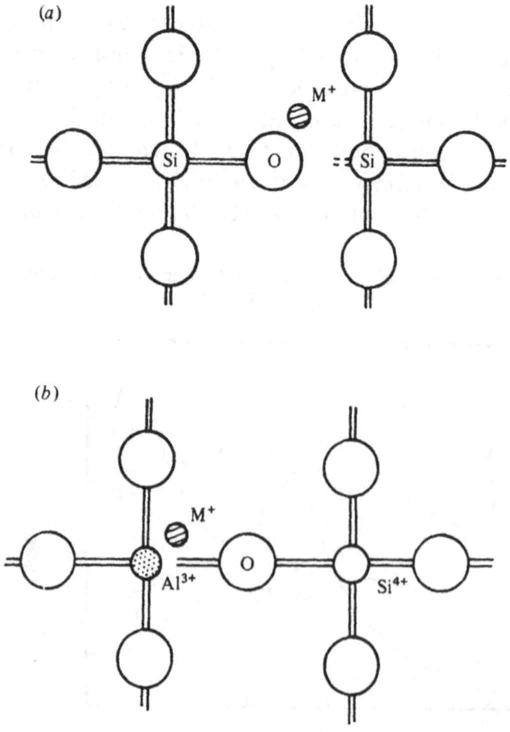

The addition of Al into quartz and silica containing alkali impurities is known to decrease the optical absorption related to non-bridging oxygen (e.g., Sigel, 1973/74). The interpretation of this is that because Al is trivalent, then any Al^{3+} which substitutes for Si^{4+} needs to be charge compensated by the presence of a monovalent ion (although Al bonded to only three oxygen atoms, AlO_3, has been reported, e.g., Brower, 1978). Thus, the alkali ions, which can easily diffuse through the open channels of the c-axis, associate with Al^{3+} and are therefore unable to act as network modifiers. The model for the Al–alkali centre (shown in figure 5.18b) was established by O'Brien (1955) following paramagnetic resonance studies by Griffiths, Owen & Ward (1954). A review of the literature concerning Al defects in α-quartz is given by Weil (1975). The picture which has emerged is that the trivalent Al^{3+} achieves its charge compensation by the nearby location of a Na^+, Li^+ or K^+ interstitial ion in the open c-axis channel. Alternatively, charge compensation can take place with a H^+ ion. Thus, several different Al centres exist, depending upon the compensator and upon the exact site chosen by the monovalent ion. Dielectric loss measurements reveal a range of possibilities, depending on the ion (Stevels & Volger, 1962).

The presence of Al shows up in the optical absorption spectrum of irradiated quartz in the form of the so-called A-bands (figure 5.17), responsible for irradiated quartz's smoky colour. Nassau & Prescott (1975) resolved these bands into three Gaussian-shaped components, peaking at 620 nm (1.85 eV; A_1), 480 nm (2.55 eV; A_2) and 355 nm (2.85 eV; A_3) although the A_1 and A_2 bands were recognized earlier by Mitchell & Paige (1956). The absorption is due to light-induced transfer of holes from one O^{2-} site to an equivalent one (Schirmer, 1976). Detailed electron spin resonance and dielectric relaxation measurements have shown these bands to be related to an $(Al^{3+})^0$-centre – i.e. a substitutional Al^{3+} ion which has trapped a hole. Double-hole centres may also be observed (Nuttal, Weil & Claridge, 1976). The location of the hole means that the charge compensation provided by the monovalent ion is no longer required. Thus, if the irradiation is performed at a high enough temperature, the monovalent ion will diffuse along the c-axis channels to become localized elsewhere in the lattice. Electron paramagnetic resonance data from Mackey (1963) suggest that the substituted Ge^{4+} may provide a localization site for the monovalent ion but in Ge-free crystals other sites must exist. If the temperature is too low, however, alkali ion diffusion does not take place and A-band coloration does not occur.

The general behaviour of the alkali ions in quartz can be followed by electrical measurements (particularly radiation-induced conductivity:

e.g., Hughes, 1975; Pfannhauser, Hohenau & Gross, 1978; Barthe, Barthe & Portal, 1980; Jain & Nowick, 1982). Electrolytic sweeping experiments have proved to be extremely useful in the elucidation of the irradiation behaviour of quartz. Application of a high field, parallel to the c-axis, at sufficiently high temperatures (400–550 °C) results in the removal of alkali ions at the electrodes (King, 1959). If the sweeping is performed in air, or in a hydrogen atmosphere, the alkali ions are replaced by protons. If it is performed in a vacuum it is thought that holes replace the charge compensators (King & Sander, 1975). Removal of H^+ ions by sweeping is confirmed by monitoring the infra-red absorption bands. When H^+ ions act as the Al^{3+} compensators, the oxygen–hydrogen bonds so introduced give rise to characteristic infra-red absorption (Wood, 1960). These absorption bands can be removed by sweeping in a vacuum (Kreft, 1975).

Recently, by employing sweeping techniques in conjunction with electron spin resonance, optical and infra-red absorption, *thermal con-ductivity* and *acoustic loss*, Martin & colleagues have confirmed the behaviour of the monovalent ions during irradiation and have presented a reasonably complete picture of the process, including a description of the temperature dependencies of ionic diffusion (Halliburton *et al.*, 1979; Sibley *et al.*, 1979; Doherty *et al.*, 1980; Koumvakalis, 1980; Halliburton *et al.*, 1981; Jalilian-Nosraty & Martin, 1981). The dielectric properties can be explained via a similar picture (Fontanella *et al.*, 1979; Jain & Nowick, 1982).

It should be noted, for both quartz and silica, that although the measured electron spin resonance signal and the A-bands are obviously related to Al, no clearcut correlation between them and the Al concentration has been found (Sigel, 1973/74; van den Brom & Volger, 1974). This may be because the Al^{3+} ion has to be in a silicon substitutional position in order to give rise to these signals. Mitchell & Paige (1956) suggested that some aluminium may be in interstitial positions. Brown & Thomas (1960) noted, from their crystal growth experiments, that at low Al contents the impurity is taken up uniformly, but interstitially. At higher Al contents, substitution for Si takes place, but this occurs preferentially on certain growth centres. Cohen (1960) made similar conclusions and also remarked that precipitation of Al could take place by slow cooling from above the α–β phase transition. The precipitate grows at the temperature of the transition and is an alkali aluminium silicate, isomorphous with β-quartz.

The poor correlation between the A_1 and A_2 absorption bands and the Al content has led some workers to suggest that these bonds are related to a different impurity altogether. Fe^{3+} is a likely candidate (van den Brom & Volger, 1974). This impurity is said to play an important role in the coloration of amethyst (Lell *et al.*, 1966). Ge, like Si, is tetravalent and enters the SiO_2 lattice substitutionally for Si. Its higher electron affinity makes it a possible electron trap, with the hole trap consisting of the Al centres. Diffusion of the alkali to the Ge sites (at temperatures > 200 K) in order to stabilize the trapped electron was suggested by Mackey (1963). In Ge-free specimens, however, hole trapping at Al still takes place meaning that other sites must be acting as the complementary electron traps. Ti^{4+} (present in appreciable quantity in rose quartz) is a possibility. So too are alkali atom clusters and atomic hydrogen.

5.3.3 Luminescence

Before discussing the possible origins of thermoluminescence from quartz and silica, it is helpful to refer to recent data concerning other types of luminescence emission from these materials. Various types have been reported (photo-, cathodo-, radioluminescence, ion beam induced) and although a complete consensus of opinion regarding the origins of the emissions is still lacking, some useful information has emerged. Sigel (1973/74) examined blue luminescence emission following a pulse of 2 MeV electrons at 4.2 K for both crystalline quartz and fused silica. The spectra are shown in figure 5.19. The source of the emission (~ 460–480 nm; 2.6–2.7 eV) has been suggested to be the relaxation of an exciton formed during the electron pulse (Sigel, 1973/74; Greaves, 1978; Trukhin & Plaudis, 1979). The exciton is believed to be an electron–hole pair localized at a broken Si—O bond. This is

Figure 5.19. Spectra of the low-temperature (4.2 K) emission of electron-pulse-irradiated crystalline quartz (squares) and fused silica (circles). (After Sigel 1973/74.)

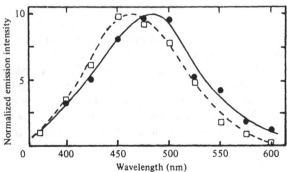

supported by the observation that blue luminescence is induced during E'-centre (215 nm (5.8 eV) absorption) annihilation (Sigel, 1973/74; Greaves, 1978). It is interesting to observe that the intensity of the luminescence is not proportional to the rate of E'-centre annihilation, implying that a significant non-radiative pathway for exciton relaxation exists in competition with the luminescent relaxation. A similar situation exists for the alkali halides (see section 5.1). Blue emission is also observed during ultra-violet excitation at both 4.2 K (Sigel, 1973/74) and 77 K (Kristianpoller & Katz, 1970) but not at room temperature.

Mattern, Lengweiler & Levy (1975) note that for quartz the radioluminescence maximum shifts from the blue emission at 85 K to an emission at 380–390 nm (3.2–3.3 eV) at 210 K. This was studied further by Alonso *et al.* (1983) who resolved the blue emission into three bands (at 440 nm, 425 nm and 380 nm). Each of these emissions was thermally quenched below room temperature. The authors suggest that the 380 nm emission results from electron recombination with unstable holes trapped at Al ions charge compensated by alkali ions.

Another major emission band is often observed at 640–670 nm in silica. This has been seen during photoluminescence (Sigel, 1973/74), radioluminescence (Sigel *et al.*, 1981; Sigel & Marrone, 1981), cathodoluminescence (Jones & Embree, 1978) and ion implantation (Chandler *et al.*, 1979; Jaque & Townsend, 1981). It has been variously assigned to different impurities such as Na^+, H^+ or OH^-, or to non-bridging oxygen introduced by irradiation, or, in silica fibres, by fibre-drawing (Sigel *et al.*, 1981; Sigel & Marrone, 1981). In the photoluminescence experiments, this emission was induced only by irradiating with ultra-violet light of wavelength 257 nm (Sigel, 1973/74) but it can also be induced by γ, electron or neutron irradiation (Sigel & Marrone, 1981).

Jones & Embree (1976) observe an emission at 260 nm (4.77 eV) in α-quartz, and 290 nm (4.28 eV) in amorphous silica and in the SiO_2 layer of metal oxide–silicon devices. The same emission is also observed in silica by Chandler *et al.* (1979) and by Jaque & Townsend (1981) and is related to an oxygen vacancy by Jones & Embree. Turner & Lee (1966) observe an intense emission at 408 nm (3.04 eV) in silica. This has been correlated by the authors to a Ge absorption band at 242 nm (5.13 eV). Jones & Embree (1978) also relate emission at 435 nm in crystalline quartz to Ge. It is interesting to note that Kristianpoller & Katz (1970) reported the blue emission discussed above only in their Ge-doped specimens. Finally, Treadaway, Passenheim & Kitterer (1975) reported an intense radioluminescence emission at 185 nm (6.7 eV) in both α-quartz and fused silica.

5.3.4 *Thermoluminescence; samples irradiated below room temperature*

Medlin (1963) has studied the glow-curves below room temperature for several types of natural and synthetic quartz crystals following X-irradiation at low temperatures. An intense peak at 165 K (heating rate not given) was correlated by Medlin to Ti^{4+}. Other peaks, between 90 and 300 K did not appear to be related to impurities. The emission spectrum of the 165 K peak gave a maximum at 380 nm (3.3 eV) whilst all the other peaks up to room temperature exhibited a complex emission spectrum, with ~ 5 emission maxima between 400 nm and 700 nm. Mattern *et al.* (1975) observe a single Gaussian emission peak at 380 nm for a glow peak at 185 K in γ-irradiated quartz, whilst Malik, Kohnke & Sibley (1981) note 450 nm emission for a glow peak in the region of 115–145 K (11 K min^{-1}) and 380 nm emission for several peaks between 145 and 270 K in different types of crystalline quartz. The glow-

Figure 5.20. The glow-curves from SiO_2 below room temperature exhibit a wide variety of shapes. Examples are given here of crystalline quartz (*a*: from Malik *et al.*, 1981; *b*: from Schlesinger, 1965) and of fused silica (*c*: from Kristianpoller & Katz, 1970).

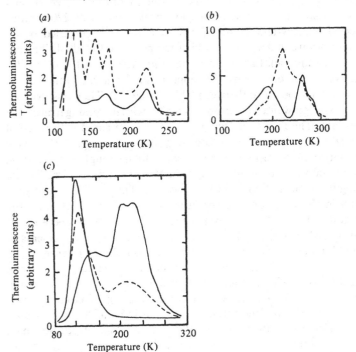

curves themselves exhibit a wide variety of shapes depending upon the sample. Some examples are shown in figure 5.20.

Schlesinger (1965) used the phototransfer technique (section 4.3.2) to suggest that all of the peaks below room temperature in both Ge-free and Ge-doped quartz are electron traps (except one at 257 K which is a hole trap). When combined with the results of Malik *et al.* (1981), which suggest that the glow peaks between 185 and 270 K are related to an Al–alkali centre (a hole trap), then one is forced to the conclusion that thermoluminescence in this region, emitting at 380 nm, is due to the recombination of electrons, released from traps, with holes at Al–alkali centres. This is clearly consistent with Alonso *et al.'s* (1983) interpretation of the radioluminescence data. Malik *et al.* further suggest that the glow peak between 115 and 145 K is due to the release of an electron from a hydrogen atom recombining with a nearby Al–hole centre producing 450 nm emission.

5.3.5 *Samples irradiated at room temperature*

An interesting observation made by Malik *et al.* (1981) which is consistent with their interpretation of the thermoluminescence mechanism is that the intensities of the glow peaks in the 180–270 K range decrease if the low-temperature (95 K) irradiation is preceded by an irradiation above 200 K (above which temperature the alkali ions are known to be mobile). Earlier, Durrani *et al.* (1977b) had performed a similar experiment on α-quartz, but concerning the thermoluminescence above room temperature (between 513 and 693 K). They observed an increase in the glow intensity from this region with increasing irradiation temperature. The results of Durrani *et al.* are compared with those of Malik *et al.* in figure 5.21. Both sets of workers give the same interpretation of the results, namely, that at high enough temperatures, the alkali ion is able to diffuse away from the Al site during irradiation, resulting in the formation of a $(Al^{3+})^0$-centre. The increase in the glow intensity with irradiation temperature is thus consistent with the suggestion that the thermoluminescence above room temperature is related to $(Al^{3+})^0$-centres. This view is supported by measurements of the emission spectra (Batrak, 1958; Schlesinger, 1964; Durrani *et al.*, 1977a; Levy, 1979). The emission for all of the peaks above room temperature is ~ 450–470 nm, which is close to the 450 nm emission reported by Malik *et al.* (1981) for recombination of electrons with $(Al^{3+})^0$-centres during thermoluminescence at 115–145 K. Furthermore phototransfer studies (Schlesinger, 1964; Bailiff *et al.*, 1977a; Bowman, 1979) indicate that all of the glow peaks above room temperature are from electron traps. Thus, the implication is that the $(Al^{3+})^0$-centre is acting as the recombination site for electrons released from traps. Durrani *et al.* (1977*a*) suggested a mechanism to explain the changes in the glow-curve shape at high radiation doses, on this basis (see figure 4.3*a*).

It should be recalled that luminescence in this spectral range is also expected from exciton annihilation (Sigel, 1973/74; Greaves, 1978; Trukhin & Plaudis, 1979). Also, there is evidence of more than one emission peak in the thermoluminescence above room temperature (David, Sunta & Ganguly, 1977*a*; Levy, 1979). Nevertheless, it is unlikely that the thermoluminescence above room temperature is caused by the exciton, because both Mattern *et al.* (1975) and Chandler *et al.* (1979) report that this luminescence quenches between 130 and 300 K, described by an expression of the form of equation (2.10) or (2.11), with an activation energy of ∼ 0.1 eV. Alonso *et al.* (1983) also note thermal quenching in this spectral region, but they resolve the emission into

Figure 5.21. (———) Decrease in intensity (normalized) of the thermoluminescence (TL) from the 185-270K region (TL(I)) for two specimens (● and ■) of quartz with increasing irradiation temperature. (–·–) Decrease in the intensity of the acoustic loss peak due to Na⁺. (---) Complementary increases in concentrations of Al—OH centres and (Al³⁺)⁰ centres. Data from Malik *et al.* (1981). Increase in thermoluminescence from the 513-693K region (TL(II)) with irradiation temperature (data from Durrani *et al.*, 1977*b*).

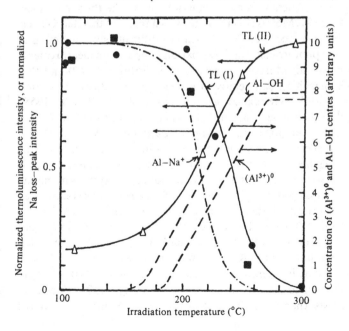

several components and arrive at activation energies of 0.21 eV, 0.43 eV and 0.35 eV for the 440 nm, 425 nm and 380 nm emissions respectively.

A cautionary note ought to be made at this point. Most published accounts of emission spectra from luminescence studies report similar emission wavelengths. Some authors resolve these into components; others do not. Thus, one cannot be certain that the emission of, say 450–470 nm reported by Durrani *et al.* (1977a) for the luminescence above room temperature is the same as that observed by Malik *et al.* (1981) for the glow peaks between 115–145 K. Similarly, the 380 nm emissions observed by Alonso *et al.* (1983) cannot be said with certainty to be the same emission as that observed by Mattern *et al.* (1975). The emission spectra from SiO_2 varies considerably from samples from different origins (e.g., David *et al.*, 1977a) and, for any one sample, contains several closely spaced bands (Levy, 1979). It would be unwise at this stage to be categorical about the thermoluminescence emission processes in these materials.

In contrast to the alkali halides, there have been very few detailed correlations between colour-centre annealing and thermoluminescence in SiO_2. An often-reported observation is that the smoky coloration can be bleached, but the thermoluminescence is unaffected (Chentsova & Butuzov, 1962; McMorris, 1971). This is clearly difficult to reconcile with the assertion made above that the $(Al^{3+})^0$-centre (which causes the smoky coloration) acts as a recombination site in the thermoluminescence emission. McMorris (1971) suggests as a result of this that Ge acts as the luminescence centre. This in turn has difficulties when viewed in the light of the phototransfer and irradiation temperature measurements already mentioned. Moreover, Chentsova & Butuzov (1962) find that ultra-violet bleaching can remove the Ge-related optical absorption band, but not the Ge-related glow peaks (300 °C). Batrak (1958), however, clearly observes annealing steps in the intensities of the A_1 and A_2 absorption bands (figure 5.22) in α-quartz and Lee (1969) notes that in some quartz specimens, but not all, the smoky coloration could be thermally bleached without the stimulation of thermoluminescence. It seems that in some specimens, strong non-radiative pathways dominate the recombination process. (The existence of such pathways is supported by the observation of thermal quenching between 100 and 200 °C (Wintle, 1975a).) It is possible to speculate that these non-luminescent sites may be associated with Fe^{3+}, a well-known luminescence 'killer' (Medlin, 1968, and section 4.5.3). It was noted earlier that Fe^{3+} is a possible candidate for some of the smoky coloration in quartz, acting in a similar fashion to Al^{3+} (van der Brom & Volger, 1974). Marfunin (1979) suggests that both the holes trapped at Al centres, and the electrons trapped at Ge (or Ti) centres act as the recombination sites, but are activated in different temperature regimes.

Several papers have dealt with the relationships between impurities and specific glow peaks, but as yet no unambiguous correlations have emerged. Batrak (1958) and Chentsova, Grechushnikov & Batrak (1957) have used electrodiffusion techniques to show that Na^+ and Li^+ play an important part in the thermoluminescence production. They conclude that the Li and Na ions replace each other in the lattice, but give different trap depths (although the main glow peaks appear in the same temperature region: ~ 300–350 °C). Chentsova & Butuzov (1962), Schlesinger (1964) and McMorris (1971) have indicated the importance of Ge as an electron trap in the thermoluminescence near 300 °C while Böhm, Peschke & Scharmann (1980) suggest that Ge is related to a peak at 57 °C. Ichikawa (1967) records relationships between Al, Li and Na, and several of the glow peaks in natural quartz. One peak (at ~ 380 °C; $75 °C \, min^{-1}$) is thought to be related to an oxygen vacancy and Al. Medlin (1963) tests the effect of the introduction of several impurities into the quartz lattice, and concludes that, apart from Ti, impurities play no role in the thermoluminescence of natural quartz – a somewhat

Figure 5.22. The 308 °C ($0.06 °C \, s^{-1}$) glow peak in smoky quartz and the thermal annealing steps in the A_1 (630 nm) and A_2 (450 nm) absorption bands (after Batrak, 1958).

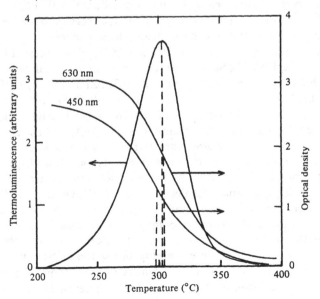

surprising result. Hensler (1959) looked at Al- and Li-doped specimens, and concluded that the thermoluminescence at 280 °C (4 °C min^{-1}) is a result of oxygen vacancy–oxygen interstitial recombination. Arnold (1960, 1973) uses ion implantation to alter the defect state in silica and concludes that a glow peak at 90 °C is common to all silica and is related to an oxygen vacancy; the 140 °C peak is associated with a large structural defect; a peak near 180 °C is probably linked with Ge; and a peak near 320 °C involves Al.

The glow-curves themselves show an enormous variety of shapes. Most published curves are from geological or pottery quartz and the geological specimens in particular exhibit a wide range of glow peak distributions. David *et al.* (1977*a*) record the glow-curves from several different natural quartz specimens and emphasize that it is difficult to compare the published thermoluminescence from different specimens because of differences in heating rate, pre-irradiation treatment, impurity content, formation conditions, etc. All these factors seriously affect the glow-curve shape. Sankaran *et al.* (1982) compile a table to describe a variety of glow-curve features, and in figure 5.23 glow-curves from several SiO_2 specimens (including silica) are illustrated for comparison. The differences between the various samples also show up in thermally stimulated conductivity measurements. Some specimens give no observable thermally stimulated conductivity (Schlesinger, 1964) whilst others give thermally stimulated conductivity peaks corresponding to the observed glow peaks (Medlin, 1963; Böhm *et al.*, 1980).

The effects of various pre-treatments upon the thermoluminescence from quartz has been studied in detail in an extensive series of papers by David and colleagues (David *et al.*, 1977*a,b*, 1978, 1979; David, 1981; David & Sunta, 1981*a,b*). Similar measurements are also dealt with by other authors (e.g., Fleming & Thompson, 1970; Zimmerman, 1971*a*; Kaul, Ganguly & Hess, 1972; Shekhmametev, 1973; Durrani *et al.*, 1977*a*; Levy, 1979; McKeever *et al.* 1983*b*). Unfortunately, these measurements add little to an understanding of the SiO_2 defect structure (the topic of this chapter) because their interpretation requires a previous knowledge of the defects involved, as was discussed generally in chapter 4. However, it may be of interest to discuss one observation which gives a little information concerning the defects in quartz. David *et al.* (1977*b*) report that pre-irradiation annealing of a quartz specimen to temperatures beyond the α–β transition point (573 °C) produces profound changes in the thermoluminescence sensitivity. McKeever *et al.* (1983*b*) have investigated this further by examining the changes in the radioluminescence and thermoluminescence intensity as a function of a pre-irradiation thermal anneal. Up to ~ 450 °C the sensitivity changes appear to be the result of an increase in the recombination centre concentration. (It is to be noted that this temperature region is that which is normally employed during alkali ion sweeping (King, 1959) supporting the view that $(Al^{3+})^0$-centres act as the luminescence sites.) Beyond the $\alpha-\beta$ transition point, however, the massive increase in the

Figure 5.23. Examples of glow-curves above room temperature from various forms of SiO_2. (a) Natural Brazilian α-quartz; 600 krad γ, 3.6 °C s^{-1} (Durrani et al., 1977a). (b) Natural pink quartz; 2.5×10^5 rad, 25 °C min^{-1} (David & Sunta, 1981a). (c) Ge-doped Clevite α-quartz; 10 min X-rays, 10 K min^{-1} (Schlesinger, 1965). (d) Smoky quartz and citrine; X-irradiated, 20 K min^{-1} (quartz), 3.5 K min^{-1} (citrine) (Maschmeyer et al., 1980). (e) Fused silica; 56 hours X-rays, various times after irradiation at 150 °C, 2 °C min^{-1} (Yokota, 1953). (f) Silica fibre optic; 7 Gy γ, 1 °C s^{-1} (West & Carter, 1980). All thermoluminescence scales in arbitrary units.

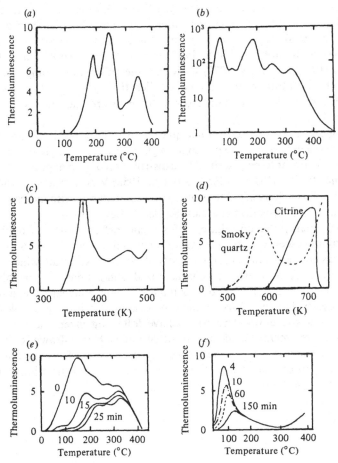

sensitivity of the 350 °C region of the glow-curve appears to be related to the formation of a metastable defect cluster. It should be recalled that cooling through the α–β transition can induce precipitation of an alkali-aluminium silicate phase (Cohen, 1960) and this may be the defect which is being observed.

A final comment should be made on the kinetics of the thermolumin-escence production. The lack of a thermally stimulated conductivity signal observed in some samples (Schlesinger, 1964) or for some peaks (Medlin, 1963) may indicate a close spatial correlation between traps and recombination sites. Low-temperature charge transfer between defects supports this view (Medlin, 1963; Schlesinger, 1965). Such correlations would favour first-order kinetics and this appears to be the case for all of the glow peaks examined. Isothermal decay curves (e.g., Hwang, 1972) and computerized curve fitting (Levy, 1979) both indicate monomole-cular kinetics. Apparent non-first-order, or variable kinetics (Sunta & Kathuria, 1978) may be the result of complex peak overlap (Hwang, 1972; David, Kathuria & Sunta, 1982). It is interesting to note that both Hwang (1972) and Levy (1979) show that the half-lives of some of the higher temperature peaks can be smaller than those of the lower temperature peaks at certain storage temperatures (as discussed in section 3.3.5).

The glow-curves from crystalline quartz can be described by assuming mono-energetic traps. As expected for a glass, however, the glow-curves from silica appear to be described by a distribution of trap depths and/or frequency factors (Yokota, 1953; Kikuchi, 1958; West & Carter, 1980). Isothermal bleaching of colour centres in these materials also indicates a distribution of activation energies (Swyler & Levy, 1976).

In summary, the correlation of specific defects with particular thermoluminescence peaks in SiO_2 presents immense difficulties. The problems are the result of the complexity of the defect structure which, as yet, has only been superficially characterized. Unfortunately, thermolu-minescence alone adds little to our understanding. Furthermore, until the necessary detailed information becomes available, it is impossible to predict how the glow-curve will behave following given thermal or radiation treatments. Once again, attention is being drawn to the problems which arise when there is a complex array of defects and one is attempting to produce correlations between defect concentrations and glow peak intensities.

6 Thermoluminescence dosimetry (TLD)

The application of the phenomenon of thermoluminescence to the measurement of absorbed radiation dose has progressed a great deal since the initial work by Daniels and colleagues referred to in chapter 1. Several thermoluminescent phosphors are now used routinely in many dosimetric applications. This chapter deals with the characteristics of thermoluminescence dosimetry (TLD) materials and several of their uses (environmental monitoring, personal dosimetry and medical applications). To begin with, it is useful to examine what properties are looked for when assessing the quality of a new TLD material.

6.1 General requirements for TLD materials

The selection of a phosphor for a thermoluminescence dosemeter requires a precise knowledge of the particular application being considered. Broadly, the applications may be conveniently listed as *personal dosimetry* (dose estimate in body tissue) and *environmental monitoring* (dose estimate in air), with special considerations for medical applications and reactor dosimetry. Clearly there are large areas of overlap between each usage. Modern dosemeters are manufactured with specific applications in mind, for instance, measurement of the low-energy X-ray or β dose absorbed by the epidermis at a depth of $5\,\mathrm{mg\,cm^{-2}}$ ($\sim 50\,\mu\mathrm{m}$) below the surface of the skin, or assessment of the dose-equivalent[†] from a mixed neutron field absorbed by body tissue at a depth of $1000\,\mathrm{mg\,cm^{-2}}$. To make these measurements within an acceptable uncertainty limit, correct choice of dosemeter is essential.

Several recommended standards have been suggested for dosemeter performance, depending upon the application. The International Standards Organisation (ISO, 1982) is presently attempting to establish performance requirements for environmental and personal dosemeters in order to satisfy the recommendations of the International Commission on Radiological Protection (ICRP, 1977) on dose limitations. Other recommended standards for TLD phosphors are being

---

[†] Dose-equivalent (H): defined (in rems (or sieverts)) as $H = D\bar{Q}N$. It is the dose D (in rad (or gray)) modified by the average quality factor $\bar{Q}$ for the radiation and other quality factors included in N (ICRUM, 1973).

discussed by the Commission of European Communities (CEC, 1975), the Physikalisch Technische Bundesanstalt (PTB, 1976), the American National Standards Institute (ANSI, 1978) and the Health Physics Society Standards Committee (HPSSC, 1977). Test procedures for dosemeter performance can be found by referring to these documents. The British Calibration Service publications 0803 and 0823 (BCS, 1977) are also useful. Additionally, the International Electrotechnical Commission is currently examining possible standards for TLD readout units, but this work is not yet finalized and is not dealt with further in this chapter. Further discussions on equipment may be found in chapter 9.

The performance of a dosemeter is assessed by examining properties such as linearity, dose range, energy response, reproducibility, stability of stored information, isotropy, effect of environment on dosemeter performance, batch inhomogeneity, and others. Several of these properties are now discussed in some detail before progressing to a discussion of individual phosphors.

6.1.1   *Dose response*

An obviously desirable property of a TLD detector is that it exhibits a linear relationship between thermoluminescence intensity (I) and absorbed dose (D). From a discussion of the phenomenon of supralinearity in chapters 3 and 4 it is clear that many (if not most) thermoluminescent materials exhibit a non-linear growth of glow intensity with absorbed dose over certain dose ranges. Examples were given in figures 3.27, 4.1 and 4.3 . Both supralinearity and the sublinear growth approaching saturation can lead to problems of under- or overestimation, respectively. (Here, saturation refers either to all traps full or to the onset of appreciable radiation damage.) The occurrence of non-linear regions in the dose response curve of a detector does not preclude its use in TLD, but it does require careful calibration and correction from which additional errors may accrue. The correction factors (and, therefore, the errors) may be particularly large when the dosemeter is used in the sublinear region, approaching saturation. Supralinearity and saturation can both be affected by previous exposure to radiation and by thermal treatments (see the references given in chapter 4) so re-use of a dosemeter may present problems such that on second use, the phosphor may exhibit a different dose response. In order to overcome these sensitization and supralinearity problems, many phosphors are required to undergo complex annealing procedures to 're-set' the original properties. Annealing procedures are discussed in section 6.1.4. The character of the supralinearity and sensitization is also dependent on linear energy transfer (LET[†]) with, for example, the onset of supralinearity occurring at different doses for different LET.

The lowest dose which can be detected by the material corresponds to that thermoluminescence intensity which is significantly greater than the intensity from an unirradiated sample. For standardization purposes this is taken to be three times the standard deviation of the zero dose reading. This minimum detectable dose is variable in that it depends not just upon the intrinsic sensitivity of the material, but also upon the size of the detector, the sensitivity of the readout equipment and, as with the other dose response characteristics, it varies with previous radiation exposure and thermal treatment. For example, the zero dose can increase because thermoluminescence can often be regenerated following α-irradiation and readout without re-irradiating the sample (Petridou, Christodoulides & Charalambous, 1978). Neutron irradiation of ^{6}LiF produces tritium by the reaction ^{6}Li$(n,\alpha)^3$H, so that a time-dependent build-up of the zero dose reading occurs (due to the tritium activity) proportional to neutron fluence and storage time (Piesch, Burgkhardt & Sayed, 1978). The 'self-dose' (from the radioactive content of the material) can also give an appreciable background signal, especially when the detector is used in long-term monitoring (Nambi, 1982).

An additional important property required of a radiation detector is that its response is independent of dose rate. It was discussed in chapter 3 how, theoretically, a dose-rate-dependent response can be possible in certain dose rate ranges and for certain doses (viz., section 3.4). The dose rate independence of several detector materials has been tested at very high dose rates, using X-ray pulses. For example Karzmark *et al.* (1964), Tochilin & Goldstein (1966) and Goldstein (1972) have observed dose-rate-independent effects in LiF between 5×10^2 rad s^{-1} and 1.5×10^{10} rad s^{-1}. At high doses (> 50 krad) loss of sensitivity is observed beyond 1.5×10^{10} rad s^{-1} (Goldstein, 1972). Goldstein also notes a dose-rate-independent response from BeO up to 5×10^{11} rad s^{-1}, and from Li$_2$B$_4$O$_7$ up to 10^{12} rad s^{-1}, even with high doses. Generally, dose-rate-dependent effects have not yet been demonstrated to be a problem in TLD.

6.1.2 *Energy response*

The intensity of thermoluminescence emitted from a material is proportional to the amount of energy initially absorbed by that material.

[†] LET: defined as dE/dl, where dE is the energy loss by a charged particle travelling a distance dl, due to collisions.

It thus becomes important to assess the variation in the material's absorption coefficient with radiation energy. For photon irradiations (γ-, X-rays), this assessment is made via the photon energy response $S(E)$ of the system, defined as the ratio of the mass energy absorption coefficient for the material $(\mu_{en}/\rho)_m$ to the mass energy absorption coefficient of a reference material (air or tissue) $(\mu_{en}/\rho)_{ref}$, where μ_{en} is the linear absorption coefficient and ρ is the density. Thus:

$$S(E) = (\mu_{en}/\rho)_m/(\mu_{en}/\rho)_{ref}. \qquad (6.1)$$

The term $(\mu_{en}/\rho)_m$ often refers to a compound or mixture of materials consisting of several elements, thus the law of mixtures applies, namely:

$$(\mu_{en}/\rho)_m = \sum_i(\mu_{en}/\rho)_i W_i, \qquad (6.2)$$

where W_i is the fraction by weight of the ith element.

The mass energy absorption coefficient is dependent upon the energy loss processes occurring during energy absorption (pair production, Compton scattering, photoelectric interaction). The amount of each process taking place depends upon the energy of the absorbed radiation, the material's isotopic content and its atomic number Z. For example, for photon energies < 15 keV, the photoelectric effect dominates, whilst for 15 keV $< E < 10$ MeV, Compton scattering dominates in low Z materials. For high Z specimens, the photoelectric effect remains dominant up to $\sim 10^2$ keV. Thus, a dosemeter which contains a mixture of high and low Z detectors may be used to determine the quality of the radiation (Cameron & Kenney, 1963; Gorbics & Attix, 1968: Spurný, Milu & Racoveanu, 1973) although this application has suffered from the lack of suitable high-Z TLD materials (Spurný, 1980).

The photon energy response is often defined with respect to the response from a given energy, often ^{60}Co γ-rays of energy 1.25 MeV. Thus, a *Relative Energy Response* (RER) is defined as:

$$(RER)_E = S(E)/S(1.25 \text{ MeV}\,^{60}\text{Co}). \qquad (6.3)$$

Plots of $(RER)_E$ versus E(MeV) for LiF, CaF_2 and for a theoretical material of $Z_{eff}^\dagger = 55$ are given in figure 6.1. For dosimetry, it is obviously desirable to have a detector which exhibits a constant response over a wide range of energies, therefore low Z materials are preferred. Additionally, for assessment of the dose delivered to body tissue, so-called 'tissue equivalent' dosemeters are desirable. Lithium borate is one such phosphor with $Z_{eff} = 7.4$, equal to that of tissue.

If necessary, the energy response can be modified by the appropriate

---

†Z_{eff} is the effective atomic number of the compound.

use of filters. For instance, if LiF is required to measure the dose from photon energies of 30 keV, the overestimation in this energy region can be reduced by inserting a filter (say 1.5 mm Al) between source and detector. Szabó, Félszerfalvi & Lénárt (1980) give some examples of appropriate filters for LiF and $CaSO_4$. Pradhan & Bhatt (1979) show that unfiltered $CaSO_4$ is accurate below 80 keV, but when filtered with 0.55 mm Sn and 0.35 mm stainless steel, it becomes accurate above 80 keV. The best results (in terms of producing a flat energy response) are obtained by using multi-element filters of different thicknesses. Normally the selection of the filters is purely empirical, but it is possible to calculate the optimum filter characteristics beforehand (Bacci *et al.*, 1982). A filter which combines energy compensation with a correction for directional dependence has recently been described by Trousil & Kokta (1982).

The energy dependence of a phosphor can also be altered by suitable pre-irradiation treatments (Mayhugh & Fullerton, 1974; Lakshmanan & Bhatt, 1980) and supralinearity is dependent upon photon energy.

The response to β-particles is more complex and measurement of the absorbed dose is not straightforward. β-particles travel a whole range of distances into a medium, depending on their energy and thus the energy response depends heavily on the physical form of the detector, particularly its thickness, or (in the case of powder) the grain size. In general, the relative response of a TLD material increases with β energy, levelling off when the β-particles' range exceeds the grain size (Greitz & Rudén,

Figure 6.1. Relative energy response versus photon energy for LiF ($Z_{eff} = 8.14$), CaF_2 ($Z_{eff} = 16.3$) and a theoretical material of $Z_{eff} = 55$. The reference material is air. (After Spurný, 1980.)

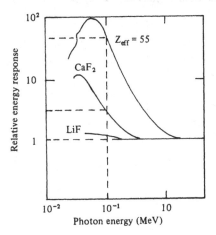

1972; Becker, 1973). Figure 6.2 gives some experimental examples for LiF:Mg,Ti. Similar discussions are given by Liu & Bagne (1974) whilst Honda *et al.* (1980) give comparable results for LiF:Mg, Cu. Analogous relationships for low-energy electrons are discussed by Lasky & Moran (1977).

Calculations of the energy absorbed using Bragg–Gray Cavity Theory (e.g., Nakajima, 1969) are complicated because of the large grain size or thickness of practical TLD detectors with respect to the β-particle range (Chan & Burlin, 1970). Similar problems arise with low-energy X-rays (Greening, 1972). O'Brien (1977) suggests the use of Monte Carlo calculations as an alternative to simple Cavity Theory.

The problem of grain size has been investigated by several researchers. It has been shown that the sensitivity decreases as the grain size decreases, with some authors reporting an optimum grain size for maximum sensitivity (Endres, Kathren & Kocher, 1970; Nakajima, 1970; Yamaoka, 1978). The discussion of these effects centres on two points: one is a lower intrinsic sensitivity of the surface regions of the grain (Zanelli, 1968) possibly related to oxygen or OH^- ions at the surface (Lasky & Moran, 1977; Driscoll & McKinlay, 1981; it should be noted that OH^- ions remove Mg from the thermoluminescence process in LiF by forming $Mg(OH)_n$ complexes (Wachter, 1982)). The second explanation has centred upon an alteration of the energy response by grains of different sizes. Burlin *et al.* (1969) and Chan & Burlin (1970) use

Figure 6.2. Relative thermoluminescence (TL) response as a function of mean β energy in LiF:Mg,Ti. A, powder, average 0.15 mm grain size. B, 1 mm LiF/Teflon rods at the surface of a vessel containing the β-emitting solution. C as B, but submerged in the solution. (After Becker, 1973.)

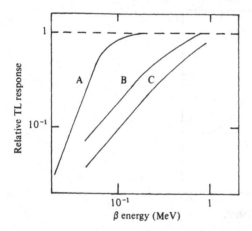

Cavity Theory to estimate the energy response for TLD powders of grain diameters ranging from 1 to 100 μm. The energy deposited in a grain depends upon the balance of the energy leaving the grain, relative to the energy entering and remaining within. This in turn depends upon energy of the incident photons, and upon the relative values of Z_{eff} for the material and the surrounding medium. For photon energies > 0.15 MeV, Compton processes (independent of Z) dominate and the energy flow into and out of the grain is balanced. Thus, no dependence on grain size is expected. At lower energies, photoelectrons generated by the radiation leave the material and penetrate the surrounding air, whilst some of those generated in the air penetrate the material. Because TLD materials have a larger Z_{eff} than air, more energy is carried out via the photoelectrons than is carried in and an energy loss results, the magnitude of which increases with decreasing grain size. Burlin and Chan and colleagues calculated the energy loss for LiF and CaF_2 and the results are shown in figure 6.3.

The neutron response of TLD materials is rather complex and calculation of the neutron dose poses some problems. Since neutrons have no charge, their detection relies upon the effects of the secondary particles produced during neutron irradiation. For thermal neutrons, in order to obtain a high yield for such particles, isotopes with a high reaction cross-section are incorporated into the detectors. For example, LiF and $Li_2B_4O_7$ respond to slow neutrons via reactions with ^{6}Li and ^{10}B. The reactions are:

$$^6_3Li + ^4_0n \rightarrow ^4_2He(\alpha) + ^3_1H \tag{6.4a}$$

Figure 6.3. Thermoluminescence (TL) response of LiF and CaF_2 grains in air as a function of photon energy, relative to the response of LiF in LiF and CaF_2 in CaF_2, respectively. (After Burlin *et al.*, 1969; Chan & Burlin, 1970.)

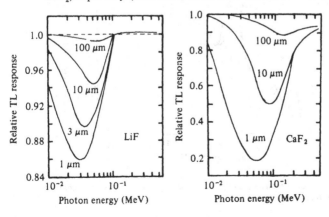

and

$$^{10}_{5}\text{B} + ^{1}_{0}n \rightarrow ^{4}_{2}\text{He}(\alpha) + ^{7}_{3}\text{Li}. \tag{6.4b}$$

The Harshaw Chemical Company produce TLD detectors of $^{6}\text{LiF:Mg,Ti}$ which exhibit a sensitive thermal neutron response via reaction (6.4a). Harshaw also manufacture a detector enriched in ^{7}Li (see table 6.1) which has a very low sensitivity to thermal neutrons. A wide range of sensitivities has been noted by Ayyangar *et al.*, (1974) for several common TLD materials, including LiF, $\text{CaSO}_4\text{:X}$ (X = Dy, Tm), $\text{CaF}_2\text{:Y}$ (Y = Dy, Mn), $\text{Li}_2\text{B}_4\text{O}_7\text{:Mn}$, $\text{MgSiO}_4\text{:Tb}$ and BeO.

The thermoluminescence response to fast neutrons arises as a result of the production of light-nuclei (recoil protons) and of heavy charged particles (fission fragments). Several authors have calculated the neutron energy response of a variety of TLD materials and have compared those calculations with experimental data. The authors include Tanaka & Furuta (1974, 1977), Horowitz & Freeman (1978) and Henniger, Hübner & Pretzsch (1982). The results of Tanaka & Furuta (1974) are shown in figure 6.4. The calculations are performed by taking account of the kerma[†] of the secondary particles (recoil protons, fission fragments) and the calculation of the thermoluminescence response (in terms of ^{60}Co equivalent rad neutron^{-1} cm^{-2}) requires the inclusion of the dependence of the thermoluminescence sensitivity upon LET. It is to be noted that in general, the fast neutron sensitivity is much lower than the

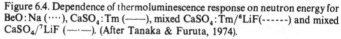

Figure 6.4. Dependence of thermoluminescence response on neutron energy for BeO:Na (·····), CaSO_4:Tm (——), mixed CaSO_4:Tm/^{6}LiF(------) and mixed CaSO_4/^{7}LiF (—·—). (After Tanaka & Furuta, 1974).

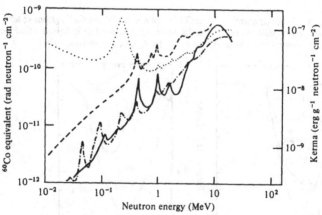

[†]Kerma: the energy transferred as the kinetic energy of the secondary charged particles neutron^{-1} cm^{-2}.

thermal neutron sensitivity. Further details are given by Douglas (1981). Carlsson & Alm Carlsson (1970) discuss the proton energy dependence of LiF.

The thermoluminescence which results from neutron irradiation stems entirely from the effects of the secondary charged particles. Radiation of this nature deposits energy in the lattice in dense, localized volumes along the track of the particle. In these volumes the thermoluminescence traps may become saturated and thus the overall sensitivity of the material is reduced. In this way high LET irradiation is seen to be much less efficient at inducing thermoluminescence than low LET radiation. For example, 4 MeV α-particles are only 5–50% as efficient as β-particles (Zimmerman, 1972, 1977b). Thus, it is generally found that as the LET increases, the thermoluminescence efficiency decreases. This trend can be seen over a wide range of LET (for protons and for α-particles) in figure 6.5 where experimental data of Jähnert (1972) and of Mukherjee (1981) are illustrated over the LET range of 0.02 to 300 keV μm^{-1}. Fain, Montret & Sahraoui (1980), Hübner, Henniger & Negwer (1980) and Horowitz & Kalef-Ezra (1981) discuss the LET dependence for heavy charged particles.

It has been noted that supralinearity and sensitization (cf. chapter 4) are both dependent on the LET of the irradiation (Suntharalingam & Cameron, 1969), the supralinearity being less for high LET. The causes of this dependence have been extensively discussed in the literature

Figure 6.5. Response of ^{7}LiF as a function of linear energy transfer. (Solid circles, Jähnert, 1972; Open circles, Mukherjee, 1981).

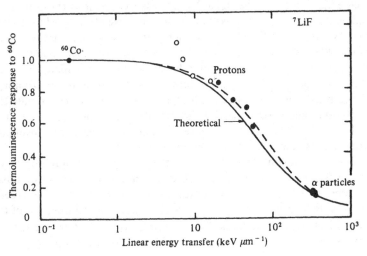

(Larsson & Katz, 1976; Waligórski & Katz, 1980; Lakshmanan & Bhatt, 1981; Lakshmanan *et al.*, 1981*a,b*).

6.1.3   *Fading and stability*

An important consideration in the choice of a TLD detector is how stable the signal is in the environment in which the dosemeter is operated. Thus, it becomes necessary to assess if the trapped charge within the material can be lost (before readout) by heat (thermal fading), light (optical fading) or any other means (anomalous fading).

The principles of thermal fading have essentially been dealt with in the discussion of the isothermal analysis of the thermoluminescence glow-curves. The main point is that if the trap depth E is too small then severe fading of the signal will occur, both during irradiation and between irradiation and readout. For dosimetry purposes, therefore, it is desirable for the detector to be characterized by a glow-curve with a peak at around 200–250 °C. This temperature range usually ensures that the trap depth is large enough ($E > kT$) for no appreciable trap emptying to have taken place, but also it is low enough such that interference from the black-body background signal is negligible.

As discussed in section 3.3.5, signal loss can occur via defect reactions as well as by simple trap emptying. In order to assess these effects properly, thermal decay experiments are required. Calculating half-lives for the peaks based on calculated E and s values is only of limited use.

Occasionally, a detector may be required to operate in a high-temperature environment. In these cases, it is necessary to select particularly high temperature peaks. This will also help to overcome accidental erasure of the signal by careless treatment of the dosemeter badge. (For instance, it is useful to relate the case of the workmen who, each evening, hung their overalls in a boiler house so that they would be warm when donned the following morning. The TLD badges, still attached to the overalls, thus lost the signal during the night that they had accumulated in the day.) The presence of deep traps in a phosphor, if not emptied during primary readout, may be utilized to re-estimate the thermoluminescence signal by the mechanism of phototransfer. The ability of a TLD detector to be re-read is a major advantage in dosimetry work and is discussed more fully in section 6.3.

The fact that some traps can be emptied by optical stimulation raises the question of optical fading. A dosemeter which is continuously exposed to sunlight, or to fluorescent lamps or to other energetic artificial light sources, may lose part of its signal by stimulation of the trapped charge by photons. The principles of optical trap emptying have already been discussed (section 4.3). For dosimetry applications, a material is normally tested by exposing an irradiated specimen to a light source of known wavelength (or spectrum) and, after a set time, comparing its thermoluminescence signal with that of a similar sample stored in the dark. Ideally, no optical effects are wanted. (Conversely, a high optical sensitivity is desirable in some dating applications – see chapter 7, section 7.5.1.)

Underestimation of the dose delivered may also result if the signal fades anomalously (section 4.4). Anomalous fading is much more difficult to detect than either thermal or optical fading because it generally occurs much more slowly. Possibly because of this it has not yet been demonstrated to be a problem with modern TLD materials.

6.1.4   Annealing procedures

Many of the properties discussed so far can be altered by thermal pretreatment. It has been discussed (section 4.2) how thermal annealing can change the defect distribution of a specimen and thereby alter its sensitivity. From the discussion of the defects responsible for thermoluminescence in LiF it is easy to see why the standard annealing treatment for this material (400 °C for 1 hour followed by 80 °C for 24 hours) has the effect that it does. This can be most easily understood from an ionic conductivity plot (cf. figure 5.4; good examples for LiF:Mg,Ti are given by Vora, DeWerd & Stoebe, 1974). The heat treatment has the effect of producing large numbers of trimer complexes which in turn give rise to a large peak 5 (the so-called 'dosimetry peak' in LiF). This is desirable because peak 5 has benign dosimetric properties (stable, moderately sensitive, reasonable linearity range). At the same time the lower temperature peaks are reduced in size, thereby effectively isolating peak 5 from undesirable interference. This can be further helped by a post-irradiation annealing of 100 °C for 10 minutes. This completely removes many of the unwanted, lower temperature peaks.

Following this annealing procedure for each irradiation has the effect of re-establishing the defect equilibrium, thereby allowing re-use of the material. It is interesting to note that according to the present state of knowledge concerning the nature of the defects in LiF (chapter 5) it should be possible to re-establish the same defect equilibrium by using a lower annealing temperature, say 250–300 °C, instead of 400 °C. If this is done, however, the same dosimetric properties are not observed for LiF. For example, the supralinearity and sensitization properties are altered (see, for example, the recent work of Lakshmanan, Chandra and colleagues). It appears that 400 °C is necessary if reproducibility is wanted. One possible cause of this is that the higher temperature also empties the higher temperature (deeper) traps which may be acting as competitors to the dosimetry traps. The strength of the competition will depend on the extent to which the competitors are filled. Annealing at 400 °C reverses the effect of the irradiation, but prolonged annealing can cause permanent reduction of the sensitivity so that annealing for too long is undesirable (Piesch, Burgkhardt & Hofmann, 1976; Piesch *et al.*, 1978). Optimal annealing conditions for several phosphors are discussed by Burgkhardt, Herrera & Piesch (1978*a*,*b*).

As if the above annealing procedure for LiF is not complex enough, it has also been established that the cooling rate from these high annealing temperatures is an important variable (Mason, McKinlay & Clark, 1976). Once again, this can be understood from a consideration of the defect structure of LiF (Dhar *et al.*, 1973; Bradbury *et al.*, 1976). Even the heating rate during readout is an important factor, the dosimetry peak being suppressed at very high heating rates (Taylor & Lilley, 1982*c*).

The annealing procedure for LiF has been discussed here in order to illustrate the complex nature of the process and how its behaviour can be understood in the light of the known defect properties of the material. For practical dosimetry it is clear that a complicated procedure such as this is a major disadvantage in the routine operation of the detector. It is mainly for this reason that the search for alternatives to LiF has been underway for two decades. Some progress has been made and now phosphors are available which do not require extensive annealing treatments. Sodium-stabilised LiF (Portal *et al.*, 1971) is one such example. For those materials which do not need thermal pretreatments, rigorous control of the procedure is essential. The different annealing procedures needed to give acceptable properties for various phosphors are given in table 6.1.

6.1.5   *Other factors*

There are several other characteristics which ought to be considered during detector selection. Among these may be included environmental effects. For example, a dosemeter should not undergo any physical or chemical changes during its use. This may necessitate encapsulation in specially designed holders which will also help to avoid exposure to high humidity (which can enhance fading and reduce sensitivity) and corrosive agents.

The above effects have a significant bearing on the re-usability of TLD materials, as does repeated thermal cycling (Martensson, 1969; Wald, DeWerd & Stoebe, 1977) and continual handling of the material (Niewiadomski, 1976*b*; Cox, Lucas & Kaspar, 1976; DeWerd, 1976).

A related phenomenon also relevant to the accuracy of measurement is 'spurious' thermoluminescence – that is, glow which does not result from irradiation of the specimen. Spurious glow may have several causes. One possibility is that charge may be unintentionally transferred from deep traps to the dosimetry traps during exposure to light (Niewiadomski, 1976*b*). An alternative which is often discussed is *tribothermoluminescence* – namely, the trapping of charge in the thermo-luminescence traps following mechanical effects upon the specimen (friction, grinding, sieving, pounding; see Walton, 1977). Many phos-phors exhibit this phenomenon which can cause serious errors if overlooked. It is obviously more of a problem with powdered samples than with single crystals or extruded specimens. Although the precise cause is not understood, it has been realized for some time that the tribo-signal can be suppressed or eliminated by performing the readout in an oxygen-free atmosphere (e.g., Schulman *et al.*, 1960; Nash, Attix & Schulman, 1965). It appears that the presence of oxygen is required for the tribo-signal to be significant. For this reason, thermoluminescence readout is normally performed in an argon or nitrogen atmosphere. A third possibility for spurious signals is chemiluminescence produced during readout (probably by oxidation of the surface impurities). An inert atmosphere also helps to reduce this problem, as does careful cleaning and handling procedures. 'Self-dose' is also an important consideration in certain applications (Nambi, 1982). Finally, spurious thermoluminescence arising from electrostatic charge build-up has recently been reported for several common dosemeters (Douglas & Budd, 1981).

If large numbers of dosemeters are to be used it is necessary to have an assessment of how reproducible the detectors are, from sample to sample, within the same batch. Batch inhomogenities can cause several problems which necessitate either the selection of individual specimens by the manufacturer, or the calibration of each detector by the user. Problems met with include variations in γ-sensitivity, LET response and neutron sensitivity (e.g., Horowitz *et al.*, 1979*a,b*; Douglas, 1981; Piesch, 1981*a,b*). Sensitivity variations for TLD detectors from the same batch may be as much as $\pm 10\%$. Individual selection can reduce this to $\pm 1\%$.

6.2 Specific examples

Table 6.1 lists some of the main properties of popular TLD materials at present in use. Other materials have been suggested and, in some cases, utilized (e.g., LiF:Ti (Mieke & Nink, 1979); NaCl (Heywood & Clarke, 1980*a*); MgO (Carter, 1976; Las & Stoebe, 1981); sulphates of Cd, Sr and Ba (Dixon & Ekstrand, 1974)) whilst at each Solid State Dosimetry Conference, a range of new materials is highlighted (e.g., CdI_2, $CeCl_3$ and $BaPt(CN)_4$; Spurný, 1980). Even DNA has been

Table 6.1. *Characteristics of popular TLD materials*

Phosphor	Commercial source	Glow peak (°C)	Approximate emission maximum (nm)	Z_{eff}	Ratio of TL response 30keV/^{60}Co	Sensitivity (1)	Linear range (rad) (2)	Saturation level (rad) (3)	Thermal fading	Optical fading	Pre-annealing procedures
LiF:Mg, Ti (4)	Harshaw (5)	210(6)	425(7)	8.14	1.3	1.0	5×10^{-3} to 10^2	10^5	~5–10% per year	NA	400°C, 1 hour 80°C, 24 hour (8)
LiF:Mg, Ti,Na (9)	D & CEC (10)	220	400(11)	8.14	1.3	0.5(12)	†	†	NA	NA	Generally unnecessary; possibly 500°C, 0.5 hour (11)
LiF:Mg, Cu,P (13)	—	232	310 410	8.14	1.3	25	10^{-4} to 10^3	$>10^4$	NA (14)	†	~250°C, 10 min
Li$_2$B$_4$O$_7$: Mn (15)	Harshaw (16)	210	600	7.4	0.9	0.4	10^{-2} to 300	3×10^6	5% in 60 days	Sensitive	300°C, 15 min
Li$_2$B$_4$O$_7$: Cu (17)	Panasonic (18)	205	368	7.4	0.8	8	10^{-2} to 10^5	$>10^5$	25% in 60 days	10% in 3 hours	†
MgB$_4$O$_7$: Dy/Tm:X (19)	—	210	480 570	8.4	1.5	7	10^{-3} to 10^2	10^3	10% in 60 days	Sensitive	600°C, 1 hour
Mg$_2$SiO$_4$: Tb (20)	DNT (21)	200	380 to 400		4–5	40–100	10^{-3} to 4×10^2	$>10^5$	NA	Sensitive	500°C, 3 hours

CaSO$_4$:Dy (22)	Harshaw	220	480 570	15.3	10–12	30	10^{-4} to 3×10^3	10^5	7–30% in 6 months	30% in 5 hours	400°C, 1 hour
CaSO$_4$:Tm (22)	Matsushita (23)	220	450	15.3	10–12	60	10^{-5} to 3×10^3	10^4	7–30% in 6 months	30% in 5 hours	400°C, 1 hour
CaSO$_4$:Mn (24)	Harshaw	80	500	15.3	10–12	60	10^{-5} to 3×10^3	10^4	10% per month	†	†
CaF$_2$ (natural) (25)	MBLE (26)	260	380	16.3	13–15	20 to 50	10^{-3} to 5×10^3	10^4 to 10^5	NA	Sensitive	†
CaF$_2$:Dy (27)	Harshaw (28)	200 240	480 575	16.3	13–15	16	10^{-3} to 10^3	10^6	25% in 4 weeks	Sensitive	80°C, 10 min (27) 600°C, 2 hours (11)
CaF$_2$:Mn (29)	Harshaw (30)	300	500	16.3	13–15	10	10^{-3} to 10^3	10^5	7% in 24 hours	†	Unnecessary
BeO (31)	BW (32)	180 to 220	330	7.13	1.4	3	10^{-2} to 50	5×10^5	7% in 2 months	50% in 1 hour	600°C, 15 min
Al$_2$O$_3$ (33)	CEC (10)	250	425	10.2	~3.5	5	10^{-2} to 10^2	10^5	5% in 1 month	Sensitive	†

(1) **Response** to ^{60}Co γ dose, relative to **LiF:Mg,Ti**.
(2) Lower limit set by minimum detectable dose. Upper limit set by onset of supralinearity.
(3) Level at which all traps are full, or at which appreciable radiation damage occurs.
(4) Cameron et al. (1968).

(Notes continued overleaf)

Table 6.1 (Contd.)

(5) Harshaw Chemical Co., Ohio, USA. Sold as:(i) TLD-100, containing Li in natural isotopic abundance (^{6}Li – 7.5%; ^{7}Li – 92.5%); (ii) TLD-600 (^{6}Li – 95.62%; ^{7}Li – 4.38%); (iii) TLD-700 (^{6}Li – 0.07%; ^{7}Li – 99.3%). Patent number: 1 059 518 (1967), London Patent Office.

(6) 'Peak 5'; heating rate $1.0\,^{\circ}\text{C}\,\text{s}^{-1}$. The other glow peak temperatures do not have heating rates quoted and are therefore approximate.

(7) Townsend et al. (1983b). Other emission maxima are approximate.

(8) Usually followed by a post-irradiation anneal of $100\,^{\circ}\text{C}$ for 10 min.

(9) Portal et al. (1971). Patent number 7103757 (1971), French Patent Office.

(10) Desmarquet & Carbonisation Enterprise et Ceramique, France. Sold as: (i) PTL-710 (Li in natural isotopic abundance, see (5) above; (ii) PTL-716 (^{6}Li enriched); (iii) PTL-717 (^{7}Li enriched).

(11) Portal (1981).

(12) Driscoll (1977). Quoted in McKinlay (1981).

(13) Nakajima et al. (1978).

(14) At room temperature; 5% after 30 days at $50\,^{\circ}\text{C}$.

(15) Schulman, Kirk & West (1967).

(16) Sold as TLD-800.

(17) Takenaga, Yamamoto & Yamashita (1980).

(18) National Panasonic/Matsushita (see 23), Japan.

(19) Prokic (1980). X is a co-activator, identity protected by Patent.

(20) Hashizume et al. (1971).

(21) Dai Nippon Toryo Co., Chigasaki, Japan.

(22) Yamashita et al. (1968, 1971).

(23) Matsushita, Osaka, Japan.

(24) Bjarngard (1967).

(25) Schayes et al. (1967).

(26) Manufacture Belge de Lamps et de Electronique, Belgium (now discontinued).

(27) Binder & Cameron (1969).

(28) Sold as TLD-200.

(29) Ginther (1954); Ginther & Kirk (1957).

(30) Sold as TLD-400.

(31) Tochilin et al. (1969).

(32) Thermalox 995: Brush Wellman Inc, Ohio, USA.

(33) Rieke & Daniels (1957).

† No available data, or no specific recommendations given.

suggested (Dolecek & Appleby, 1974). The table is designed as a quick reference for material selection. The listed commercial source is generally not the only place from which the material can be obtained. Alternative suppliers of the materials can be found.

The temperature of the main glow peak used in dose measurements is only important because it is desirable for this not to be too low so that the peak is unstable at operating temperatures, nor is it to be too high so that there is interference from black-body radiation. Otherwise it has no significance. The numbers quoted serve only as a guide because the temperature depends strongly upon heating rate. Likewise, the emission wavelength is best in the blue part of the spectrum in order to coincide with the maximum sensitivity of most photomultiplier tubes. Emission in the red is an unattractive property because the sensitivity of most tubes is poor in this spectral region and black-body emission is at the same wavelengths.

The effective atomic number Z_{eff} is an indication of the material's energy response. For dose measurement it is useful to compare these figures with that for air, namely $Z_{eff(air)} = 7.64$, and for tissue $Z_{eff(tissue)} = 7.42$. The ratio of the response at $30\,keV/^{60}Co$ is also listed. The sensitivity of the detector is given normalized to LiF for a given dose of ^{60}Co γ-rays. The sensitivity can, of course, be altered by radiation and thermal treatments and the sensitivities quoted correspond to unused specimens following the recommended annealing treatment.

The linear range is taken to be that region of the dose response curve which extends from the minimum detectable dose up to the onset of supralinearity. Both of these features are dependent upon the pretreatment, and supralinearity also depends upon the LET of the irradiation. The minimum detectable dose is a flexible parameter and the figure given in the table must be taken as approximate. Similarly, the saturation level depends on LET and sensitization and, because of the sublinear region, is difficult to quote exactly. Thus, estimates only are given. Strictly, the saturation level corresponds to all traps being full, but here it may also refer to the onset of radiation damage (cf. figure 4.3).

The details of thermal fading have been taken from reports of the degree of signal loss observed at a given temperature in a given time. The figures are not calculated from E and s estimates. Unless noted, the storage temperature is room temperature (although the exact value is not always known). The optical fading characteristics depend upon the wavelength and the intensity of the light. Some workers use daylight, others use fluorescent light, others use more sophisticated artificial lamps. Where detailed figures have been given in the literature, these are quoted in the table; otherwise simply the fact that the material is optically sensitive is noted.

Finally, the details of the recommended annealing procedures are given. Even here, there is little consensus, many slightly different treatments being recommended in the literature. Where possible, manufacturer's procedures are listed.

Not all of the important properties of a detector material can be summarized in table form. Therefore, each of the tabulated materials are now discussed in more detail.

6.2.1. *Lithium fluoride, LiF*

Already, during the course of this book, many pages have been devoted to LiF and this material is in many ways the 'standard' thermoluminescence phosphor. It has been used throughout to illustrate many of the principles and features of thermoluminescence, and a repetition of that information would be inappropriate here. Nevertheless, there are several additional points which can be made regarding the application of LiF to dosimetry.

LiF is a useful material in dose measurement for several reasons, including its general resistance to corrosion and wear. It is barely soluble in water and treatment with strong acids (e.g., fluoboric acid) is normally required before significant chemical attack is observed. It can be produced in several versions, being manufactured by Harshaw as TLD-100, TLD-600 and TLD-700 and is available as powder, extruded chips and ribbons, single crystals and Teflon-impregnated discs.

A sodium-stabilized version is also available (LiF:Mg,Ti,Na) made with 200 ppm magnesium fluoride and 1–2% by weight of sodium fluoride (Portal *et al.*, 1971). ^{6}Li- and ^{7}Li-enriched versions of this can also be obtained. (This material, known as PTL-710, 716 or 717, is manufactured by D & CEC; see table 6.1.) Its main advantage over the TLD-100 (600 and 700) versions is its improved thermal stability.

It should be noted that the thermal fading rates of the dosimetric peak in TLD-100 observed in practice do not generally correspond to those calculated from E and s determinations. There is quite a wide range of observed fading rates, as can be seen by reference to the several books and review articles on TLD where fading characteristics are listed (e.g., Cameron *et al.*, 1968; Becker, 1973; Horowitz, 1981; McKinlay, 1981; Oberhofer & Scharmann, 1981; Pradhan, 1981). In general, the rates vary from ~5% in one year to 15% per month at ambient temperature. The calculated fading rate (using the E and s data of McKeever (1980*a*,*b*), Taylor & Lilley (1978) and Fairchild *et al.* (1978*b*)) is ~10–15% per year. The probable cause of these discrepancies has already been discussed, namely the clustering and precipitation of divalent impurities during isothermal storage (cf. section 3.3.5 and chapter 5). After rigorously controlled annealing procedures (which govern the defect distribution), a fading rate of some 5–10% per year appears to be a representative value.

The sensitivity quoted in table 6.1 for the TLD specimens is for a material which has undergone the standard annealing treatment. For those applications where the measurement of very low doses is required, the sensitivity can be improved by adopting the treatment recommended by Mayhugh & Fullerton (1975) which combines the usual sensitization procedure suggested by Cameron *et al.* (1968) (namely, pre-exposure with a dose $> 10^3$ rad, followed by an anneal at 300 °C for ~ 1 hour) with ultra-violet exposure during the 300 °C annealing period. A smaller minimum detectable dose is then obtained.

The high-temperature peaks in LiF have been little discussed so far, yet they can be quite useful in certain dosimetric applications. In particular, there is a peak at ~ 275 °C which is normally quite small for low-dose, low-LET irradiations, but increases in size (relative to peak 5) as the dose increases. It is found to be especially sensitive to high-LET

Figure 6.6. Neutron-irradiated TLD-600, showing a large peak at ~ 275 °C compared with peak 5 (at ~ 210 °C). The specimen has been annealed at 400 °C for 1 hour, followed by a slow cool to room temperature. (After Townsend *et al.*, 1983*b*.) Compare with figure 2.14.

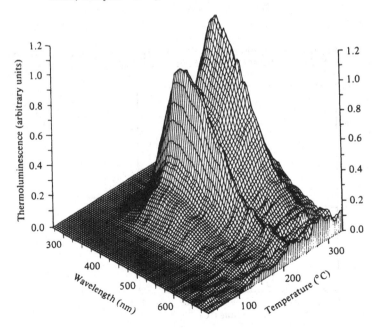

radiation (e.g., α-particles). Figure 6.6 shows an isometric plot for LiF TLD-600 irradiated with thermal neutrons. The high capture cross-section of ^{6}Li for slow neutrons results in the production of high-LET α-particles which induce a large 275 °C peak relative to peak 5. The increase in the ratio of the 275 °C/210 °C peak height has been used by many workers as an indication of neutron dose (e.g., Nash & Johnson,

Figure 6.7. Photon energy response curves for several thermoluminescent phosphors. A, CaF_2; B, $CaSO_4$; C, Al_2O_3; D, LiF; E, $Li_2B_4O_7$; F, BeO (from Becker, 1973); G, data points for Mg_2SiO_4 (from Lakshmanan & Bhatt, 1980); H, MgB_4O_7 (from Prokic, 1980).

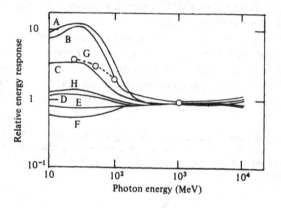

Figure 6.8. Comparison of A, LiF TLD-100 glow-curve (400°C, 1 hour) and B, LiF: Mg,Cu,P (250°C, 10min; curve from Nakajima *et al.*, 1978).

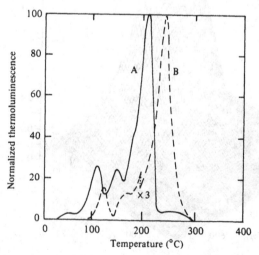

1977; Budd *et al.*, 1979). The fact that the 275 °C/210 °C peak–height ratio increases with imparted dose has been exploited by Jones & Martin (1968) to extend the range of γ-dose measurement up to $\sim 10^7$ rad.

The effective atomic number of LiF (8.14) is close enough to the value for tissue to make it almost tissue-equivalent. Of the materials in table 6.1, only beryllium oxide and lithium borate have lower values. Figure 6.7 shows the photon energy responses for the materials listed.

Nakajima *et al.* (1978) describe the preparation and properties of a highly sensitive LiF dosemeter, incorporating Mg, Cu and P as dopants. Mg and Cu doping alone gives the 'usual' emission near 410 nm, but the addition of P increases emission at 340 nm. When properly heat treated (250 °C, 10 min) the phosphor exhibits sensitivity comparable to $CaSO_4$, CaF_2 and Mg_2SiO_4:Tb. The heat treatment stabilizes the glow-curve structure and removes problems due to tribothermoluminescence. The glow-curve for LiF:Mg,Cu,P is compared with that from TLD-100 in figure 6.8.

6.2.2 *Lithium borate*, $Li_2B_4O_7$

As part of a search for a tissue-equivalent TLD material, Schulman, Kirk & West (1967) prepared specimens of $Li_2B_4O_7$, doped with manganese. The effective atomic number of this material (7.4) gives it an extremely good energy response, shown for pure $Li_2B_4O_7$ in figure 6.7. An example of a glow-curve from $Li_2B_4O_7$:Mn is shown in figure 6.9. The low-temperature glow peak is unstable at room temperature but the high-temperature peak is sufficiently well separated to allow its use in TLD. Brunskill (1968) developed the preparation of this material to produce a useful, tissue-equivalent detector (Langmead & Wall, 1976) but its main disadvantage stems from the fact that it emits at ~ 600 nm, which unfortunately corresponds to the most insensitive region of the response of the majority of photomultiplier tubes. The situation can be improved by careful selection of a tube with an extended red sensitivity (such as can be obtained using a GaAs photocathode).

None the less, owing to the unsuitability of the 600 nm emission for most thermoluminescent readout equipment (which means that, typically, $Li_2B_4O_7$:Mn has a sensitivity which is only $\sim 40\%$ that of LiF) much effort was put into finding an alternative activator to Mn, in order to produce a blue-emitting phosphor. Thus, Thompson & Ziener (1973) determined that silver-doped $Li_2B_4O_7$ emitted at 290 nm, but only a marginal improvement in sensitivity resulted. Rare earth doping (Rzyski & Morato, 1980) improves linearity, but not sensitivity. Greatest overall improvement results from the addition of copper. Takenaga, Yamamoto & Yamashita (1977) prepared samples of $Li_2B_4O_7$ doubly doped with

Ag and Cu, with optimum concentrations of 0.02% by weight for each dopant. More recent developments have determined that material singly doped with Cu produces the largest sensitivity enhancement (Takenaga *et al.*, 1980). The preparation of $Li_2B_4O_7$:Cu is discussed by Pradhan, Bhatt & Vohra (1982). $Li_2B_4O_7$:Cu is found to emit at 368 nm and has a sensitivity several times that of LiF. A recent comparison of Mn-, Cu:Ag- and Cu-doped $Li_2B_4O_7$ specimens was carried out by Wall *et al.* (1982) who concluded that the Cu:Ag system was the least attractive and that Cu-doped material exhibits far better sensitivity than the Mn-doped specimens, but worse fading. Similar results are given by Lakshmanan *et al.* (1981c) who also warn about possible batch-to-batch variations in dosimetric properties.

Fading can be a particular problem with $Li_2B_4O_7$. Mn-activated material exhibits very little fading in dry conditions either in the dark or

Figure 6.9. Glow-curves from $Li_2B_4O_7$:Mn showing fading (at room temperature) of the low-temperature peak. A, 3.4 min after irradiation; B, 8.6 min; C, 27.3 minutes; D, 69.2 minutes; and E, 960 min. (After Schulman *et al.*, 1967.)

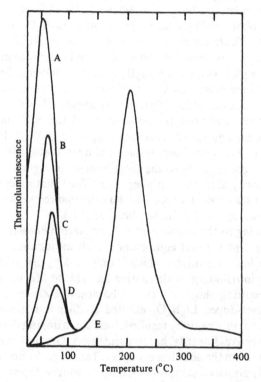

when exposed to light. With the introduction of Cu, however, severe fading is observed when illuminated by either fluorescent or tungsten light. This problem can only be overcome by keeping the phosphors in the dark. Although optically resistant, $Li_2B_4O_7$:Mn exhibits a large increase in its thermal fading rate (and a decrease in its sensitivity) when exposed to humid atmospheres (Mason *et al.*, 1974). Below $\sim 30\%$ relative humidity a small ($\sim 5\%$) loss of sensitivity occurs; above 30% humidity, a more rapid decline in sensitivity with time is seen. Cu-activated materials exhibit greater moisture resistance with a stoichiometric matrix composition (i.e., $1Li_2O:2B_2O_5$) showing the minimum water intake (Takenaga *et al.*, 1980). Adding silica (0.25% by weight) also improves the moisture resistance (Bøtter-Jensen & Christensen, 1972). Furetta & Pellegrini (1981) note an *increase* in the glow peak height with storage time at ambient temperature. The cause of this (which is reminiscent of similar effects in alkali halides, e.g., Inabe & Takeuchi, 1978) is unknown.

Not only is the material's reaction to high humidity affected by stoichiometry and doping, but also the sensitivity and the energy response can both be tailored by suitable adjustments to the preparation procedure. Jayachandran (1970) and Thompson & Ziener (1972, 1973)

Figure 6.10. Variation in response with photon energy for different lithium borate materials: A, $Li_2B_4O_7$:Cu,Ag; B, $Li_2B_4O_7$:Mn; C, $Li_2B_4O_7$ (pure); D, $Li_2B_4O_7$:Cu (from Wall *et al.*, 1982). E ($3Li_2O.3BeO.8B_2O_3$):0.5% $MnCl_2$ (from Becker, 1972).

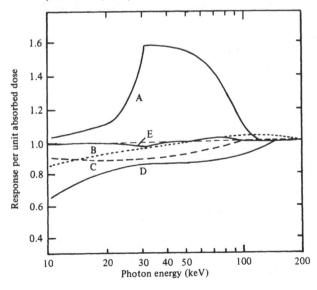

showed how the energy response depends on the dopant concentration whilst Christensen (1968) demonstrated that an improved energy dependence can be obtained, at lower energies, by preparing material with a higher borate content. Preparation in a reducing atmosphere results in higher sensitivity. DeWerd, Kim & Stoebe (1976*b*) describe the preparation of both glassy and crystalline $Li_2B_4O_7$:Mn, in stoichiometric and non-stoichiometric compositions. Although the glassy material is less sensitive (non-stoichiometric glass shows the higher sensitivity) it exhibits greater linearity (up to $\sim 10^5$ rad). The non-linearity in crystalline $Li_2B_4O_7$ appears to be related to the growth of a second peak in the high-temperature region. The glasses exhibit just one, broad peak.

The different energy responses that can be obtained from $Li_2B_4O_7$ are shown in figure 6.10 which illustrates the effect of the different dopants and of the stoichiometry. Horowitz *et al.* (1979*a*) warn that the LET response of $Li_2B_4O_7$ is highly sensitive to ppm dopant concentration variations, giving rise to large batch-to-batch differences.

6.2.3   *Magnesium borate, MgB_4O_7*

This is the most recently introduced TLD detector to be discussed in this book, having been introduced by Prokic in 1980. The material was developed by the Böris Kidric Institute for Nuclear Sciences (Belgrade) and is doubly activated with either Dy or Tm, plus one other component, the identity of which is unspecified. Its sensitivity is ~ 7 times (although Barbina *et al.* (1981) report 4 times) greater than TLD-100 and its glow-curve is displayed in figure 6.11. Its tissue

Figure 6.11. Normalized glow-curves for A, MgB_4O_7: Dy (from Barbina *et al.*, 1982); B, Mg_2SiO_4:Tb (from Nakajima, 1972).

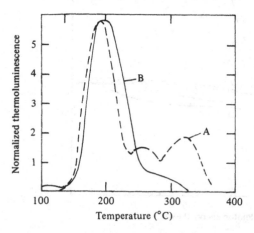

equivalence is only marginally worse than that of LiF (see figure 6.7).

Initial recommendations (Prokic, 1980) did not specify any thermal annealing treatment, but later investigations noted that maximum sensitivity could be achieved following a pre-irradiation anneal of 500 °C for 1 hour (Barbina *et al.*, 1981). The same authors later modified this to 600 °C for 1 hour (Barbina *et al.*, 1982). Thermal fading is at an acceptable level, but Driscoll, Mundy & Elliot (1981) note an unacceptable level of light-induced fading on the low-temperature side of the dosimetry peak. They recommend a short post-irradiation treatment of 160 °C for approximately 16 s to overcome this.

A potentially important problem noted by both Barbina *et al.* (1981) and by Driscoll *et al.* (1981) is the high level of variability within a batch. Unless this can be improved, individual calibration of the detectors is necessary and this may prove to be a hindrance which may prevent the material from being widely used in routine dosimetry.

6.2.4 *Magnesium orthosilicate*, Mg_2SiO_4

In 1971 Hashizume *et al.* described the preparation of a promising new phosphor, Mg_2SiO_4. This material appears in nature as forsterite (iron free) or olivine, but was manufactured as a TLD material with terbium as dopant. In 1975 Jun & Becker remarked that suprisingly little attention had been given to this material. Unfortunately, since that date little more has been reported which suggests that this phosphor has not fulfilled its early promise.

The glow-curve of Mg_2SiO_4:Tb is shown in figure 6.11 with the dosimetry peak at ~ 200 °C. The fading characteristics of this peak are a little complex in that, upon storage, a 10% rise in output is observed during the first few hours, although this can be stabilized by a thermal treatment of 40–50 °C for 30 min (Jun & Becker, 1975). Overall, however, excellent thermal stability is exhibited.

The energy dependence is quite large which necessitates the use of filters (Nakjima *et al.*, 1971). Embedding the material in Teflon also help to reduce the energy dependence. As with other phosphors, the energy dependence can be improved by sensitizing the phosphor by pre-exposure to γ radiation and annealing at 300 °C for 1 hour (Lakshmanan & Bhatt, 1980). A problem with sensitized Mg_2SiO_4 is that a substantial increase in the background signal is observed (Jun & Becker, 1975). The material is also highly light sensitive and may be used as an ultra-violet dosemeter (see section 6.3 for ultra-violet effects).

For completion, the material produced by Bhasin, Sasidharan & Sunta (1976) ought to be mentioned. This version of Mg_2SiO_4:Tb has a very different glow-curve to that produced by the Dai Nippon Toryo Co.

(Japan) in that its main dosimetry peak is at $\sim 300\,°C$. It is reported to be insensitive to room light, but strongly affected by 254 nm ultra-violet light. With a sensitivity some 80 times that of TLD-100, it appears to be one of the most sensitive phosphors available.

6.2.5   *Calcium sulphate, $CaSO_4$*

The application of $CaSO_4$ doped with Mn or with other rare earths (notably Dy, Tm or Sm) began in the late 1960s to early 1970s (for example, Bjarngard, 1967; Yamashita *et al.*, 1968, 1971; Becker, 1972). The mechanism of thermoluminescence production in these materials appears to be related to the emission of light corresponding to the fluorescent emission of the activator material (Nambi *et al.*, 1974; see also section 2.1.4). For rare-earth-doped $CaSO_4$, the glow-curve shape is remarkably constant, independent of the rare earth ion, with the dominant glow peak for dosimetry appearing at $\sim 220\,°C$. The Mn-doped material exhibits its main peak at $\sim 80\,°C$ (figure 6.12) a consequence of which is that this glow peak exhibits severe thermal fading. Hence, despite the fact that $CaSO_4$:Mn is one of the most sensitive dosemeters available, its use is limited to short-duration experiments. The higher stability of the rare-earth-activated detectors make these more popular (Burgkhardt *et al.*, 1978*a,b*). As with LiF, a wide range of fading rates have been reported for these materials, as discussed recently by Nambi (1982).

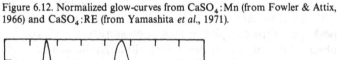

Figure 6.12. Normalized glow-curves from $CaSO_4$:Mn (from Fowler & Attix, 1966) and $CaSO_4$:RE (from Yamashita *et al.*, 1971).

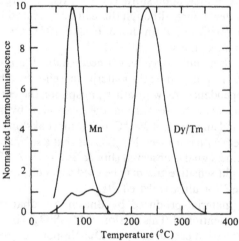

The supralinear response of $CaSO_4$:Dy is in part related to the growth of new peaks on the high temperature side of the main dosimetry peak (Lakshmanan *et al.*, 1978). The growth of these peaks undoubtedly has a large bearing on the observation of complex and variable kinetics for this material with increasing dose (Srivastava & Supe, 1979; Furetta & Gennai, 1981). To some extent these problems can be offset by sensitizing the material with a pre-exposure of 100 rad and a post-irradiation anneal of $\sim 600\,°C$ for 1 hour (Lakshmanan *et al.*, 1978).

Sensitization also helps to reduce the energy dependence of $CaSO_4$:Dy (Shinde & Sastry, 1979; Pradhan & Bhatt, 1982). With an effective atomic number of 15.3, $CaSO_4$ exhibits a strong photon energy dependence and attempts to overcome this have included embedding the material in Teflon (Pradhan & Bhatt, 1982) and the use of filters (Yamashita *et al.*, 1971; Pradhan & Bhatt, 1979; Szabó *et al.*, 1980). Morato *et al.* (1982) have recently proposed the use of cold-pressed pellets of $CaSO_4$:Dy with NaCl as a binder. With suitable filters, this system can produce virtual energy independence between 20 keV and 1.25 MeV. Pradhan & Bhatt (1977) observe that the energy response to β-rays can be improved by mixing $CaSO_4$:Dy with graphite. The neutron response varies markedly, depending upon dopant. Dy-doped material is found to have the highest neutron sensitivity because of the neutron reaction with ^{164}Dy (Yamashita *et al.*, 1971).

The use of particularly sensitive $CaSO_4$:Dy or $CaSO_4$:Tm phosphors co-activated with an unspecified second dopant has been described by Prokic (1980). Improvements in the preparation of 'standard' $CaSO_4$: Dy were described by Prokic (1978) and by Azorin, Salvi & Moreno (1980).

6.2.6 *Calcium fluoride, CaF_2*

A significant advantage of CaF_2 is that it is naturally occurring (as the mineral fluorite) and can thus be cheaply obtained in developing countries. A standard natural fluorite used to be marketed by MBLE, Belgium, although this is now discontinued. One major problem of using natural fluorite is the complexity of its glow-curve (Schayes *et al.*, 1967). The natural mineral contains many activators which appear to be predominantly rare earths (Becker *et al.*, 1973; Sunta, 1979). Synthetic rare-earth-doped calcium fluoride detectors were introduced by Binder & Cameron (1969; CaF_2:Dy) and by Lucas & Kaspar (1977; CaF_2:Tm). Ginther & Kirk (1957) introduced CaF_2:Mn. The glow-curves for MBLE fluorite, CaF_2:Dy and CaF_2:Mn are shown in figure 6.13.

The fading of the glow peaks in CaF_2 is suprisingly high considering their estimated half-lives from E and s calculations (Adam & Katriel,

1971) and several fading rates have been reported (Nambi, 1982). Considering the processes occurring during isothermal annealing in alkali halides (cf. LiF) it is relevant to recall that impurity clustering has also been observed to influence thermoluminescence in CaF_2 (Bangert *et al.*, 1982*b*).

A frequently noted problem with CaF_2 is the build-up of a significant response within the phosphor during storage periods. This is possibly due to 'self-dose' (Brinck *et al.*, 1977; Nambi, 1982).

The ultra-violet response of CaF_2 can be enhanced severalfold by annealing at 900°C in air, while the same treatment significantly reduces the sensitivity to ionizing radiation. Pradhan & Bhatt (1983) examine this further and relate the increased ultra-violet sensitivity to surface oxidation and speculate that the reduction in the γ-response may be related to a reduction of the charge state of the rare earth impurities before irradiation. The production of thermoluminescence in rare-earth-doped CaF_2 is believed to involve a reaction of the type $RE^{3+} + e^- \rightarrow RE^{2+}$ during irradiation and $RE^{2+} + h^+ \rightarrow RE^{3+}*$ during heating, where $RE^{2+}*$ is an excited state which decays to the ground state with the emission of luminescence, as discussed in section 2.1.4.

Figure 6.13. Normalized glow-curves of A, MBLE CaF_2 (natural); B, CaF_2:Mn (from Fowler & Attix, 1966); C, CaF_2:Dy (post-irradiation anneal of 115 °C for 10 min, from McCurdy, Schaiger & Flack, 1969).

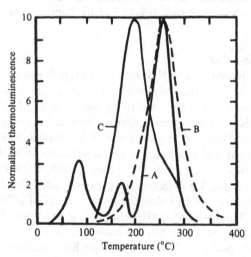

6.2.7 *Beryllium oxide, BeO*

Tochilin, Goldstein & Miller (1969) and later Scarpa (1970) looked at the potential usefulness of BeO as a TLD material, with further examinations of this phosphor being carried out at later dates by Gammage and colleagues (Gammage & Garrison, 1974; Crase & Gammage, 1975; Gammage & Cheka, 1977) and by Busuoli *et al.* (1977) and Moos (1979). The material is seen to have some rather interesting dosimetric properties including an LET response which increases with increasing LET (as opposed to the usual decrease), a low thermal neutron response (but a higher fast neutron sensitivity) and good tissue equivalence ($Z_{eff} = 7.13$). To offset these features, the material also exhibits strong light sensitivity, small linear range, a complex glow-curve structure and (in powder form) high toxicity.

The glow-curve shape depends critically upon the dopants (e.g., Al^{3+}, F^-) and the preparation conditions. Crase & Gammage (1975) report that a commercial form of BeO called Thermalox 995 (with an emission wavelength of 265 nm) is suitable for dosimetry, although quite a wide variety of other forms of BeO (some suitable for TLD, some not) also exist. For example, glow-curves from hot pressed and sintered BeO are shown in figure 6.14. These materials were obtained from the

Figure 6.14. Glow-curves for hot pressed and sintered BeO from Consolidated Beryllium, UK. (Reproduced with permission of T.A. Lewis, CEGB, Berkeley, UK.)

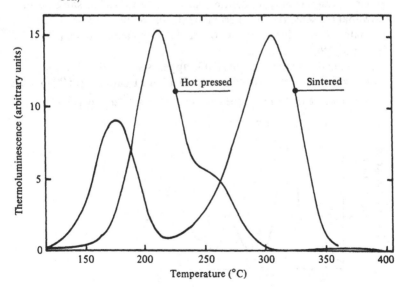

same manufacturer (Consolidated Beryllium Ltd, UK) but exhibit entirely different properties, the high-temperature peak of the sintered material displaying the more favourable dosimetric behaviour. Increasing the zero dose level and decreasing sensitivity with re-use appear to be features of most BeO types and individual calibration is a necessity.

6.2.8   *Aluminium oxide, Al_2O_3*

The dosimetric properties of this material were first described by Rieke & Daniels (1957) with a later investigation of its TLD behaviour by McDougall & Rudin (1970). The material used by McDougall & Rudin was nominally pure Al_2O_3, but it probably contained a few ppm of Cr^{3+} because the thermoluminescence emission was at 650 nm – i.e., the same as that from ruby. Buckman (1972) points out that the emission from Cr^{3+}-free Al_2O_3 is predominantly at ~ 410 nm, which is a much more desirable wavelength for TLD. Mehta & Sengupta (1976a,b, 1977) describe the preparation of an Al_2O_3 phosphor for TLD which emits at 420 nm. The glow in this material is believed by these authors to be related to the presence of Si and Ti. Osvay & Biró (1980) discuss a Mg- and Y-doped Al_2O_3 phosphor, but no emission wavelengths are given.

The material studied by Mehta & Sengupta is described as being a sensitive γ detector, but is insensitive to α-particles. Al_2O_3 also has a response which is only moderately sensitive to LET. The same authors show that the sensitivity to thermal neutrons can be increased by mixing the material with Dy_2O_3 or by covering it with cadmium foil (Mehta & Sengupta, 1981).

The main dosimetry peak in Al_2O_3 appears at $\sim 250\,°C$ but the glow-curve itself is quite complex (figure 6.15). Flame treatment (to 2000 °C) is

Figure 6.15. Thermoluminescence from α-Al_2O_3 (after Portal *et al.*, 1980).

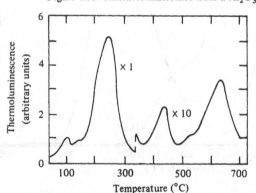

seen to produce large increases in sensitivity and to introduce new, high-temperature peaks into the glow-curve (Mehta & Sengupta, 1976a). In particular, a peak at 625 °C is produced which appears to have reasonable dosimetric properties (Portal, Lorrain & Valladas, 1980). However, there does not appear to be a large enough advantage to be gained in using this peak to offset the problem of having to subtract the black-body background signal during routine dosimetry.

6.3 Ultra-violet effects and dose re-estimation

Generally, the thermoluminescence response of dosemeter materials to ultra-violet rays is of two distinct types, namely, the glow which can be stimulated by illuminating 'virgin' dosemeters with ultra-violet light, and that which can only be stimulated after firstly irradiating the specimen with ionizing radiation (γ, X-ray, β, etc.). Since the ultra-violet wavelengths usually employed are not energetic enough to cause band-to-band ionization, the mechanism responsible for the thermoluminescence production following illumination of a virgin sample may be due to the creation of excitons and their subsequent decay to electrons and holes. On the other hand, the thermoluminescence which can only be induced by ultra-violet light if the sample has been previously irradiated is probably due to the phototransfer of charge from deep traps (section 4.3.2). A convenient example is the ultra-violet response of MgO. In this case phototransfer effects were ruled out by Kirsh et al. (1977) as being the cause of ultra-violet-induced thermoluminescence, but instead they preferred an exciton mechanism.

All of the phosphors mentioned in section 6.2 exhibit ultra-violet responses and have been investigated as possible ultra-violet dosemeters. $Li_2B_4O_7$, $CaSO_4$, Mg_2SiO_4 and Al_2O_3 all appear to respond to ultra-violet irradiation without the need for previous γ exposure (e.g., Buckman, 1972; Chandra, Ayyangar & Lakshmanan, 1976; Mehta & Sengupta, 1978; Lakshmanan & Vohra, 1979; Lakshmanan et al., 1981c) but additionally, all display phototransfer properties (cf. chapter 4). For example, the phototransfer of charge from deep traps to the dosimetry traps in $CaSO_4$:Dy was suggested by Caldas & Mayhugh (1976) to be a convenient method of estimating high exposure levels (i.e., doses > the saturation dose of the dosimetry traps).

In addition to the above mentioned materials, phototransfer is also observed in LiF, CaF_2 and BeO (MgB_4O_7 has yet to be fully investigated). LiF is the most widely studied and for this material it has been found that the efficiency of phototransfer can be increased by elevating the temperature (to 80 °C) at which the phototransfer takes place (Mason et al., 1977). These studies have arisen from the possibility of using the phototransfer phenomenon to re-estimate the dose absorbed by the material. Such confirmation measurements may be necessary because of, say, instrument malfunction during readout resulting in the loss of the signal without a record being made. Alternatively, there may be a suspicion of an overexposure in personal dosimetry, which needs checking. Doses as low as several hundred mrad may be checked in this way.

Aside from phototransfer, it has been suggested that photostimulated luminescence may be used to re-estimate the dose. Here, the ultra-violet light optically releases charge from the deep traps, but instead of reading the phototransferred thermoluminescence, the luminescence intensity resulting from the recombination of the released electrons at luminescence sites is recorded. For a given ultra-violet exposure, the luminescence intensity is proportional to the original imparted dose. This method is described for $CaSO_4$:Dy by Pradhan & Bhatt (1981).

An entirely different method of dose re-estimation is suggested by Oliveri, Fiorella & Mangia (1979), also for $CaSO_4$:Dy. These authors estimate that the trapping sites in this material are distributed in energy. Thus, by partially reading out the peak (up to a certain temperature) only 'one piece' of the distribution is being monitored, with the rest of the trapping states remaining unaffected. To re-estimate the dose, the specimen is read out again, but to a higher temperature, thus sampling traps not yet emptied. Clearly a different calibration curve is required for each temperature reached.

The preferred method of dose re-estimation, however, is still phototransfer. A potential problem arises when sensitized phosphors are used (section 6.2.1). Here the sensitizing pre-dose gives the deep traps a charge population before the exposure to be measured takes place. Thus, the phototransferred charge is proportional not to the exposure but to the pre-dose plus exposure (Bartlett & Sandford, 1978). However, Charles, Khan & Mistry (1980) argue that, for LiF, if the Mayhugh & Fullerton (1975) ultra-violet sensitization procedure is adopted, re-estimation can still take place. This is because the traps emptied by the ultra-violet exposure are not the same as those which cause the sensitization. Douglas *et al.* (1980) further illustrate that the phototransfer efficiency increases with LET by an amount which varies from batch to batch. This is consistent with the idea that the deep traps are multi-hit centres which are more sensitive to high LET radiation than to γ photons.

6.4   Personal dosimetry

6.4.1   *Introduction*

Personal dosimetry is concerned with assessment and limitation of the radiation dose delivered to individuals. The aim is to restrict the dose absorbed by a person to below the ICRP recommended limits. These limits correspond to the maximum dose-equivalent values which are not expected to cause injury to the person during his working lifetime, the chief problem at low doses being carcinogenesis.

The tissues at risk during the irradiation of an individual include the skin, and deeper lying material including the gonads, lungs, thyroid, bone surface and bone marrow stem cells, and the breasts in females. In practice, by using a 'skin' dosemeter (i.e., one which estimates the dose delivered to a depth of $5-10\,\mathrm{mg\,cm^{-2}}$ ($50-100\,\mu\mathrm{m}$)) and a 'body' dosemeter (i.e., one which estimates the dose at a depth of $300-1000\,\mathrm{mg\,cm^{-2}}$), an estimate can be made of the dose delivered to the skin and to the male gonads (between $\sim 400-1000\,\mathrm{mg\,cm^{-2}}$). Such a dosemeter will also provide an overestimate of the dose to the bone marrow ($\sim 2000\,\mathrm{mg\,cm^{-2}}$), to the female gonads ($\sim 7000\,\mathrm{mg\,cm^{-2}}$) and to other deeper lying organs. With respect to all of these tissues, the aim of radiation protection is to limit the probability of stochastic effects to acceptable levels (i.e., as low as possible, but less than the ICRP limits), and to prevent non-stochastic effects by keeping dose levels below all known thresholds.

6.4.2   Materials

TLD dosemeters used in personal dosimetry may be classified according to their primary function, namely:

> Skin dosemeters: dosemeters worn on the body for the purpose of assessing skin dose due to photons or electrons.
>
> Body dosemeters: dosemeters worn on the body for the purpose of assessing body dose due to photons or electrons.

For skin dosemeters, the absorbed dose is to be measured at a depth of (typically) $7\,\mathrm{mg\,cm^{-2}}$. For body dosemeters, the depth is typically between 300 and $1000\,\mathrm{mg\,cm^{-2}}$ and therefore deals only with penetrating radiation (X-rays $> 15\,\mathrm{keV}$, γ-rays). Additionally, extremity dosemeters are required in personnel monitoring where it is believed that the above dosemeters will not give a true indication of the dose received by body extremities (e.g., the fingers). For more detailed information about the dose received, neutron detectors may be required, along with detectors which enable not just the dose to be calculated (i.e., non-discriminating detectors), but are also able to provide data on the quality and direction of the radiation (i.e., discriminating detectors). Thus, a dosemeter package might contain both γ and neutron-sensitive detectors, along with a series of filters to give information on the energy of the radiation. (An alternative to filters might be to use a tissue-equivalent material (e.g., LiF, $Li_2B_4O_7$) and an energy-dependent material (e.g.,

CaF_2).) Such a dosemeter would provide for an assessment of the dose, type and energy of the radiation being monitored.

A problem arises with β-dosimetry. Here the β-particles may not penetrate the gonads or bone marrow, but can affect some deeper tissues in the eye, contributing (over a period of years) to the possibility of lens clouding. Several reports have been published on attempts to manufacture ultra-thin detectors for β-dosimetry. These include deposition of fine grains onto discs (Zimmerman, 1972), graphite loading to limit the thermoluminescence output to the surface grains (Pradhan & Bhatt, 1977), grains impregnated into polytetrafluoroethylene (PTFE) and bonded on to a PTFE substrate (Charles, 1977), LiF grains incorporated in a polyethersulphone matrix (Lowe, Lakey & Yorke, 1979) and diffusion of boron into LiF to a depth of up to $2 \, mg \, cm^{-2}$ (Christensen & Mäjborn, 1980). Ion implantation to the required depths has not yet been tried, but it appears to be a possibility (e.g. Bangert *et al.*, 1982*a,b*).

Low-energy photons may also cause problems because of the detector energy response, detector thickness and shape, and self-absorption in the dosemeter container (badge). High Z_{eff} materials and thick detectors should be avoided in the low-energy range. For high-energy photons ($> 1 \, MeV$) the response of most detectors decreases. The use of tissue-equivalent filters of the required thickness ($\sim 3 \, mg \, cm^{-2}$) can correct the energy response for tissue-equivalent detectors.

With the proliferation of nuclear reactors, the detection of neutrons in personal dosimetry has taken on increased importance. There are several possibilities for the measurement of neutron doses, depending upon the energy and upon the nature of the radiation field (i.e., pure neutron, or mixed γ/neutron) and recent reviews by Lakshmanan (1982) and by Douglas (1981) detail the principles involved. For the present purposes we may note that $Li_2B_4O_7$ and LiF have excellent thermal neutron sensitivities and therefore are popular detector materials. In a mixed γ/neutron field the combined use of neutron-sensitive ^{6}LiF and neutron-insensitive ^{7}LiF is able to separate the two radiation types. Alternatively, the high sensitivity of $CaSO_4$ to charged particles may be exploited by mixing the phosphor with a neutron-sensitive compound (e.g., a ^{6}Li compound). Examples are given by Ikeya, Ishibashi & Itoh (1971) and Iga *et al.* (1977).

Fast neutrons may be detected by neutron activation. Here the activity of certain radioactive isotopes, created within the material during neutron irradiation, doses the phosphor to produce a thermoluminescence signal. For example, the fast neutron reaction $^{32}S(n,p)^{32}P$ is utilized by Pradhan *et al.* (1978) in $CaSO_4$:Dy, mixed with sulphur, to give a sensitive neutron detector. ^{32}P is a β-emitter and the post-neutron accumulation of thermoluminescence gives a measure of the ^{32}P content, and, therefore, of the initial neutron dose.

The most popular method of fast neutron detection is albedo dosimetry. Albedo neutrons are those which re-emerge from the body after undergoing moderation by multiple scattering. The backscattered neutrons can then be detected by thermal neutron detectors. The correction factors for converting neutron flux into dose, and the fraction of incident neutrons which emerge as thermal neutrons, are both energy dependent and so the response of an albedo dosemeter is also energy dependent. Therefore, additional detectors to provide correction for the incident neutron energy spectrum are sometimes used. Piesch (1977) and Douglas & Marshall (1978) discuss albedo dosemeter design and performance, whilst Douglas (1981) reviews the general field of neutron dosimetry using TLD.

In general, most of the phosphors listed in table 6.1 find usage in personal dosimetry, but tissue-equivalent materials are generally preferred. The requirements suggested by the various standards organizations differ slightly and the reader is referred to the documents mentioned earlier in section 6.1 if further details are required.

6.4.3 *Practical application*

In personal dosimetry, the dosemeter has to be worn by the user and so a TLD 'badge' is required which may be clipped to the user's clothing, worn as a ring, or fixed to a finger. The badge design is such that it can hold the required detectors (be they in powder, chip, rod or PTFE-disc form) along with any filters which may be necessary. (The badge plastic itself may act as a filter.) In order to ensure reproducibility of dose measurements, careful control over the manufacture of the badge is essential (i.e. composition, dimensions, etc.). Examples of several TLD badges, along with their dosemeter inserts, are shown in figure 6.16.

Many dosemeter manufacturers have designed automated TLD readout systems for the rapid and reliable handling of large numbers of dosemeters. Many are linked with computers for rapid manipulation, storage and retrieval of the relevant data (user's name, dose estimate, previous dose history of the patient, previous history of the dosemeter including dose received and thermal treatments carried out, etc.). The National Radiological Protection Board (NRPB) in the UK operates a large-scale automated system designed to deal with a high throughput of personal dosemeters using LiF:Mg, Ti PTFE-discs (Dennis, Marshall & Shaw, 1974). The most recent estimate is that 320 000 dose assessments are made with this system each year, and this number is increasing at the rate of about 55 000 per year. In addition, 47 000 extremity dosemeters

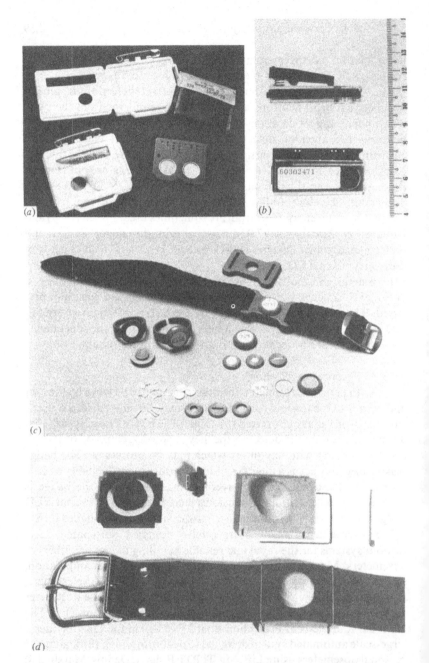

Figure 6.16. Examples of thermoluminescence dosemeters and their respective badges. (a) National Radiological Protection Board (NRPB), showing the holder (top left), the insert cover (top right), the insert itself consisting of two LiF polytetrafluoroethylene discs (bottom right) and the complete arrangement

are processed. Duftschmid (1980) describes a similar system used in Austria, and Julius (1975) discusses the introduction of a similar service in the Netherlands. Other radiation protection services use smaller-scale, purpose-built TLD readers available from a variety of manufacturers and using a range of dosemeter materials (see chapter 9). For a recent comparison of many of the TLD systems commercially available, see Julius (1981) and Duftschmid (1981).

6.5 Environmental monitoring

6.5.1 *Introduction*

Environmental monitoring is concerned with the measurement of the natural radioactivity arising from radioactive materials present in

Figure 6.17. Genetically significant dose received by the population of Great Britain. (After Wall *et al.*, 1981.)

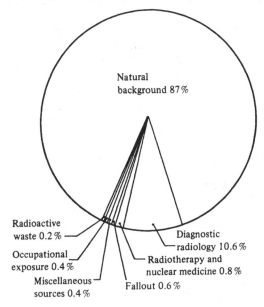

Natural background 87%

Radioactive waste 0.2%

Occupational exposure 0.4%

Miscellaneous sources 0.4%

Fallout 0.6%

Diagnostic radiology 10.6%

Radiotherapy and nuclear medicine 0.8%

Figure 6.16 (*Contd.*)
(bottom left). Reproduced with permission of NRPB. (*b*) National Panasonic, UD-800 series, four-element badge. Several variations are available based on $Li_2B_4O_7$:Cu and $CaSO_4$:Tm. Reproduced with permission of National Panasonic (UK) Ltd. (*c*) An extremity dosemeter based on BeO discs and LiF chips. Reproduced with permission of H.W. Julius, Radiolgische Dienst TNO. (*d*) A neutron albedo dosemeter using LiF. Reproduced with permission of E. Piesch, Karlsruhe Nuclear Research Centre.

the surroundings and from cosmic rays. (When entering the earth's atmosphere cosmic rays consist of 92% protons, 7% α-particles and 1% heavy nuclei. Few of these primary particles survive at sea level where secondary cosmic rays consist of 75% muons and 25% electrons and positrons. The filtering action of the atmosphere means that the cosmic ray dose rate at sea level is less than at high altitude. For example, at the equator the mean dose rate to the gonads at sea level is 23 mrem year^{-1}, whilst at an altitude of 3000 m it is 56 mrem year^{-1} (Martin & Harbison, 1979). This difference is not trivial and aircraft crews may accumulate a significant additional radiation dose due to cosmic rays during their working lifetime.)

The measurement of radiation from manmade sources is also encompassed. An example of recent public concern is the monitoring of radiation levels released from nuclear power stations (e.g., Brinck *et al.*, 1977; Oatley, Hudson & Plato, 1981). Another well-documented example of environmental dosimetry is the measurement of the natural

Table 6.2. *Typical annual gonad dose per year due to natural and manmade radiation in Great Britain.* (*Data from Martin & Harbison, 1979*)

Source	Dose (mrem year^{-1})
Natural	
Gamma	50
Carbon 14	1
Radon and decay products	4
Potassium 40	20
Cosmic rays	50
Sub-total	125
Manmade	
Diagnostic radiology	7
Therapeutic radiology	3
Isotopes in medicine	0.2
Radioactive waste	0.2
Fallout	0.7
Occupational exposure	0.5
Other sources	0.4
Sub-total	12.0
Total	137

radioactivity from soil and pottery in thermoluminescence dating (dealt with in the next chapter). With the advent of manned and unmanned space flight, assessment of the background radioactivity in space has also become desirable.

Figure 6.17 illustrates the different sources of radioactivity to which the population of Great Britain is exposed (from Wall *et al.*, 1981). A table of the typical annual gonad doses due to natural and manmade sources is given in table 6.2.

6.5.2 *Materials*

The detector materials used in the environmental dosemeters require a different set of performance characteristics to those used in personal dosimetry. The dose levels to be measured are much lower (requiring high sensitivity) the measurement periods are much longer (from months to years, requiring high thermal and optical stability) and the dosemeter often spends considerable periods exposed in the open (requiring high moisture and chemical resistance). An environmental dosemeter may be defined as one which is placed in the environment for the purpose of assessing the absorbed dose in air due to photons and muons. In general, its performance characteristics will be different to those of personal dosemeters, except perhaps for batch homogeneity, stability and effect of climatic changes.

The major requirements of an environmental dosemeter (high sensitivity and low fading rate) are exhibited by many of the materials listed in table 6.1, although optical effects and long-term instability make the application of some over very long periods rather difficult. CaF_2 and $CaSO_4$:Dy are most often used, but attention must be drawn to the degree of self-irradiation (which must be minimal) and to the energy dependence of the material. (Fortunately with natural radioactivity, the proportion of energy below 100 keV is only a few per cent, but with manmade sources considerable low-energy radiation is present.) Filters are required to remove the energy dependence. Protection from moisture and from sunlight is necessary with some materials. In all, many effects of the undesirable features of the detector materials can be minimized by appropriate dosemeter design.

6.5.3 *Practical application*

The location of the dosemeter in the field is of importance. Placing the dosemeter on dense objects (buildings, trees) or on the ground can cause shielding and directional anomalies which ought to be avoided. Suspending the dosemeter ~ 1 m above the ground is widely practised.

Clearly the encapsulation of the detector(s) in a badge, as in personal dosimetry, is not required in environmental monitoring. Nevertheless, the dosemeter needs to be durable and easily identifiable, and easily detectable in the field. Moisture and light protection, plus the necessary filters, also need to be incorporated. A popular dosemeter arrangement is to seal the detector in a bulb which is filled with inert gas and which includes a small heater element for use during readout. An example of such a system is illustrated in figure 6.18. The outer shield provides the necessary energy compensation for the CaF_2 phosphor which is mounted on the electrical filament. Bulb dosemeters of this sort (which can also be used for personal dosimetry) are manufactured by several commercial concerns, along with suitable TLD reader systems. More details can be found in chapter 9 and in the article by Julius (1981).

Piesch (1981*b*) suggests a calibration and measurement procedure for environmental dose assessment using TLD. His recommendation is outlined in figure 6.19. Here the original batch of dosemeters is given the normal pre-irradiation annealing treatment appropriate to the phosphor being used. The batch is then divided into five groups. The first group (of at least 10) is to be used for calibration. These are stored in a shielded container until they are given their calibration irradiation in the

Figure 6.18. Harshaw bulb dosemeter for personal and environmental dosimetry, to be used with CaF_2:Mn or CaF_2:Dy. The phosphors are mounted on a heater filament and sealed in a glass bulb. The bulb is shown in its position in a TLD reader and in its holder. (Reproduced with permission of Harshaw Chemical Co., Ohio, USA.)

laboratory. The second group (minimum of 10) are to be used to assess the zero dose level. Thus, these are given no irradiation at all, but are stored in the shielded container. The three other groups are to be transported to the field site. Of these, one group is given a test irradiation in the laboratory before being transported to the field where they are sealed in a shielded container. These are used to assess the degree of fading. A second group is also transported to the site and shielded. These will be used to assess the dose accumulated during transit. (This may be significant if the dosemeters are carried by aircraft. Goedicke, Slusallek & Kubelik (1983) use $CaSO_4$:Dy to estimate that doses of several tens of mrad can be accumulated.) The final group is used for the actual environmental dose assessment. After the measurement period, the dosemeters are transported back to the laboratory and are each given the same post-irradiation annealing treatment as is given to those dosemeters used for the calibration and zero dose assessment. Following this treatment, the thermoluminescence from all the samples is read out on the TLD apparatus. This lengthy but careful procedure provides all the data necessary to correct for fading and for any signal which arises from a source other than that which is desired to be measured.

Several interlaboratory test and calibration programmes have been initiated for the purpose of examining the overall reproducibility and accuracy of TLD environmental monitoring. The results of recent intercomparisons are described by Burgkhardt, Piesch & Seguin (1980), Gessel & de Planque (1980) and Piesch (1981b).

Figure 6.19. Calibration measurement procedures for environmental TLD (as suggested by Piesch, 1981b). See text for description.

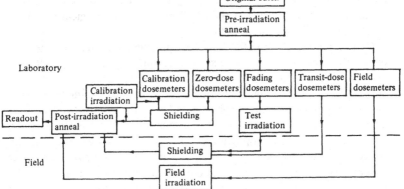

6.6   Medical applications

6.6.1   *Introduction*

The application of TLD in medicine for the purpose of clinical absorbed dose estimation has been carried out successfully for many years since the first proposal made by Daniels in 1954 (see chapter 1). The major usage of TLD in medicine is in the measurement of the dose received by the patient during radiotherapy and diagnostic radiology, and in beam calibration and uniformity checks.

The aim of radiotherapy treatment is to irradiate cancerous tissue with a precise dose of radiation, which has been previously prescribed by a physician, whilst at the same time ensuring that the dose delivered to surrounding, healthy tissue is kept to a minimum. The delivered dose may be calculated by the hospital physicist, or using *transit dosimetry* or more advanced *computerized tomography*, but the advantage of TLD is that the delivered dose can actually be measured either indirectly with a 'phantom' or directly with *in vivo* techniques (e.g., Thomasz *et al.*, 1980; Velkley, Cunningham & Strockbine, 1980). An accuracy in the dose assessment of $\pm 5\%$ is required because errors greater than this can adversely affect the result of the treatment.

The International Atomic Energy Agency (IAEA) in cooperation with the World Health Organization (WHO) currently operate a TLD postal dosimetry service (based on LiF) for radiotherapy practitioners. The work stems from concern at the IAEA about the accuracy of the dosimetry of radiotherapy units in use around the world. Before the service began, a survey showed that approximately one-third of the therapy units available lacked any sort of radiation measurement instrument, or, where one was available, it had not been recently calibrated. The service began in 1966 and its findings are regularly published, the latest data being given by Eisenlohr & Jayaraman (1977). A similar survey organized by the US Centres for Radiological Physics is described by Wochos *et al.* (1982).

Of the genetically significant dose-equivalent delivered to the UK population from manmade sources, as illustrated in figure 6.17, $\sim 90\%$ is from diagnostic radiology and it has become desirable to limit the dose delivered to an individual from diagnostic X-rays to a minimum. Once again computerized tomography provides a complex method of dose assessment, but TLD enables useful comparisons and checks to be made.

In both of the above applications TLD is ideally suited, and in many cases, alternative measurement methods are simply not available. TLD dosemeters are small, nearly tissue-equivalent and sensitive. They can be conveniently attached to, or internally located in, the patient and are easily retrievable.

6.6.2   *Materials*
   An important requirement of a TLD material when used in clinical dose assessment is tissue-equivalence. Thus, the photon energy and LET dependence of the materials need to be considered. The radiation energies used in radiotherapy and radiology can vary from 10 keV X-ray to 60 MeV ^{60}Co photons. Electron energies may be as high as 40 MeV. For measurements *in vivo* it is not feasible to surround the detector with appropriate filters so tissue-equivalent detectors are necessary. BeO is not favoured owing to its potential toxicity and as a result LiF and (especially) $Li_2B_4O_7$ are preferred. The ability to alter the energy response of $Li_2B_4O_7$:Mn by controlling its stoichiometry and dopant concentration (as discussed in section 6.2.2, cf. figure 6.10) is particularly useful in radiology where the X-ray energies used could result in large errors if consideration was not given to the material's energy response. Generally, $Li_2B_4O_7$:Mn powders are preferred in this role because of their cheapness.
   LiF and $Li_2B_4O_7$ are able to cover the rather wide dose range experienced in medical applications (10^{-3} to 10^2 rad in radiology; up to 10^3 rad in radiotherapy), although at the high dose end of the range correction for supralinearity may be necessary. For radiography, $CaSO_4$:Dy may be useful because of its high sensitivity (e.g., Jain *et al.*, 1979). Although energy dependent, its response in the X-ray energy range used for radiology (~ 15–35 keV) is approximately constant, overreading by a factor of ~ 10 (Pradhan, 1981).
   The forms of the dosemeters, especially in radiotherapy, generally tend to be chips or rods, although powders (less popular now) and PTFE-impregnated samples are also used. When selecting the dosemeter form, the precise application needs to be considered (i.e., external use, insertion in catheters, will it be in contact with body fluids?, will it need sterilizing?, etc.). Often the smaller the physical size of the detector the better, commensurate with the requirements for electronic equilibrium. This is particularly so if dose gradients are to be measured and this requires a reasonably sensitive dosemeter. The standard geometries of the solid LiF TLD detectors produced by Harshaw are well suited to these applications (rods: $1 \times 1 \times 6$ mm (16 mg); chips: 3.2 mm $\times$ 3.2 mm $\times$ 0.089 mm (25 mg)).

6.6.3   *Practical application*
   TLD is used in medical dosimetry in two regimes, namely *in vivo* and phantom measurements. The purpose of *in vivo* dosimetry is to verify the dose delivered to a patient during radiotherapy or diagnostic radiology and can be conveniently subclassified into four different kinds of measurement. There are: (i) Entrance dose measurement – used to

assess the source output, the dose distribution across the patient, and the position of any shielding. (ii) Exit dose measurement – to check the dose absorbed by the patient's body. (iii) Intracavitary dose measurement – to measure directly the dose delivered to a particular body cavity. (iv) Spared organ dose measurement – to assess the dose delivered to shielded organs. Entrance and exit doses can be measured by attaching the dosemeters, sealed in opaque polythene sachets, to the incident and reverse surfaces of that part of the patient which is being irradiated. Wax or perspex is used as a build-up material. Intracavitary measurements are made by sealing the dosemeter in a catheter which is then inserted into position within the patient. This kind of measurement is almost exclusive to TLD.

On the basis of the findings of the IAEA–WHO radiotherapy

Figure 6.20. (*a*) Anthropomorphic 'Rando'® phantom developed by Alderson Research Laboratories. The phantom is built in sections, each one of which contains numerous locations, in a grid pattern, for TLD phosphors. Reproduced with permission of Radiation Components Ltd, Stamford, Conn., USA. (*b*) Plastic resin phantom used by McKlveen (1980*a*,*b*) to determine X-ray exposures to patients during dental radiography. Each section of the phantom contains locations for TLD phosphors. Reproduced with permission of J.W. McKlveen, Arizona State University, USA.

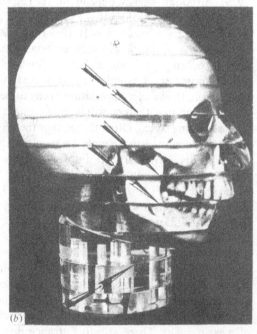

(*a*)   (*b*)

dosimetry service, Eisenlohr & Jayaraman (1977) conclude that the participants who use 'in phantom' measurements for radiotherapy beam dosimetry have a better control over the dose administration than those who measure doses in air. Measurement of the dose distribution in a phantom is a necessary condition in order to assess correctly the dose delivered to a patient. As recently pointed out by Johns (1981), there is considerable backscattered radiation incident on a dosemeter when used *in vivo* and both the energy and direction of this radiation differ from that of the incident radiation. As a result, it is necessary to simulate these effects by using the dosemeter in a phantom. Examples of materials used in simple-geometry phantoms are water or polystyrene. Illustrations of the use of such phantoms to calculate depth–dose profiles for radiotherapy units are given by McKinlay (1981).

The greatest benefit from using phantoms can be achieved, not with simple, homogeneous water or water-equivalent material, but by constructing an anthropomorphic phantom designed to mimic the human body in its reaction to radiation (e.g., Shrimpton, Wall & Fisher, 1981). Modern anthropomorphic phantoms contain human skeletons and are built up with organ and muscle tissue-equivalent materials (resin, synthetic rubber, liquids, gels). Also included are channels simulating airways, sinuses, etc. Examples of anthropomorphic phantoms for various applications are shown in figure 6.20.

Anthropomorphic phantoms can be used in both radiotherapy and in diagnostic radiology to assess realistically the dose delivered to the various parts of the body by positioning them in the radiation beam in the same way as a patient would be positioned. The dose to specific organs can then be determined from TLD detectors located within.

6.6.4 *Specific examples*

It is instructive to illustrate the principles of *in vivo* and phantom TLD measurements in both radiotherapy and in diagnostic radiology by discussing some specific examples of cases where TLD has proven useful, if not invaluable. Recent uses of $Li_2B_4O_7$:Mn in either powder form (McKinlay, 1981) or disc form (Marinello, Barret & Le Bourgeois, 1980) have been described for the treatment of the rare skin disease mycosis fungoides (cancer of the surface lymph cells). Treatment of this condition requires partial or whole body radiation, normally with electrons, in order to irradiate the outer few millimeters of the skin with the minimum of penetration to inner organs. It is difficult in irradiations of this nature to calculate the dose delivered to the patient, and TLD provides a suitable method to assess this and its distribution over the body.

The dosemeters are wrapped in black polythene sachets and fixed to the skin of the patient at the required locations. Marinello *et al.* (1980) describe a pulsed electron irradiation treatment for this disease. The use of TLD is able to account for the non-uniformity of the delivered dose caused by variations in shape, orientation and size of the body. As pointed out by Marinello *et al.*, measurements of this nature could hardly be carried out using alternative dosimetry methods.

A second medical application in which TLD has proved to be of high value is in treatment of carcinoma cervix. The radical treatment of this disease requires a high dose and therefore a high morbidity in adjacent organs is inevitable. Thus, there is an important need to restrict unwanted reactions, whilst at the same time controlling the disease. There have been several publications in which the dose delivered to the patient is assessed using intracavitary *in vivo* TLD techniques (e.g., Joelsson *et al.*, 1972; Heywood & Clarke, 1980*b*; Björnsson & Sorbe, 1982). The usual treatment of this disease involves the insertion of a radioactive source (e.g., radium 226 or radon 222) into the uterus and vagina (e.g., Brand & Kerr, 1982). The dose delivered to the surrounding tissue can be calculated by taking a radiograph of the area (with the source in position) and noting the various distances of particular tissues from the source. However, by using thermoluminescence dosemeters, the delivered doses can actually be measured by sealing the dosemeters in appropriate catheters and inserting them into the desired locations (femoral arteries, rectum, bladder, etc.). The position of the catheters can be noted by the use of lead markers which show up in the radiograph.

A problem often encountered in this technique is that the dosemeters have to be inserted in the operating theatre along with the radioactive source. This not only presents a radiation hazard to the hospital staff, but also means that when the patient is returned to the ward for the irradiation period (24–36 hours) the subsequent relaxation and change in position of the patient means that the dosemeter-to-source distances will have altered since the time the radiograph was taken. To counter this, an after-loading system was developed in 1972 whereby a uterine/vaginal applicator is inserted whilst the patient is in the theatre, but the source is not inserted into the applicator until the patient has settled down in the ward. The source (^{137}Cs) can be removed at any time to allow staff and visitors to attend the patient. After a prescribed period, the dosemeter catheters are removed and the actual dose delivered to the various organs can be determined. This system has resulted in a significant decrease in the incidence of morbidity. Figure 6.21 shows a radiograph of the dosemeter catheters and uterine/vaginal applicator in position. The differences between the measured and calculated doses found by Heywood & Clarke (1980*b*) using this system demonstrate the advantages to be gained from using TLD.

In diagnostic radiology, phantoms are often used to assess the dose delivered to a patient with a view to minimizing the dose to individuals by this treatment. Wall *et al.* (1981) discuss how the use of diagnostic X-rays has increased by several per cent over the last few years. Several examples of the use of TLD materials in phantoms can be found in which dose assessments have been carried out. For example, Wall, Green & Veerappan (1979*a*) used a Rando phantom (see Shrimpton *et al.*, 1981) and $Li_2B_4O_7$ dosemeters to estimate that the maximum dose delivered to a patient's skin during brain and body scanner usage was between 3 and 5.6 rad. A survey by Fitzgerald *et al.* (1981) of 61 National Health Service centres in Britain to assess techniques and exposures from mammographic radiology units utilized TLD dosemeters and a breast phantom representing 50% fat and 50% water, homogeneously mixed. The survey was carried out to provide data for consideration of the balance between benefit and risk during screening tests for breast cancer using mammography (Mooney, 1981). The measurements indicated that the dose to the skin surface per centre per patient varied between 0.54 rad

Figure 6.21. Lateral radiograph illustrating the uterine/vaginal applicator and the lead markers of the rectal and bladder catheters. (After Heywood & Clarke, 1980*b*.)

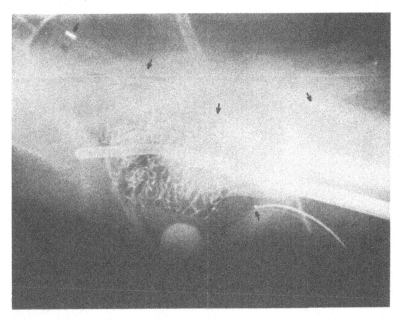

and 3.26 rad. Watson (1977) measured skin doses up to 10.7 rad using TLD in a similar survey of New Zealand mammography units.

McKlveen (1980*a, b*) utilized both *in vivo* and phantom TLD measurements to determine the X-ray dose given to dental patients during diagnostic and preventative dentistry. An elastic mask to be worn by the patients and the phantom (shown in figure 6.20) were instrumented with TLD dosemeters and both patient and phantom were subjected to full-face diagnostic dental X-rays, using a variety of X-ray units. Locations of interest included the skin surface, eyes, upper and lower teeth, and the thyroid gland. In general, doses of 100 mR were common, and with full-face irradiation from a periapical unit, doses of up to 6 R were sometimes measured. Similar findings were reported by Wall *et al.* (1979*b*) who conclude that if the incidence of diagnostic radiology is kept to less than one per year for adolescents, and less than one in every three to five years for adults, there will be minimal risk of somatic injury.

7 Thermoluminescence dating

7.1 General

In its simplest form, the equation for calculating the age of a specimen using thermoluminescence is as given in equation (1.8), rewritten here in a slightly modified fashion thus:

$$\text{Age} = \frac{\text{natural thermoluminescence}}{\text{(thermoluminescence per unit dose)} \times \text{(natural dose rate)}}. \quad (7.1)$$

Here the natural thermoluminescence is the thermoluminescence accumulated by the specimen during its lifetime and the thermoluminescence per unit dose is the sensitivity of that specimen for exhibiting thermoluminescence following a given absorbed dose of radiation. Strictly, the 'age' being determined is the length of time that the sample has been subjected to irradiation following its last heating. The radiation originates from the local surroundings and from the specimen itself, and the 'last heating' may correspond to the firing of pottery (e.g., for archaeological pot sherds) or the geological formation of the material (e.g., lava). Alternatively, the age may represent the time since formation of a sedimentary deposit (e.g., ocean sediments, stalagmites) where the initial formation did not involve high temperatures.

The application of thermoluminescence to the dating of archaeological and geological specimens has experienced a rapid expansion since the first suggestion by Daniels and colleagues in 1953. This has been particularly true since the late 1960s, since which time techniques to arrive at absolute dates for pottery from archaeological sites have been developed to a high degree of sophistication, utilizing the so-called *fine-grain* and *quartz-inclusion* methods. The introduction of the *pre-dose* and *phototransfer* techniques served to enhance the reputation of thermoluminescence as a powerful means of archaeological age determination.

The above techniques have their main applications in the dating of pottery, but thermoluminescence has also been developed to date flints, rocks, lavas and several other non-pottery materials with an acceptable degree of reliability. More recently, the advent of sediment dating using thermoluminescence has opened up a wide range of potential applications.

A development which stems directly from the application of thermoluminescence in age determination has been the use of electron spin resonance, proposed and developed for this purpose primarily by Ikeya and colleagues (e.g., Ikeya, 1978*a,b*; Miki & Ikeya, 1978) and now finding increased popularity. Here, the trapped electron population is assessed via the electron spin resonance signal strength, rather than the glow peak intensity. All other details remain essentially the same.

In this chapter, techniques for both pottery and non-pottery dating will be discussed, along with the principal problems which are encountered. The use of thermoluminescence in checking the authenticity of antique ceramics will also be described.

7.2   Techniques in pottery dating

7.2.1   *Introduction*

The discovery by Grögler *et al.* (1960) and by Kennedy & Knopff (1960) of thermoluminescence from ancient pottery made way for the possibility of calculating the age of the pottery (since firing) using relationship (7.1). Pottery consists of a baked clay matrix containing small inclusions of up to a millimetre in diameter. The inclusions (mainly quartz and feldspar) are found to be far more sensitive (as regards the

Figure 7.1. Thermoluminescence (bright regions) from coarse-grained Roman pottery from Alcester, England. The slices have been irradiated with 9.5 MeV protons to a dose of $\sim 10^6$ rad. The light emission is photographed using a polaroid camera (focal length 1.9). Reproduced with permission of J.H. Fremlin, University of Birmingham, UK.

production of thermoluminescence) than the matrix material. This was unequivocally demonstrated by Fremlin & Srirath (1964) who photographed the light emission from a sample of pottery during thermoluminescence. An example is shown in figure 7.1 in which it can be seen that the inclusions are glowing more brightly than the clay matrix. Separation of these grains from the matrix results in a higher light yield per unit weight of material, but estimating the radiation dose absorbed by the inclusions is not straightforward.

The clay matrix typically contains 2–3 ppm U, 10–12 ppm Th and 1–2 % K. Quartz, on the other hand, (the most common inclusion material) contains only ~ 1 ppb of these materials (e.g., Fleming, 1966; Fleischer, Price & Walker, 1975). The total natural dosage recieved by the pottery is a combination of α-, β- and γ-radiation (plus a small amount from cosmic rays) and can be estimated using the dose rate data of Bell (1979), given here in table 7.1. The calculations indicate that the α activity contributes $\sim 40\%$ to the total delivered dose (taking account of the lower efficiency of αs in inducing thermoluminescence compared with γs or βs – Aitken & Fleming, 1972). Since most of this dose is delivered from outside the quartz grains, account must be taken of the attenuation of the α dose with depth. Bell (1980) has recently revised earlier calculations by Howarth (1965) on the α dose attenuation in quartz grains and concludes that, for grain diameters of 1–8μm, the average attenuation is $\sim 5\%$, resulting in only a 2% reduction in the total dose delivered. In contrast, grains $> 100\,\mu$m will receive very little α dose in their inner volume. Beta-particles penetrate $\sim 500\,\mu$m whilst a γ photon has a range

Table 7.1. *Dose rate (mrad/year) for 1 ppm by weight of parent. (From Bell, 1979.)*

	α	β	γ
Thorium series			
No thoron loss	73.8	2.86	5.14
100% thoron loss	30.9	1.03	2.08
Uranium series			
No radon loss	278.3	14.62	11.48
100% radon loss	126.3	6.09	0.56
Natural potassium			
1% K_2O	—	68.93	20.69
Natural rubidium	—	4.64	—

of $\sim 30\,\text{cm}$. Realization of this has resulted in the development of two different methods of dating pottery, namely, the fine-grain and the quartz-inclusion methods.

7.2.2   Fine-grain dating

The fine-grain dating technique was described by Zimmerman (1967, 1971c). Approximately 1 mg of $1-8\,\mu\text{m}$ grains are deposited on to aluminium discs by crushing the pottery sample and allowing the grains to settle in acetone. The grains themselves are not necessarily just quartz, and a study by Singhvi & Zimmerman (1978) using cathodoluminescence indicated that many fine-grain discs are dominated by feldspars (K-feldspar or plagioclase). Smaller amounts of apatite and zircon are also present. Several discs are prepared together and are used to measure the natural thermoluminescence, the thermoluminescence sensitivity per rad for both α and β radiation and growth curves to check and correct for supralinearity (see section 7.3.3). The age is then determined using equation (7.1). It is convenient to define an 'equivalent dose' (ED) as

$$\text{ED}_{\text{FG}} = \frac{\text{natural thermoluminescence}}{\text{thermoluminescence per unit dose}}, \qquad (7.2)$$

where the subscript FG refers to the fine-grain technique. It is assumed that the equivalent dose is equal to the archaeological dose N and so equation (7.1) becomes:

$$\text{Age} = \text{ED}_{\text{FG}}/\text{dose rate}. \qquad (7.3)$$

To evaluate the dose rate, the α-, β-, γ- and cosmic-ray contributions need to be included. Zimmerman (1972) discusses the relative efficiencies of high linear energy transfer (LET, see previous chapter) α-particles at producing thermoluminescence in quartz compared with β-particles. If the relative efficiency is described by a 'k-factor', defined, thus:

$$k = \frac{\text{thermoluminescence per unit dose of } \alpha \text{ radiation}}{\text{thermoluminescence per unit dose of } \beta \text{ radiation}}, \qquad (7.4)$$

then it can be said that α-particles have a typical k value of ~ 0.1. Thus, by taking account of this, equation (7.3) can be rewritten

$$\text{Age} = \text{ED}_{\text{FG}}/(kR_{\alpha} + R_{\beta} + R_{\gamma} + R_{\text{c}}) \qquad (7.5)$$

where R_{α}, R_{β}, R_{γ} and R_{c} are the dose rates from α-, β, γ- and cosmic-ray irradiations respectively (discussed further in section 7.4).

Huxtable (1978) gives a useful description for the steps which are necessary in order to arrive at fine-grain dates for pottery samples. She includes techniques which are required to overcome spurious signals and unwanted fading effects (discussed further in section 7.3). Additional detail is given by Fleming (1979*a,b*).

7.2.3 Inclusion dating

As discussed in the above sections, the small grains (1–8 μm) which are used in the fine-grain technique receive the full α dose (except for $\sim 2\%$). In contrast, the inner portions of the larger grains receive no α contribution at all. Thus, if the outer part of these grains (which have received the full α dose) are removed, there remains a collection of grains for which the contribution from α irradiation can be entirely ignored. This is the principle of the quartz-inclusion technique introduced by Fleming (1966, 1970) and further described by Fleming (1978, 1979*a*). The extracted grains are washed in 1N HF acid for several minutes in order to etch away the α-irradiated regions. This removal of the outer portion also helps to increase the grain transparency by elimination of the more opaque, contaminated surface regions. Furthermore, the glow-curve from etched quartz is seen to be rather different from that of untreated material because the surface contamination provides traps and luminescence centres not found in virgin quartz. When properly carried out, the quartz-inclusion method can reduce the α dose to $\sim 1/100$ that of the combined β and γ doses. Thus, the α dose can be ignored and the age equation becomes:

$$\text{Age} = \text{ED}_{\text{INCL}}/(R_\beta + R_\gamma + R_c). \tag{7.6}$$

An examination of equations (7.5) and (7.6) will reveal that by combining the fine-grain and quartz-inclusion techniques, the environmental contribution from the γ-, β- and cosmic rays may be eliminated, thus:

$$\text{Age} = (\text{ED}_{\text{FG}} - \text{ED}_{\text{INCL}})/kR_\alpha. \tag{7.7}$$

This development is often called *subtraction dating* and it requires a high level of accuracy in the measurement of the equivalent doses.

A similar technique is described by Mejdahl (1983; Mejdahl & Winther-Nielsen, 1982) for feldspar inclusions, whilst Zimmerman (1971*d*, 1979*b*) described a rather different inclusion method which relies upon the extraction of those grains which display a high level of internal radioactivity. For example, pottery zircon has a very high (50–3000 ppm) uranium content compared with the clay matrix. As a result, the rate of external irradiation of these grains is very small compared with the high internal dose rate (e.g., ~ 50 rad year^{-1}). The zircon grains may be separated by a flotation technique and each grain is individually assessed for its uranium and thorium content (by *fission-track* mapping –

see Fleischer *et al.*, 1975) and for its α sensitivity. (This can create problems in that each grain has to be handled separately.) The natural α dose is evaluated from a measurement of the natural thermoluminescence and the age of the specimen is calculated from:

$$\text{Age} = \text{natural } \alpha \text{ dose}/\text{natural } \alpha \text{ dose rate.} \tag{7.8}$$

Zimmerman (1979*b*) describes the detailed experimental procedures required for this method and discusses some of its major problems, including that associated with 'zoning'. In some grains the uranium and the thermoluminescence sensitivity are inhomogeneously distributed (zoned) and show an anti-correlation. This may result from long-term irradiation damage and means that the regions with the more intense thermoluminescence have received less dose than the grains as a whole, and this in turn results in an underestimate of the age. Simple corrections for this do not appear feasible (Sutton & Zimmerman, 1976; Zimmerman, 1979*b*) and examination of the thermoluminescence zone by zone appears necessary (Templar & Walton, 1983).

7.2.4   Pre-dose dating

The pre-dose technique gets its name from the fact that the archaeological age of the specimen is related, not to the natural thermoluminescence intensity, but to the *sensitivity* of a particular glow peak found in all pottery quartz. The sensitivity of the peak is found to be dependent upon the amount of radiation previously received (i.e., the 'pre-dose') which, in the case of an archaeological specimen, is the archaeological dose.

As with the quartz-inclusion method, the pre-dose technique is concerned with those large quartz grains (90–150 μm) which do not possess an α dose contribution (although acid etching is not required because of the observation that the efficiency of α radiation in inducing the pre-dose effect is negligibly small (Aitken & Fleming, 1972). Martini, Spinolo & Vedda (1983*a*) examine this more closely to show that the strength of the pre-dose effect decreases markedly with increasing linear energy transfer and α-particles are shown to be only 1/100 as efficient as β-particles.) Unlike the quartz-inclusion (and the fine-grain) techniques, however, pre-dose dating does not concern itself with the measurements of the time-dependent accumulation of thermoluminescence, but rather with the time-dependent increase in the *efficiency* of a particular trapping site at producing thermoluminescence. A prerequisite of normal thermoluminescence dosimetry and dating is that there is no significant loss (fading) of trapped charge occurring during the irradiation period. In dating applications, the irradiation periods are extremely long (many thousands of years) so that even small fading rates cannot be tolerated. For this reason conventional dating procedures deal only with the high-temperature portion of the glow-curve. (This topic is dealt with more thoroughly in section 7.3.1.) Thus, a quartz specimen which has been slowly irradiated at ambient temperatures during antiquity displays no low-temperature thermoluminescence. On the other hand, if that specimen were to be irradiated within a very short period in the laboratory to the same level of dose, the glow-curve would display a peak at $\sim 110\,°C$ (at a heating rate of $\sim 20\,°C\,\mathrm{min}^{-1}$) which is not observed in the natural glow-curve. It appears that the natural fading rate from the traps in question is such that there is no charge left to contribute to thermoluminescence (at $110\,°C$) during readout in the laboratory. The traps are said to be unstable, with a half-life of a few hours at ambient temperatures.

At first sight, it would not be expected that the $110\,°C$ peak could be of any benefit to the thermoluminescence dating, yet it is exactly this peak which is exploited in the pre-dose method. The pre-dose effect, as described by Fleming (1973), can be best explained by reference to figure 7.2. A fresh specimen of pottery quartz is given a small test dose of 1 rad and the thermoluminescence monitored in the usual fashion. This produces a small $110\,°C$ peak of size S_0, representative of the initial sensitivity of the sample. The specimen in then heated to $500\,°C$ before being cooled to ambient temperature and re-irradiated, again with 1 rad. This gives an enhanced signal of S_N. A second heating to $500\,°C$ and 1 rad irradiation is also seen to give the same sensitivity, denoted here as S_N'. If the specimen is now given a larger dose of β irradiation (β) and is again heated to $500\,°C$, then, on subsequent re-irradiation with the 1 rad test

Figure 7.2. Procedure for the determination of the thermluminescence sensitivity of the $110\,°C$ peak in quartz after various pre-treatments. (From Fleming, 1979b.)

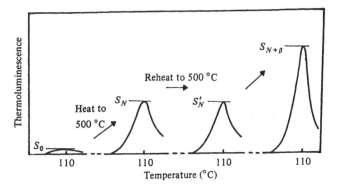

dose, a sensitivity value of $S_{N+\beta}$ is obtained. The important observations here are that: (i) $S_N = S_N'$; this implies that heating to 500 °C *alone* does not cause any increase in sensitivity. (ii) The growth S_N' to $S_{N+\beta}$ must be related to the 'pre-dose', β. (iii) The growth S_0 to S_N must relate to the pre-dose delivered before the specimen was received by the laboratory – i.e., the natural dose, N. Thus, if the increase from S_N' to $S_{N+\beta}$ due to β is directly proportional to the increase from S_0 to S_N due to N, then we may write:

$$N = \beta(S_N - S_0)/(S_{N+\beta} - S_N'). \tag{7.9}$$

If equation (7.9) is to be used to date the pottery specimen it must be assumed that the sensitivity S_0 is the original sensitivity of the pot when it was removed from the kiln. This in turn assumes that the kiln firing removed the pre-existing 'memory' of the 110 °C peak and re-set the pre-dose sensitization to zero. This point will be raised again after a discussion of the mechanism by which the pre-dose effect occurs.

The currently accepted model for the pre-dose effect was established by Zimmerman (1971a). She noted that not only did the thermoluminescence sensitivity increase after pre-dose activation, but that the radioluminescence yield was also enhanced. Furthermore, the thermally stimulated exoemission sensitivity was decreased. An additional observation was that irradiation with ultra violet light of wavelength 230–250 nm could remove the sensitivity enhancement, but further heating and irradiation treatments would restore it. Thus, sensitization of the 110 °C peak can be achieved only by *both* heat treatment (to ~ 500 °C) *and* irradiation; the sensitivity increase is proportional to the dose received; the effect can be reversed by ultra violet irradiation; and it is the luminescence centres, not the traps, that are being affected.

On the basis of these observations and conclusions, Zimmerman

Figure 7.3. Zimmerman's (1971a) model for the pre-dose effect in quartz. For explanation see text.

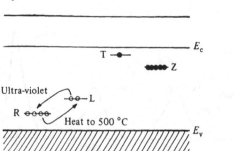

suggested a mechanism for the sensitization based on the transfer of charge from one recombination site to another. The main elements of Zimmerman's model are displayed in figure 7.3. These consist of two electron traps, one of which (T) corresponds to the 110 °C peak, while the second (Z) is included for charge neutrality. Also, there are two hole traps which act as recombination centres. In particular, holes trapped at L act as luminescent sites for the 110 °C emission. Holes trapped at R function as non-luminescent sites and are termed reservoir centres. During natural irradiation in antiquity, electrons populate Z, but T is too shallow to allow for appreciable long-term charge accumulation, as previously discussed. It is also assumed that the probability of hole trapping at R is greater than that at L because of either a larger capture cross-section, or a larger concentration of available traps, or both. The hole population in R is thus proportional to the natural dose N.

Irradiation in the laboratory with a small test dose serves to put a few electrons in T, and holes in L. Heating to over 110 °C releases the electrons, enabling them to recombine at L which in turn gives a thermoluminescence peak of height S_0. Increasing the temperature beyond 500 °C transfers holes from the reservoir into L. This transfer probably takes place via the valence band, although this is not explicitly shown in figure 7.3. The electrons in the deep centre Z remain unaffected. Thus, when the specimen is cooled to room temperature, a significant concentration of holes now exists in L such that, after a subsequent test irradiation, there is a much larger number of holes in L with which the electrons in T can recombine. (Note that after 1 rad, the number of electrons in T is very much less than the number of holes in L.) When the specimen is now heated, a larger 110 °C peak, of height S_N is obtained. Thus, the pre-dose (in this case the natural dose) populates the R centres, whilst thermal activation (to ~ 500 °C) transfers holes from R to L. The number transferred depends upon the number already in R, i.e., the size of the pre-dose, and thus the sensitization observed (i.e., $S_N - S_0$) is also dependent on the pre-dose. The model also explains the enhancement of radioluminescence, but not of thermally stimulated exoemission. Irradiation with ultra-violet light is assumed to reverse the effect by simply transferring holes from L back to R. (Not all of the sensitivity increase can be removed, indicating a residual hole concentration in luminescent sites (Bailiff, 1979).)

It is straightforward to see that the assumption made earlier concerning the interpretation of S_0 can easily be satisfied with this model. It is only required that the kiln firing temperatures are enough to empty all traps so that the pre-dose effect starts afresh when the pottery is made. The model proposed appears to be the only one so far suggested that is able to explain all of the experimental observations, although direct confirmation is lacking. Some much-needed support has recently emerged from combined thermally stimulated conductivity and thermoluminescence measurements by Böhm *et al.* (1980). Here, the changes in the relative positions and sizes of the thermally stimulated conductivity and thermoluminescence peaks with thermal treatments are exactly those predicted using Zimmerman's model. Some minor modifications have been suggested by Chen (1979) and although the essential elements of the model remain unchanged, the modifications are worthy of a little more discussion. Chen points out that the requirement that the concentration of electrons in T is less than that of holes in L, conflicts, at first sight, with the observation that the sensitivity is dependent upon the number of trapped holes. This is because one would normally expect the thermoluminescence intensity to be proportional to the smaller of the two concentrations (i.e., trapped electrons and trapped holes). However, if one includes an extra competing centre, this conflict is removed. The extra centre may be an electron trap or a non-luminescent recombination centre (e.g., R), but in either case it competes with the holes in L for the electrons released from T. In this case, the thermoluminescence intensity will depend both on the number of trapped electrons and the number of trapped holes, as required.

Further minor modifications are suggested by David (1981) who deals with the effects on the pre-dose properties of heating beyond the α–β and tridymite phase changes (573 °C and 870 °C respectively). These modifications do not alter the important features of the model.

Since the transfer of holes from R to L is stimulated by heat, it might be expected that the rate of transfer is thermally activated according to equation (1.1). Such behaviour is indeed observed and is most clearly described by a so-called *thermal activation curve*. An example of this curve for a quartz specimen is shown in figure 7.4. An Arrhenius plot for these data indicated that R is distributed in energy between 1.34 and 1.55 eV above the valence band. If the activating temperature is increased beyond 500 °C, a decrease in the sensitivity enhancement is often observed ('thermal deactivation'). An example was given by Aitken (1979) in which deactivation was observed for a sample given a β dose on top of its natural dose, but not for an as-received specimen. The cause of this remains unclear. It is to be noted that the activation curve may vary from sample to sample and needs to be determined for each specimen individually. For some samples ambient activation may have occurred in antiquity.

Aside from the general problems (discussed later) which are met with in all thermoluminescence dating methods, there are additional difficulties to confront when using the pre-dose technique. For example, a worrying observation recently reported concerns a small increase in the sensitivity of the 110 °C peak following heating (above 400 °C) alone (David *et al.*, 1977b; McKeever *et al.*, 1983b). The cause of this is unknown but appears to be related to an increase in the luminescence probability.

Additionally, it is often found that the sensitization does not exhibit a linear increase with pre-dose. In particular, sublinearity (i.e., a less-than-linear growth) is sometimes observed, for which there may be several causes. Aitken & Murray (1976) note that the value S'_N is often less than S_N. Termed 'radiation quenching', this is caused by a significant neutralization of the holes in L either by electrons excited during irradiation or by electrons released from T during previous readout. (For this latter case, the hole concentration in L would have to be only a little larger than the electron concentration in T.) This would give rise to a non-linear relationship between dose and sensitization and Murray & Aitken suggest that correction can be made by adding the value $S_N - S'_N$ to $S_{N+\beta}$. However, this procedure is often insufficient to remove the sublinearity completely. This prompted Chen (1979) to examine in detail the effects of the saturation of both L and R centres. Chen uses a simplified growth expression, similar to equation (3.50) or (3.53), to describe the variation of sensitivity with dose, namely:

$$S = S_{\infty}(1 - \exp[-N_R/B]), \qquad (7.10)$$

Figure 7.4. Thermal activation curve of the 110 °C peak in quartz for a pre-dose of β rad (after Fleming, 1979a). The curve is obtained by heating to a given temperature after the pre-dose and then measuring the sensitivity of the 110 °C peak. A fresh sample is used for each datum point.

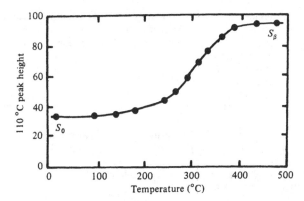

where S is the sensitivity, S_∞ is the maximum attainable sensitivity at very high dose, N_R is the number of holes in the reservoir and B is a constant. When the correct value for S_∞ is chosen, equation (7.10) will give a straight line on a plot of $\ln(S_\infty - S)$ versus dose D, assuming $N_R \propto D$. From extrapolation, the value of D at $S = S_0$ (i.e., the natural dose N) can be obtained. On the basis of this, Chen develops a procedure for calculating N no matter if it is the L or R centre, or both, which is saturating, using Fleming's so-called 'p- and q-tests'. Fleming (1973) suggests that the sensitivity enhancement rate per krad may be expressed in either of two ways. One is to take two portions of the same sample and measure the sensitivity S_N of one portion after activation to 500°C. The other portion is given a dose β and then activated to give $S_{N+\beta}$. The enhancement rate is then given by

$$p = (S_{N+\beta} - S_N)1000/\beta S_0 \tag{7.11}$$

where S_0 is the original sensitivity of the sample.

Alternatively, the *same* sample is given a dose β after its first activation.

Table 7.2. *Pre-dose characteristics as measured for various cases.* (*After Chen*, 1979.)

	Linear (R)[a]	Exponential (R)	Saturated (R)
Linear (L)[b]	Straight line[†]; correct N irrespective of β (even if different βs are used in one series) $p = q$	Straight line[†]; correct result only if $\beta = N$. $q > p$	Straight line[†] giving $N = \beta$ irrespective of real N. $q > p = 0$
Exponential (L)	Exponential growth curve[†]; correct N irrespective of β (even if different βs are used in one series). $p = q$	Exponential[†]; correct result only for $N = \beta$. $q > p$	Exponential[†]; curve yields $N = \beta$ irrespective of real N. $q > p = 0$
Saturated (L)	No result $p = q = 0$	No result $p = q = 0$	No result $p = q = 0$

[†]Growth curve of the 110°C peak sensitivity versus pre-dose.
[a]R = reservoir centre; [b]L = luminescence centre.

One then measures $\bar{S}_{N+\beta} - S_N$ and the enhancement rate is given by

$$q = (\bar{S}_{N+\beta} - S_N)1000/\beta S_0 \qquad (7.12)$$

Ideally, $q = p$ for linearity.

Chen uses these expressions to establish tests for non-linear behaviour of the L and/or R centres and describes a method by which the natural dose can be calculated. The results are summarized in table 7.2.

Bailiff (1983) introduces an additional complication by discussing the possibility of sensitization of the R centres following thermal activation of the natural dose, leading to an underestimate of the natural dose. He suggests using multiple activation and additive dose procedures to test for this.

The complete procedure for arriving at a reliable date using the pre-dose technique is quite complex if it is to deal with all of the above factors. Nevertheless the basic routine is described by Fleming (1979a) and by Aitken (1979). The latter author describes the following steps in the production of a pre-dose date:

(a)   Heat to 150 °C to remove any light-transferred thermoluminescence.

(b)   Give test dose[†] and measure S_0, heating to 500 °C.

(c)   Give test dose and measure S_N, heating to 500 °C.

(d)   Give dose, β, of several hundred rad (i.e., the total 'pre-dose' now equals the natural dose plus β).

(e)   Heat to 150 °C and check that the thermoluminescence at 110 °C has been removed.

(f)   Give test dose and measure S'_N, heating to 500 °C.

(g)   Give test dose and measure $S_{N+\beta}$, heating to 500 °C.

(h)   Give dose β again.

(i)   Heat to 150 °C.

(j)   Give test dose and measure $S'_{N+\beta}$, heating to 500 °C.

(k)   Give test dose and measure $S_{N+2\beta}$, heating to 500 °C.

Further irradiations may be made to complete a full growth curve. Measurement of the S'_N, $S'_{N+\beta}$, etc. values enables a correction for radiation quenching to be made, if necessary. Once the linear region of the growth curve has been identified (Fleming's p- and q-tests may be used) the natural dose may be calculated from equation (7.9). If saturation effects cannot be avoided, Chen's more lengthy graphical procedure may be employed.

---

† The test dose may be very much smaller than the pre-dose. Usually a test dose of 1 rad is used. If not, correction must be made for the holes that have been put into the reservoir centre.

When correctly used, the pre-dose method can calculate ages as young as 100 years, owing to the high efficiency of the sensitization mechanism (reacting noticeably to a few rad). This makes it an attractive technique in authenticity testing (Fleming, 1975a) and its use in fallout dosimetry has recently been suggested (Haskell, Wrenn & Sutton, 1980). The maximum age calculable appears to be a few thousand years.

7.2.5   *Phototransfer dating*

The use of phototransfer in thermoluminescence dating deserves some attention. Among the difficulties met with when applying the fine-grain and quartz-inclusion methods are spurious signals, anomalous fading, supralinearity and sensitization, thermal quenching and low sensitivity. The fine-grain technique is particularly susceptible to anomalous fading because of the high feldspar content of many fine-grain discs. Zircon too is plagued by this unwanted property, presenting difficulties in zircon inclusion dating. The glow from the high-temperature region of the quartz glow-curve suffers from thermal quenching and all of the specimens are affected by the problems associated with subtracting black-body background radiation from the high-temperature glow-curve. Many of these difficulties could be circumvented if it was possible to transfer the charge from the deeper lying traps to the more shallow ones, and this is exactly what is done in phototransfer dating.

Phototransfer has already been discussed in depth in section 4.3.2 and its potential usefulness in thermoluminescence dating was investigated thoroughly by Bailiff *et al.* (1977a), Bailiff (1976) and Bowman (1979). The principle is that trapped charge is transferred from the deep, donor traps to more shallow, acceptor levels by ultra-violet radiation. Quartz, fluorite, fluorapatite and zircon were investigated by Bailiff *et al.* (1977a) and the transfer characteristics, in particular the optimum wavelength, were determined. The transfer characteristics of a specimen of archaeological quartz, along with a typical quartz glow-curve showing the donor (325 and 375 °C) and acceptor (110 °C) peaks, are displayed in figure 7.5. Bowman (1979) found that the best results were generally achieved using 260 nm illumination, but in a couple of cases, 320 nm had to be used. This may possibly relate to the fact that the lower wavelength light must also be transferring holes from L to R centres, as in the pre-dose effect. This complication is not discussed in the literature and an energy level model incorporating both the phototransfer and pre-dose effects is not yet available.

The maximum transfer efficiency attainable (obtained using 260 nm) is no more than 5%, but because the thermoluminescence sensitivity of the acceptor level (i.e., the 110 °C peak) is far greater than that of the donor levels, a large 'transferred' signal is obtained.

In phototransfer dating, the signal in the 110 °C peak, for a given irradiation time, is a direct measure of the fraction of charge transferred from the deeper, donor levels. This in turn is proportional to the charge initially trapped in the donor levels, which is related to the archaeological dose. Thus, if a calibration dose is given to a previously drained sample, the natural dose can be estimated. The procedure generally adopted to obtain the archaeological dose from quartz is outlined thus:

(a) Heat to 210 °C for 1 or 2 min.
(b) Apply a test dose of 1 rad at room temperature
(c) Heat to 210 °C and measure size of 110 °C peak (X_0).

Figure 7.5. (*a*) The glow-curve from natural quartz (after 900 rad β) showing the donor and acceptor peaks used in phototransfer. (*b*) Transfer characteristic for archaeological quartz. (After Bowman 1979.)

(*a*)

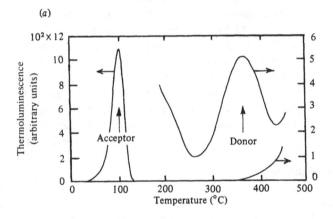

(*b*)

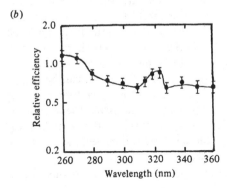

(d)   Illuminate with monochromatic ultra-violet light for 1 min at room temperature.

(e)   Heat to 210 °C and measure 110 °C peak (X_N).

(f)   Drain the donor levels (or nearly so) by prolonged ultra-violet irradiation, or heating to 500 °C (Bowman, 1979).

(g)   Repeat (b).

(h)   Repeat (c) and measure $\bar{X}_0$.

(i)   Apply a calibration dose β at room temperature.

(j)   Repeat (d).

(k)   Repeat (e) and measure X_β.

The archaeological dose N is then calculated from

$$N = \beta X_N \bar{X}_0 / X_\beta X_0. \tag{7.13}$$

Ideally, $X_0 = \bar{X}_0$ but occasionally this is found not to be true, possibly because of partial activation of the pre-dose effect during heating to 210 °C. The purpose of the warming to 210 °C is to remove the 110 °C and 210 °C peaks. This heat treatment will not affect the donor levels but removes interference from the overlapping 210 °C peak. Furthermore, if $\beta \simeq N$ (which can be ensured by trial and error) then any non-linear behaviour in the growth of the donor levels to dose N should be duplicated (or approximately so) in the application of dose β. Thus, supralinearity or sublinearity problems should be offset. As with the use of phototransfer in dose re-estimation in dosimetry applications, steps (b) to (e) and (g) to (k) can be repeated several times because only a small proportion of the donor level is depleted with each ultra-violet irradiation. Thus, several measurements can be made on one sample.

7.3   General problems

7.3.1   Fading

It hardly need be stressed that as with thermoluminescence dosimetry (TLD) (see chapter 6) dating an artefact using thermolumines-cence requires that there be no loss of charge from the trapping levels during natural irradiation. For TLD the problem is perhaps less acute than for dating because the irradiation periods are much shorter and the conditions of temperature, light, humidity, etc. can be closely monitored. In favourable cases, corrections for the signal loss caused by fading can be made. With dating applications, however, the irradiation periods are much longer and the environmental conditions are less well characterized and are probably non-constant. It thus becomes of utmost importance to eliminate fading effects.

Of the main causes of fading (heat, light and anomalous effects) possibly the easiest to circumnavigate is thermal fading. As mentioned in the preamble to the pre-dose method, the lower temperature peaks are expected to display a greater degree of thermal fading than the higher temperature peaks, for a given irradiation temperature. Thus, by selecting only those glow peaks which are stable enough to resist thermal fading, the dating can be performed with some confidence. However, the problem has to be faced of how high in temperature the glow peak has to be before thermal drainage during antiquity can be ignored. The simplest way to address this is to apply the so-called 'plateau test'. Here, the glow-curve of the natural thermoluminescence (TLN) is compared with that of the thermoluminescence following artificial irradiation (TLA) and a plot is made of the ratio TLN/TLA against glow-curve temperature. Those traps which do not have the stability required for long-term retention of the charge are indicated by a non-constant value for the ratio. The region of stable traps is indicated by a constant ratio – or 'plateau'.

In practice, samples tested in this fashion often produce poor plateaux, or sometimes fail to show a plateau at all. One reason for this may be the change in the sensitivity caused by heating the sample to record the natural thermoluminescence. To overcome this problem, the procedure most widely adopted is to irradiate a previously undrained sample and a glow-curve corresponding to the artificial-plus-natural dose $(TL(A + N))$ is recorded. The plateau test is then carried out by plotting the ratio $TLN/(TL(A + N) - TLN)$ against glow-curve temperature. Occasionally, workers multiply the ratio by the dose delivered during the artificial irradiation and produce an 'equivalent dose plateau'. From the equivalent dose, an age is often calculated and an 'age plateau' is plotted. The same information is being given in each case. Examples of plateaux typically found are given in figure 7.6. For quartz inclusions, the high-temperature 325 °C and 375 °C glow peaks have the necessary stability to retain their charge and it is this region which is exploited when dating this material.

The plateau shown in figure 7.6a is an ideal shape, but unfortunately examples like this are the exception rather than the rule. More usually, complex shapes are seen, as in figures 7.6b and especially 7.6c. The reasons for such shapes are many and varied. They include supralinearity affects, anomalous fading and spurious signals (indeed the presence of spurious thermoluminescence is often tested for by the plateau test, and specimens with poor plateaux are discarded). Furthermore, it is important to be aware of the problems associated with non-first-order kinetics which may also give rise to undesirable plateaux. With first-order kinetics, the natural and the artificial glow peaks will appear at the same temperature and so comparison of the glow-curve shapes is straightforward. With non-first-order kinetics, however, unless the natural and artificial doses are of the same level, the peaks will not be in the same position and a poor TLN/TLA plateau is inevitable. A similar problem exists when plotting $TLN/(TL(A + N) - TLN)$ where it will be impossible to get the peaks in the same position in the TLN and $TL(A + N)$ glow-curves. In this last case the problem will be less acute the smaller the dose A is compared with N.

An additional source of complexity, not readily appreciable at first sight, is the fact that the higher temperature peaks do not necessarily fade more slowly than those at lower temperature, as already fully discussed in section 3.3.5.

Aside from the thermal fading, signal loss due to optical effects is also a problem. Optical stimulation in general was discussed in section 4.3.1 and can only be guarded against in dating projects by discarding the outer 1 or 2 mm of the sample to be dated (Zimmerman, 1978). This is usually done in subdued, preferably red or yellow, light, using a diamond saw. All subsequent experiments on the specimens also ought to be

Figure 7.6. Examples of plateaux for various materials. (a) Ratio $TLN/(TL(A + N) - TLN)$ versus temperature for pottery quartz (after Aitken & Fleming, 1972). (b) Same as (a) but for a Roman clay pipe (after Stadler & Wagner, 1979). (c) Equivalent dose (ED) plateaux for various prehistoric chert artefacts (selected from Melcher & Zimmerman, 1977).

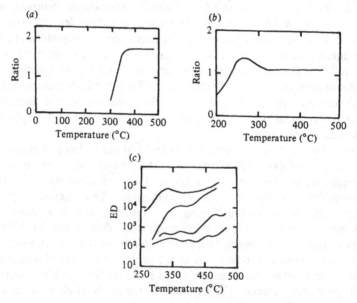

carried out in subdued lighting, unless it has been established that visible or fluorescent lighting does not affect the thermoluminescence.

The phenomenon of anomalous fading (Wintle, 1973, 1977) has been discussed in depth in section 4.4. It need only be reiterated here that quartz does not appear to suffer from the problem, whereas fluorapatite, feldspar and zircon are badly affected. Those fine-grain discs which contain mainly feldspar are therefore prone to trouble. (Feldspars are also characterized by second-order kinetics (Levy, 1979) so fine-grain discs are also likely to display problems with the plateau test.) At present the only way of detecting anomalous fading appears to be by long-term storage tests (e.g., Huxtable *et al.*, 1976) whilst the best way to overcome the problem appears to be to use phototransfer dating (Bailiff, 1976).

7.3.2 *Spurious thermoluminescence*

Spurious, or more specifically non-radiation-induced thermoluminescence, has proved to be a greater hazard to dating than to TLD applications, probably as a result of the nature of the specimens under study – i.e., fragments of ancient debris retrieved from the earth as opposed to pristine, manmade samples of uniform composition, size and structure as used in radiation dosimetry. It has already been mentioned (section 6.1.5) that the causes of spurious thermoluminescence are varied and can arise from various sources including grinding, sawing, mechani-

Figure 7.7. (*a*) Natural glow-curves for a pottery sherd (fine grain) sensitive to oxygen: (1) in still air; (2) oven flushed with a high purity gas (N_2 or Ar; flow rate 2 l/min); (3) as (2) but 5 l/min; (4) oven evacuated and then high purity N_2 or Ar at 5 l/min. (*b*) Natural glow-curves for a pottery sherd: (1A) untreated; (2A) after washing in dilute acetic acid. (1B) and (2B) are the background signals for (1A) and (2A) respectively. (After Huxtable, 1978.)

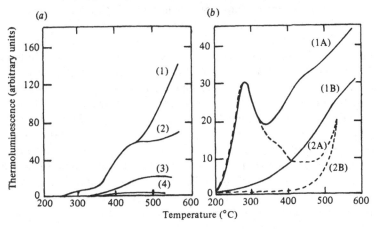

cal agitation, gas adsorption, decomposition, combustion, pyroelectric effects, etc. As related in the previous chapter, the spurious signals appear to stem from surface effects and thus it is no suprise to discover that fine-grain samples are more prone to spurious thermoluminescence than the large inclusions. Particularly bad examples are described by Huxtable (1978) and shown here in figure 7.7.

Schulman *et al.* (1960) and Aitken *et al.* (1968*b*) addressed themselves to the problem of eliminating the spurious signals from the glow-curve. Greatest improvement was found when performing the thermoluminescence readout in a dry, oxygen-free atmosphere. Figure 7.7*a* shows the behaviour of an oxygen-sensitive sample exposed to different glow-oven atmospheres and clearly there is an increase in signal strength with the presence of oxygen; but even with oxygen completely removed there is still some residual spurious signal associated with surface effects. Some of these may be eliminated by chemical treatment. For example, figure 7.7*b* shows that, in this particular instance, washing fine-grain pottery grains in dilute acetic acid has a marked beneficial effect on the removal of the spurious signal. Aitken *et al.* (1968*b*) demonstrate the effectiveness of using a reducing paint on the heating-plate in the elimination of spurious effects.

Not all materials display undesirable features to the same extent. Some quartzes are apparently free of the problem, whilst calcites and fluorites appear to be particularly bad. Generally, if adequate tests are performed and if the sample preparation is carried out carefully, spurious thermoluminescence ought not to be a barrier to obtaining a date in the majority of cases.

7.3.3   *Sensitization and supralinearity*

When considering the likelihood of thermal fading of a particular glow peak from quartz during antiquity, the plateau test will reveal that peaks at 325 °C and 375 °C both appear suitable for dating purposes. (A typical glow-curve from pottery quartz is shown in figure 7.8.) However, both peaks exhibit disturbing properties that can present difficulties – namely, they both exhibit supralinear growth (see chapters 3 and 4). The character of the supralinearity is as shown in figure 4.1, i.e., a supralinear region followed by a linear region. Figure 7.9 illustrates the problems this causes when attempting to assess the natural dose. The normal procedure is to impart to the sample varying amounts of artificial irradiation on top of the natural dose and to plot $TL(A + N)$ versus dose. If the growth is perfectly linear the natural dose N can be obtained by extrapolating to zero thermoluminescence at which point the intercept will give the required dose value. However, if the growth curve is supralinear, then the *apparent* natural dose N' will be an underestimate of the true value by an amount I_0. The problem then becomes one of estimating I_0. The procedure adopted utilizes a 'second-glow' of the material, namely, the sample is first drained of its natural

Figure 7.8. Glow-curve of a coarse-grained quartz extract from pottery irradiated with 550 rad β irradiation (curve A). Curve B is the black-body background. The glow-curve shows the 'malign' 325 °C and the 'benign' 375 °C peaks. (After Aitken & Fleming, 1972.)

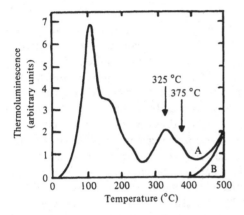

Figure 7.9. First- and second-glow method of correction for supralinearity. Curve A is the growth of $TL(A + N)$ whilst curve B – the 'second-glow' – is the growth of TLA with imparted artificial dose.

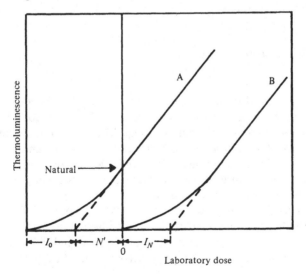

thermoluminescence by heating to 500 °C and is reirradiated in the laboratory to give TLA. A growth curve of TLA versus dose is then obtained. If there has been no change in the supralinear properties of the specimen during these irradiations and readouts, then the second-glow growth curve will display the same degree of supralinearity as the first – i.e. curves A and B in figure 7.9 will be parallel. Under these circumstances, I_0 can be estimated from the second-glow intercept I_N, where $I_N = I_0$. Thus, the true natural dose is then estimated to be

$$N = N' + I_N. \tag{7.14}$$

The parallel growth of the first- and second-glow growth curves requires that there is no change in sensitivity caused either by radiation or by heating. Unfortunately, at least in the case of the 325 °C peak in quartz, this appears not to be so. The sensitivity change means that this method of supralinearity correction cannot be applied and so the 325 °C peak is said to possess 'malign' properties. In contrast, the 375 °C peak is described as 'benign' in that the first- and second-glow growth curves are indeed parallel, at least over the dose ranges normally used. For this reason, the 375 °C peak is the one exploited in thermoluminescence dating of quartz, although care has to be taken to account for overlap of the 325 °C region. In practice the glow-curve area above ~ 380 °C is generally used (Fleming, 1979a). Occasionally, sensitivity changes in the 375 °C peak also occur, but this does not necessarily mean that $I_0 \neq I_N$ (Fleming, 1975b). This can be checked by comparing second- and third-glow growth curves. If the intercepts are seen to be equal, then there is good reason to assume that $I_0 = I_N$ and use of equation (7.14) is then justified.

7.4   Dose rate evaluation

7.4.1   *Introduction*
The natural dose delivered to archaeological and geological specimens arises from traces of radioactive substances both in the objects themselves and in the surroundings. The main contributors to the dose are isotopes of uranium, thorium and potassium, namely ^{238}U, ^{232}Th and ^{40}K. Smaller amounts of ^{235}U and ^{87}Rb are also present. The decay chains for the three dominant radioactive isotopes are given in table 7.3, along with the relevant half-lives and types of radiation emitted. As discussed in section 7.2.1 a penetration depth $> 100\,\mu$m will result in complete attenuation of α-particles, whilst the ranges for βs and γs are respectively $\sim 500\,\mu$m and ~ 30 cm. Thus, an individual quartz grain will receive most of its internal dose from α- and β-particles, whilst

the bulk of the external radiation from the surroundings will be due to γ-photons. An external contribution from cosmic rays also needs to be considered (~ 15 mrad year^{-1}). Assessment of the dose rate experienced by a sample, therefore, involves a calculation of these individual contributions (Aitken, 1978d).

In principle there are several methods by which these calculations can be performed. They either involve a direct measure of the dose rates using TLD, or they involve some means by which the individual abundances of the isotopes can be determined. Bell's data of table 7.1 may then be used to calculate the dose rates. (It should be noted that the

Table 7.3. *The uranium, thorium and potassium decay chains giving half-lives, radiations emitted and energies. Data from Littlefield & Thorley (1968).*

Isotope	Half-life	Radiation emitted (MeV)
^{238}U	4.5×10^9 years	(4.18) α, (0.045) γ
^{234}Th	24.5 days	$(0.192, 0.104)$ β^+, γ
^{234}Pa	1.14 min	$(2.32, 0.8)$ β
^{234}U	2.5×10^5 years	(4.83) α, β, γ
^{230}Th	8×10^4 years	$(4.68, 4.61)$ α, γ
^{226}Ra	1620 years	$(4.79, 4.61, 4.21)$ α, (0.188) γ
^{222}Rn	3.825 days	(5.486) α
^{218}Po	3.05 min	(5.998) α
^{218}At, ^{214}Pb	2s, 26.8 min	(6.63) α, (0.65) β, γ
^{214}Bi	19.7 min	(5.502) α, (3.15) β, (1.76) γ
^{214}Po, ^{210}Tl	1.6×10^{-4}s, 1.32 min	(7.68) α, (1.8) β
^{210}Pb	25 years	(0.025) β, γ
^{210}Bi	4.8 days	(5.0) α, $(1.65, 1.08)$ β, γ
^{210}Po, ^{206}Tl	140 days, 4.2 min	(5.298) α, (1.8) β, γ
^{206}Pb	Stable	—
^{232}Th	1.39×10^{10} years	(3.98) α, (0.055) γ
^{228}Ra	6.7 years	(0.03) β
^{228}Ac	6.13 hours	(4.54) α, $(1.5, 2.0)$ β, γ
^{228}Th	1.9 years	$(5.42, 5.34)$ α, γ
^{224}Ra	3.64 days	$(5.66, 5.44)$ α, (0.25) γ
^{220}Rn	54.5 s	(6.282) α
^{216}Po	0.16 s	(6.774) α, β
^{216}At, ^{212}Pb	3×10^{-4}s, 10.67 hours	(7.64) α, $(0.59, 0.36)$ β, γ
^{212}Bi	60.48 min	(6.054) α, (2.25) β, γ
^{212}Po, ^{208}Tl	3×10^{-7}s, 3.1 min	(8.776) α, (1.82) β, (2.62) α
^{208}Pb	Stable	—
^{40}K	1.25×10^9 years	(1.33) β, (1.46) γ
^{40}Ar, ^{40}Ca	Stable	—

effective dose rate from α-particles needs to be multiplied by a 'k-factor' of ~ 0.1 to 0.15 to account for the lower efficiency of these irradiations at inducing thermoluminescence compared with β and γ irradiations (Zimmerman, 1972).) The 'k-factor' is the thermoluminescence per rad of α-particles compared to that per rad of β-particles. Alternatively, the 'a-value' system of Aitken & Bowman (1975) may be used. These authors estimate the thermoluminescence per unit α-particle track length and evaluate an 'a-value' which is a measure of the thermoluminescence per track length to that per rad of β or γ irradiation. (This method is slowly superceding the 'k-factor' system.)

The techniques used include thick-source *α-counting* to determine ^{238}U and ^{232}Th contents, along with, for example, *flame spectrophotometry* to determine total potassium content (from which the ^{40}K concentration can be calculated). Alternatively, *γ spectroscopy* or *neutron activation analysis* (both well-established techniques) can be employed. Recently, emphasis has been laid on the need to utilize several different techniques rather than relying solely on one. Alternatives include fission track analysis, *α spectrometry, proton-induced X-ray emission* and *mass spectrometry*. Much effort has gone into estimating the relative merits of the individual techniques and a recent interlaboratory comparison of β dose rate assessments, using brick samples of known age as a standard, is one such example. Preliminary results were described by Haskell (1983). The more popular of the techniques will be described briefly here, but before doing so, some general problems met with in dose rate evaluation need to be mentioned.

Firstly, the accurate determination of the dose rate requires an assumption of secular equilibrium in the decay chains, where the daughter nuclei disintegrate and form at the same rate. This in turn requires that there is no mechanism by which any of the isotopes can be lost, by say, emanation of gas or by leaching. As discussed by Meakins, Dickson & Kelly (1978) the daughter products of the decay chains include several distinct chemical elements which possess different chemical reactivities and solubilities. As such, some may undergo exchange with their surroundings, to varying degrees, by mechanisms which are as yet improperly understood. Some recent examples of disequilibrium in non-pottery materials are noted by Wintle (1980).

The most frequent cause of disequilibrium is the loss of radon (^{222}Rn, a gas) from the uranium series. Loss of thoron (^{220}Rn) from the thorium series is also a danger, but because of its short half-life (54.5 s) it is unlikely to experience significant diffusion and will probably remain trapped for long enough to transform to ^{216}Po. In contrast, ^{222}Rn has a half-life which is sufficient (3.825 days) for appreciable gas loss to take place. The rate of loss is affected by wetness, with the escape rate normally being reduced by the presence of water because of the lower diffusion rate of Rn. As pointed out by Aitken (1978c), however, in some circumstances the presence of water can actually aid the escape by reducing the radon atom recoil length. Aitken (1976) addresses this problem in some detail and uses an α-counting system to arrive at an estimate of the loss rate by capturing the escaped gas in a gas cell. An alternative design of gas cell is discussed by Wu & Kendall (1982). The amount of radon present is then determined by α-counting. The emanation of ^{222}Rn in the laboratory, however, may be quite different from that found in the buried state. The problems of radon loss and disequilibrium in general have been discussed in several papers at the Second (Oxford, 1980) and Third (Helsingør, 1982) Specialist Seminars on Thermoluminescence Dating. As described by (for example) Martini *et al.* (1983c) the best method for establishing exactly where in the decay chain, and by how much, disequilibrium may be occurring is by α spectrometry. Here the concentrations of each of the α-emitting components in the decay chains may be arrived at by monitoring the energies of the emitted α-particles. Thus, the Th/U ratio, Rn escape and other sources of disequilibrium can be determined.

Aside from influencing the radon emanation the water content of a sample also affects the dose rate by absorbing part of the α, β and γ radiation. Thus, the dose rate is reduced with respect to its value for a dry sample. Zimmerman (1978) gives equations of the following form for correcting for the effect of water;

$$R = R_{dry} / \left[1 + H_r \left(\frac{w}{w_d} - 1 \right) \right] \tag{7.15}$$

where R is the dose rate for the as-received (wet) sample; R_{dry} is the dose rate for the sample when dry; w is the weight of the as-received sample; w_d is the weight of the dried sample. H_r is a constant whose value varies with radiation type (subscript $r = \alpha, \beta$ or γ). For α-particles, $H_\alpha = 1.5$; for β-particles, $H_\beta = 1.25$; for γ radiation, $H_\gamma = 1.14$. The presence of water in the surrounding soil will affect mainly the γ component of the dose whereas the biggest effect on the β and α doses will be from water within the specimen. Seasonal variations will complicate the dosimetry further.

7.4.2 *Thermoluminescence dosimetry*

Potentially the simplest and certainly the most direct method by which the dose rate can be estimated is actually to measure the dose accumulated by a thermoluminescence dosemeter in a known period. Essentially this is a classic problem of environmental dosimetry and, as described in the previous chapter, TLDs are ideally suited for the task. There is also the added advantage that additional complex apparatus does not have to be purchased and mastered.

The measurement of the dose rate using TLDs presents two distinct problems. The most straightforward one is to measure the γ dose delivered by the surroundings to the object at the site where it is found. The other, and more complex problem, is to determine the internal α and β dose rates within the sample itself. Mejdahl has long been an advocate of this method of dosimetry in dating and describes methods for internal β dosimetry and environmental γ dosimetry in two papers delivered at the 1978 Specialist Seminar on Thermoluminescence Dating (Mejdahl, 1978*a,b*). References to many of the early papers are given in these articles. Aitken (1978*d*) is also useful.

For *environmental* γ measurements, the TLD may be placed in a similar position to that of the find and left there, suitably enclosed (e.g., in a steel tube), for periods of up to 1 or 2 years. Such long periods enable some account to be made of seasonal variations in such things as temperature and water content. This will also automatically account for the cosmic ray dose rate. Because of the long integration times, the phosphor must exhibit low fading and low self-dose. The photon energy dependence should be similar to that of quartz (if possible), and a high sensitivity (to detect doses of a few mrad) is desirable. Several possible alternative materials are discussed by Mejdahl (1978*b*), although CaF_2 seems a popular choice (e.g., Martini *et al.*, 1983*b*; Michael, Andronikos & Haikalis, 1983). Aside from some minor problems associated with the γ energy spectrum (Aitken, 1978*d*) the important features of this rather straightforward application of environmental TLD have been discussed in chapter 6 and need not be repeated.

Estimating the *internal* dose rate with TLDs is a more complex problem and thermoluminescence dosimetry is not really suited to an assessment of the α dose. Fleming (1968) suggests a subtraction technique using two phosphors of markedly different thermoluminescence sensitivities to α irradiation but the accuracies involved are not high. As a result, TLDs are not normally used in conjunction with the fine-grain method of dating (section 7.2.2) and α-counting (see below) is preferred. For inclusion dating, TLDs may be more acceptable provided a reliable method of arriving at the β dose can be used. To do this, the α contribution to the radiation has to be filtered out. This can be achieved in several ways. Fleming (1968) uses 'desensitized' 100 μm grains of CaF_2 to accumulate a β dosage. The outer part of the phosphor is made insensitive to α-particles by annealing at 600 °C in a wet atmosphere before use, and thus, primarily, it is the β radiation which contributes to the thermoluminescence. There may be a small γ contribution (Fleming estimates 3% from U, 7% from Th and 1% from K) but this may offset the lack of consideration of the actual internal γ dose to the object itself. Mejdahl (1978b) prefers to use samples sealed (to trap radon) in polyethylene bags in order to filter out the α-particles. TLDs ($CaSO_4$) are attached to the outside. Calibration of the dosemeter is performed using a 'standard' containing known amounts of U, Th and K, mixed with $CaSO_4$ as a background material. Valladas & Valladas (1983) describe a similar system also based on $CaSO_4$, in which the detector is contained in a nylon tube which is inserted into the ceramic to be dated. Calibration is performed using standards of U, Th and K.

The biggest problem with the use of TLDs concerns variations in the K content from sample to sample (especially problematic with feldspar inclusions). Examination of table 7.3 will reveal that the β energy spectrum is strongly isotope dependent and will therefore depend upon which isotope is dominant. Shielding from α-particles results in some attenuation of the β-particles, depending on their energy. Furthermore, a γ component will always be present, the relative intensity of which also depends on the amounts of each isotope present. Bailiff et al. (1977a) conclude that the attenuation of the β-particles from the ^{232}Th series is approximately compensated for by this isotope's higher γ dose. Mejdahl's (1978b) improved calibration technique keeps errors (for quartz inclusions) to as low as $\pm 5\%$ using β TLD, and this method of dosimetry is a strong contender when using inclusion dating.

The primary requirement of the phosphor is that it is extremely sensitive so that the very low doses from the object may be detected. CaF_2 and $CaSO_4$ specimens are normally used, with $CaSO_4$ (doped with Mn) being preferred because of its lower self-dose.

7.4.3 Alpha-counting and K-analysis

The most straightforward method of determining the U and Th content of a specimen is to measure the α activity, using a scintillation counter. The simplest technique, which is adopted by most laboratories involved in α-counting, was suggested by Turner, Radley & Mayneord (1958) and consists of a ZnS scintillation screen in front of which is placed the powdered sample which acts as a 'thick source'. A photo-multiplier is used to record the scintillations. Radon from the sample is allowed to escape and Aitken's (1978c) gas cell can be used to determine the radon emanation. Despite slight revisions and adaptations this experimental arrangement, which can achieve a counts-to-background ratio of 100:1, remains popular.

Alpha-counting is the preferred technique when using fine-grain dating methods in which the major contributers to the dose are α-particles. For assessment of the β and γ doses (from U and Th) it is much less satisfactory. Firstly, the β and γ contributions from the ^{238}U and ^{232}Th decays are quite different for each chain so that variations in the U/Th ratio can have a marked effect on the dose rate. Calculation of the U/Th ratio by α-counting is difficult, and can only be crudely estimated using the 'pairs' technique in which coincident (i.e., within 0.2s) α-counts are recorded. This effectively measures the ^{216}Po activity and thereby the ^{238}Th content (Aitken & Fleming, 1972; a more promising technique has recently been noted by Martini, 1981). Because the α energy is insensitive to the series from which the α-particle originates, an estimate (usually taken to be 4:1) for the U/Th ratio is all that is required. For the β and γ doses, however, gross errors may easily be introduced. Secondly, although the α dose is acceptably insensitive to radon loss, the γ dose in particular differs greatly between the pre-radon and post-radon part of the ^{238}U series. Aitken (1978d) estimates that if radon loss is by as much as 5%, a 25% error in the γ dose will result. Thus, α-counting is only suitable for fine-grain dating where the internal β and γ activities may be ignored.

Beside the above mentioned difficulties for evaluating the β and γ doses, there are additional problems associated with measuring the α dose. The most prevalent is 'overcounting'. Recent studies (e.g., Murray, 1982; Goedicke, 1983; Jensen & Prescott, 1983; Mangini, Pernicka & Wagner, 1983) have been unable to confirm the cause of overcounting, although it appears not to be always related to disequilibrium and can be improved by careful sample preparation procedures. Further problems are associated with the degree of reflectivity of the specimen – perfectly reflecting samples can give a pulse height of twice that from a perfectly absorbing sample (Huntley & Wintle, 1978; Martini, 1981).

Alpha-counting only arrives at values for the ^{238}U and ^{232}Th concentrations. The ^{40}K contribution is determined by flame emission spectrophotometry (Suhr & Ingamells, 1966). This well-established, analytical tool enables the bulk K content to be determined, from which the ^{40}K concentration may be calculated using the known abundance of this isotope (0.011%). Alternatively, *atomic absorption* or *X-ray fluorescence* may be used.

It is an impossible task to designate a 'best' system for dose rate evaluation. For environmental dose rate assessment, TLD appears to give the greatest rewards in terms of ease of opertion, accuracy and equipment. The internal dose rate estimation is more complex. Aitken (1978d) advocates two guidelines for choosing a particular method; the first is to simulate the natural environment as much as possible and to estimate the individual α, β and γ contributions separately; the second is to use those techniques with which the experimenter is most familiar ('better the devil you know...'). Otherwise the choice is quite open and several examples of the use of each of the above techniques can readily be found in the literature.

7.4.4 Other techniques

Among the other possible techniques mentioned in the introduction to this section, α spectrometry is a potentially useful contender. This method enables the concentration of each emitting nucleus to be determined and therefore allows ^{222}Rn loss (and other sources of disequilibria) to be noted. The U/Th ratio can also be determined (e.g., Martini *et al.*, 1983*c*). Frisch ionization chambers, Si surface-barrier detectors or (especially for low levels) ionization chambers may be employed as detectors. A low count-to-background ratio needs to be improved upon for the technique to be used for low-activity specimens.

Gamma spectrometry detects γ-rays of specific energy from particular members of the decay chains and is widely used in geological applications (e.g., Løvborg, 1972). The degree of radon retention can also be estimated. High-sensitivity Ge detectors will enable small samples to be used but with the more popular sodium iodide detectors, up to 100 g of material may be required. Measurement of the 1.76 MeV γ-photon from ^{214}Bi allows both the ^{238}U concentration and the level of ^{222}Rn escape to be evaluated (by measuring the γ-count from sealed specimens). ^{232}Th and ^{40}K concentrations are determined by monitoring the 2.62 MeV γ emission from ^{208}Tl and the 1.46 MeV γ-photons from ^{40}K, respectively. The α and β dose rates are less accurately determined, especially if other causes of disequilibrium exist (Aitken, 1978*d*). The environmental γ dose may be measured in a similar fashion to γ TLD measurements, namely, by burying the detector on-site close to where the sample was found (e.g., Murray, Bowman & Aitken, 1978). There is the disadvantage that the detector cannot be left at the site for long periods, as can the TLD.

Neutron activation analysis is a well-developed technique but it cannot account for disequilibrium and will, therefore, result in large errors if any form of disequilibrium is present. Nonetheless, in favourable circumstances, its use can be attractive (Mangini *et al.*, 1983). Fission track analysis (Fleischer *et al.*, 1975) may also be usefully employed. Irradiation of the specimen by thermal neutrons induces fission of ^{235}U. The fission fragments leave etchable tracks, the concentration of which is proportional to the ^{235}U content, from which the ^{238}U concentration may be determined. Malik, Durrani & Fremlin (1973) successfully used the technique to indicate an inverse spatial correlation between thermoluminescence intensity and U distribution in some archaeological materials.

7.5   Special dating applications

7.5.1   *Sediments*

Undoubtedly the most innovative of the recent developments in thermoluminescence age determination have centred upon the dating of sediments. This has been owing, primarily, to the extensive efforts of Huntley, Wintle and colleagues following an observation by Huntley & Johnson (1976) that the natural thermoluminescence from deep-sea sediments from the North Pacific increased with core depth and who concluded that dating of the sediments by means of thermoluminescence was possible. Initially, it was assumed that the thermoluminescence stemmed entirely from the siliceous material, but later investigations revealed that this component yielded little luminescence and that in fact the glow stemmed almost entirely from small particles of silt adhering to the siliceous tests. Wintle & Huntley (1980) continued the development and described the detailed sample preparation procedure which is necessary. Essentially, the process is a fine-grain ($4-11\,\mu$m) dating method, the principles of which have already been described (section 7.2.2). There are, however, two major differences. The first concerns the event which re-set the 'thermoluminescence clock', i.e., the zeroing mechanism: the second relates to a modification to the age relationship (equation 7.5) as a result of precipitation of ^{230}Th and ^{231}Pa.

As already noted, the age normally calculated during thermoluminescence dating is the time since last heating (to $> 500\,°$C). With pottery specimens, this corresponds to the time since firing, i.e., the age of the pot. With sediments, however, the zeroing event is less obvious. Morozov (1969), Shelkoplyas (1971) and Huntley & Johnson (1976) present evidence that sunlight and mechanical grinding are able to remove a large fraction of the trapped charge within the sediments, thereby reducing the natural thermoluminescence signal. Thus, the premise is that the mechanical grinding and, particularly, the exposure to sunlight experienced by the sediment during deposition is the zeroing event. Hence, the thermoluminescence age is assumed to be the time since deposition. Thus, this dating method relies on the fact that the material possesses extreme sensitivity to light in a manner which is a disadvantage in other dating techniques, and in dosimetry (as discussed in the previous chapter, particularly section 6.1.3).

Clearly, the age of the sediment will increase with depth and this is thought to be the cause of the increase in natural thermoluminescence with depth observed by Huntley & Johnson (1976). In a series of tests on the efficiency of sunlight in removing the pre-existing thermolumines-cence, it has emerged that although the reduction in signal is initially very rapid, there is a small residual signal, I_0, which cannot be removed (Wintle & Huntley, 1979a,b, 1980). It is assumed, but cannot be proven, that this is the level of thermoluminescence within the specimen at the time of deposition. This is an important assumption because it requires that the zeroing mechanism in nature was solely by optical bleaching and that the bleaching conditions were similar to those in the laboratory. This latter requirement is unlikely (for instance, in nature the bleaching would probably have taken place in the atmosphere and, therefore, at low temperatures) and the errors introduced have not been estimated. Alternatively, if the actual zeroing mechanism is by bleaching and by other means (e.g., mechanical grinding) it has to be assumed that the laboratory ultra-violet exposure and the natural agents both reduce the trapped charge concentration to the same level – a fortuitous situation. The laboratory simulation of the zeroing process remains one of the biggest problems confronting the experimenters in this field as a recent paper by McKeogh, Rendell & Townsend (1983) demonstrates.

Figure 7.10 shows a set of natural glow-curves from a North Pacific deep-sea sediment and illustrates the removal of the natural glow by ultra-violet exposure. After ~ 1000 min, no further reduction takes place. Thus, the natural thermoluminescence is made up of two

Figure 7.10. Glow-curves showing the natural thermoluminescence from North Pacific deep-sea sediments and the effect of exposure to a sunlamp for various lengths of time (from Wintle & Huntley, 1979b).

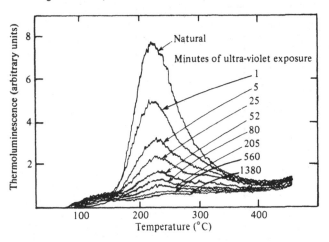

components, the residual I_0 and that induced by natural radioactivity since deposition I_D. Both components vary with glow-curve temperature and so we have an expression for the natural glow intensity I_N of:

$$I_N(T) = I_0(T) + I_D(T). \tag{7.16}$$

The problem then becomes one of isolating the component I_D and arriving at a value for the equivalent dose (ED).

Wintle & Huntley (1980) describe three methods by which ED can be estimated. (a) The sample is exposed to a sunlamp for a set period and the fraction f of I_D remaining is calculated. The sample is then irradiated with different γ doses and the dose G required to bring the thermoluminescence back to its original level is determined. The equivalent dose is then found from:

$$ED = G/(1-f). \tag{7.17}$$

It has been found that the fraction removed in a set time is independent of the level of the natural or γ dose. (b) G is determined for a series of samples which have received an additional γ dose D on top of the natural dose. A plot of G versus D is made and when extrapolated to $G = 0$, gives ED on the D-axis (figure 7.11). (c) The reduction R in the thermoluminescence, due to a standard exposure to the sunlamp, is measured for samples which have received a γ dose D on top of the natural dose. A plot of R versus D is made and extrapolation to $R = 0$ yields the value of ED on the D-axis (figure 7.11). All three methods give essentially the same

Figure 7.11. Schematic representation of the G–D and R–D methods of determining the equivalent dose for sediments (after Wintle & Huntley, 1980).

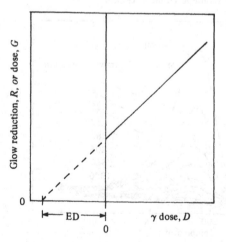

ED value, so Wintle & Huntley (1980) recommend method (c) for its simplicity.

The second major consideration concerns the dose rate determination. The contributions from internal ^{232}U, ^{232}Th and ^{40}K can be determined using α-counting and chemical analysis of K. Such a determination is subject to the usual problems associated with radon loss. Additionally there is a small problem for the ocean sediments due to ^{226}Ra–^{210}Pb exchange. Errors of $< 1\%$ have been estimated. However, before equation (7.5) can be used to determine the age a substantial modification to the equation is necessary owing to the precipitation of ^{230}Th and ^{231}Pa from the sea water. These isotopes, which decay with half-lives of 8×10^4 years and 3.4×10^4 years respectively, were incorporated into the specimen during deposition and, for young ($< 10^5$ years old) specimens, dominate the dose rate. Thus, a time-dependent dose rate contribution has to be included. The modified age equation is:

$$ED = (R_K + R_U + R_{Th})t + R_1\tau_1[\exp(t/\tau_1) - 1] + R_2\tau_2[\exp(t/\tau_2) - 1]$$
$$(7.18)$$

where R_K, R_U, R_{Th}, R_1 and R_2 are the dose rates from ^{40}K, ^{238}U, ^{232}Th, ^{230}Th and ^{231}Pa respectively, τ_1 and τ_2 are the ^{230}Th and ^{231}Pa mean-lives and t is the sample age. The α dose rate contributions include an efficiency factor (Aitken & Bowman, 1975) and all of the dose rate terms include an appropriate attenuation factor due to water (cf. equation (7.15)).

In principle, the method of dating deep-sea sediments, as described above, can be applied to any sedimentary deposit whether marine or terrestrial. Thus, the principles outlined have been applied, with varying degrees of success, to land deposits such as loess, glacial till and sand dunes (e.g., Wintle & Huntley, 1979b; Wintle, 1981; Singhvi, Sharma & Agarwal, 1982). A major simplifying factor with the terrestrial applications is that precipitation of ^{230}Th and ^{231}Pa is not a consideration and the standard age equation (7.5) can be used. Many examples of sediment dating using the techniques of Wintle & Huntley were described at the Third Specialist Seminar on Thermoluminescence and Electron Spin Resonance Dating (Helsingør, 1982). The different applications share many of the same worries, in addition to their own unique ones. Some of the major difficulties need to be mentioned.

Wintle & Huntley (1980) state that a large uncertainty in the age evaluation stems from the assessment of the water content and its effect on the dose rate. In any attempt to account for this factor the minimum requirement is that the water content of the sample is estimated. In ocean sediments a water content as high as 50–60% is often found, but it will be less abundant in terrestrial deposits, particularly from arid areas. Allowing for the variation in water content after deposition is a more difficult problem. Rendell (1983) discusses this in detail for loesses in general, and for loess from Northern Pakistan in particular. The changes in water content over millenia are complex and although information may be gained from measurement of bulk density, carbonate content and studies of the microstructure, it is difficult to arrive at a true picture. One needs to account for porosity and changes in load and lateral pressure and, especially for marine sediments, changes with depth. Calculation of the dose rate by TLD cannot make allowance for these changes, and errors of 20% or more can easily result. For these reasons, age estimates of sediments from arid locations have the most favourable chance of success.

Most attempts at sediment dating do not separate the different minerals responsible for the thermoluminescence signal. The fine-grain extracts are polymineralic mixtures, although the dominant luminescent minerals are quartz and feldspar. Several attempts to separate the components by conventional mineral separation techniques have been described (acid washing, flotation, etc; see Miallier, Sanzelle & Fain (1983) for a discussion of feldspar and quartz separation techniques). For example, Hütt, Smirnov & Tale (1979) use quartz extracts from Quaternary deposits in the Soviet Union. Others prefer to work with feldspar grains (e.g., Olsen, Thoresen & Mejdahl, 1983). Feldspar possesses several advantages with respect to quartz in sediment dating, for example the dose response of quartz from sediments appears to have several undesirable features the most important of which is that it saturates within the time since deposition (for older sediments); feldspar, on the other hand, has higher saturation doses. Also, its thermoluminescence signals appear to be more readily bleachable by sunlight. Debenham & Walton (1983) investigated the possibility of separating the two components, using their different emission spectra. As discussed in chapter 5, the emissions from quartz peak at $\sim 450\,nm$, whereas feldspar appears to emit predominantly at $400\,nm$. Debenham & Walton successfully separated the components (with filters) from polymineralic loess and found that several of the malign properties associated with quartz thermoluminescence can be avoided.

7.5.2   Stones and rocks

Dating burnt stones by using thermoluminescence follows the same principles as have already been outlined for pottery dating. Fine-grain, inclusion (quartz and feldspar) and pre-dose dating have each been utilized to arrive at an age since last heating for several varieties of stone (e.g., quartz pebbles, lava, flint). The formation periods of stalagmites and stalactites have also been estimated. Each particular application has its own difficulties and a few of those problems will be discussed in this section. It should be noted, however, that apart from speleothems (stalagmites and stalactites) and lava, the geological formation age of rock or stone is unlikely to be datable by thermoluminescence because the accumulation time for the natural radioactivity is so long that the thermoluminescence will have saturated (i.e., all traps/recombination centres full) and the glow intensity will no longer be a function of time. The thermoluminescence 'clock' is said to have 'stopped ticking'. Thus, it is normally required that a much more recent event has resulted in the heating of the stones and the time since this heating may then be calculated.

A particularly good example which illustrates several of the kinds of problem which may be encountered concerns flint, the most common lithic material used by prehistoric peoples. As discussed by Melcher & Zimmerman (1977) marked improvements in the knapping qualities result when flint is heated, and heat treatment of flint tools prior to shaping is believed to have been a widespread practice among prehistoric man. There exists, therefore, the possibility of dating the manufacture of the tool using thermoluminescence (or electron spin resonance), a particularly attractive proposal because no other absolute dating technique exists for these materials. However, the miriad of problems associated with low light levels, insufficient heating, nonuniform radioactivity, spurious glow and irreproducibility have made flint extremely difficult to work with. For example, a recent dating project on flints from Combe Grenal in France (Bowman & Sieveking, 1983) utilized samples from 64 Palaeolithic horizons, but only six datable flints were found.

Early sample preparation procedures used the grains extracted following crushing of the sample, but in doing so large spurious signals were observed, making this technique unworkable. Göksu & Fremlin (1972) and Göksu et al. (1974) proposed an alternative procedure using thin, polished slices and the use of slices has now been widely adopted, although acid etching of grains has been observed to reduce noticeably the spurious signals (Valladas, 1978). Bowman & Demetsopoullos (1980) suggest the use of the pre-dose method to date the flints in order to overcome low light levels and spurious effects. Provided saturation of centres can be avoided, the technique appears useful. A typical glow-curve from flint is shown in figure 7.12.

A major problem concerns the difficulty of normalizing the glow intensity in order to account for sensitivity variations from slice to slice.

Bowman & Seeley (1978) carried out a careful investigation of the normalization procedure and found that changes in the optical absorption characteristics, induced by the procedure itself, can give rise to serious difficulties. This, along with factors such as a non-uniform uranium content (giving non-uniform self-dose), leads to the rejection of many samples.

Many flint samples appear not to have the capacity to store enough thermoluminescence for determining ages less than ∼ 20 000 years (Göksu *et al.*, 1974). Saturation effects put an upper limit of ∼ 50 000 years on the age determination, although, on occasion, the age limit may be extended further (Wintle & Aitken, 1977). In all these instances, however, it must be assumed that the level of heating was sufficient to reduce the pre-existing thermoluminescence to zero. Melcher & Zimmerman (1977) applied the plateau test (see figure 7.6) to several flint artefacts in order to assess the extent to which the materials have been heated. Some had been sufficiently heated to be datable; others had not.

The dating of volcanic lava possesses its own unique set of problems, the most important of which is anomalous fading from the feldspathic minerals (Wintle, 1973; see also section 4.4). Valladas, Gillot & Guerin (1979) and Gillot *et al.* (1979) report that this particular problem may be side-stepped if the high-temperature (500–700 °C) portion of the glow-curve from the feldspars is utilized. The experimental difficulties associated with the advent of intense black-body radiation at these temperatures was overcome with the use of ultra-violet filters. Spurious glow is also a problem (Aitken *et al.*, 1968a) but this can be reduced by careful preparation procedures (Hwang, 1970). Chemical fractionation

Figure 7.12. Glow-curves from flint from Terra Amata. A, Natural glow; B, glow-curve following natural + artificial irradiation; C, Black-body background signal. (From Wintle & Aitken, 1977.)

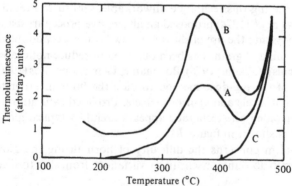

causing disequilibrium in the ^{238}U decay chain is also particularly troublesome (Nishimura, 1970).

The extraction of quartz enclaves from the lava appears a promising technique (Gillot, Valladas & Reyss, 1978) although finding suitable quartz grains may not be easy. Instead, dating stones which have been baked by the lava which has flowed over them appears to be the best alternative to dating the lava itself. This approach which utilizes conventional dating techniques on quartz extracted from the stones has been successfully employed by a number of workers (e.g., Huxtable & Aitken, 1977; Huxtable, Aitken & Bonhommet, 1978; Valladas & Gillot, 1978). One difficulty is that the thermoluminescence from the quartz is almost in saturation and therefore the natural signal is on the sublinear portion of the growth curve. Valladas & Gillot (1978) account for this by using a growth expression of the form of equation (3.50) to calculate the equivalent dose. Otherwise, the uncertainties involved in dating burnt stones are the same as those met with when dating ancient pottery (e.g., Huxtable *et al.*, 1976).

Calcite (calcium carbonate) is an attractive natural material to date using thermoluminescence because it forms cave stalagmites and stalactites when precipitated from ground water (e.g., Aitken & Bussell, 1982). Many of the problems with the use of calcite stem from sample preparation procedures and these have been discussed in depth by Wintle (1975b) and by Bangert & Hennig (1979). Clay inclusions must first be removed from natural samples to prevent contamination of the calcite glow-curve and different grinding and acid washing procedures have been suggested to avoid spurious signals. Additional difficulties stem from the spatial inhomogeneity of the thermoluminescence emission (Walton & Debenham, 1982) and the possibility of non-zero thermoluminescence in newly formed calcite (Debenham, Driver & Walton, 1982).

7.5.3 Shells, bones and teeth

Calcium carbonate is also present in shells (in fact, it is the primary constituent), occurring as either calcite or aragonite. Early studies of thermoluminescence from shells indicated that aragonite was unsuitable for dating (Johnson & Blanchard, 1967) and that the glow-curve properties of the calcite shells show a strong dependence on taxonomic group. Thin slices may be used but the organic content (conchiolin) has to be removed by treatment with ethylene diamine (Driver, 1979). The glow peak stability and growth with imparted dose needs to be examined for each individual shell.

The possibility of using the natural thermoluminescence from bone mineral salts to date bones was suggested by Jasińska &

Niewiadomski (1970) following a preliminary investigation by them of the glow-curves from several dinosaur and contemporary and fossil human bones. Difficulties arise owing to tribothermoluminescence induced by grinding the bone material and chemiluminescence associated with oxidation of the residual organic matter. Christodoulides & Fremlin (1971) described an extraction technique for the unwanted organic material (chiefly collagen; $\sim 35\%$ of the bone matter), using ethylene diamine. Rejection of small grain sizes after powdering also reduces tribo-effects. Driver (1979) examined further the preparation techniques needed to eliminate the particularly troublesome chemiluminescence and concluded that a sophisticated procedure of chemical extraction followed by exposure of the bone slices to re-excited oxygen was required to reduce the organic content to acceptable limits.

The basic properties of the glow-curve from the bone mineral have yet to be fully investigated. The luminescence material is hydroxyapatite $(Ca_5(PO_4)_3(OH))$ with a few per cent octocalcium phosphate $(Ca_8H_2(PO_4)_6.5H_2O)$. An examination of synthetic and bone hydroxyapatite by Chapman, Miller & Stoebe (1979) revealed that the most intense thermoluminescence could often only be stimulated after an initial anneal at $\sim 400\,°C$. Clearly this is unsuitable for dating. Several complex phase changes occur at these higher temperatures (e.g., the formation of pyrophosphate in bone mineral, Driver, 1979) and changes in the emission spectra have been recorded in hydroxyapatite and other forms of apatite (e.g., Lapraz, Baumer & Iacconi, 1979; Davies, 1982). These properties need to be examined further before dating bone by thermoluminescence becomes feasible.

The major inorganic component of dental enamel is also hydroxyapatite and the feasibility of using thermoluminescence to date fossil teeth has also been examined (Benko & Koszorus, 1980), with limited success.

7.5.4   Authenticity testing

According to Fleming (1979c), using thermoluminescence to detect fake 'objet d'art' is the most fully developed aspect of dating with this technique. Certainly, if one judges the degree of development from the success/failure ratio of attempts to detect forgeries among ceramic works of art, then one would be forced to agree with Fleming's assertion. Authenticity analysis of ceramics is almost exclusively the realm of thermoluminescence with its unrivalled accuracy and absolute nature (not requiring standards). The basic premise is that a forgery, no matter how fine the craftmanship, will not have absorbed enough radiation to give a strong enough natural thermoluminescence signal, commensurate with its supposed age. Thus, once the approximate dose rate (internal and external) is known, an estimate of the age of the ceramic can be made from its natural glow-curve.

For artworks older than ~ 500 years, the high-temperature region of the quartz glow-curve may be exploited, using quartz grains from a sample of the pottery extracted unobtrusively from the object (e.g., from the base; ~ 30 mg is all that is required). However, the technique really comes into its own for objects < 500 years old, which can be accurately dated using the pre-dose method. This latter period covers the Renaissance era which has attracted scores of gifted forgers. A full discussion, with illustrative examples, can be found in Fleming's books on the subject (Fleming, 1975a, 1979a). The present discussion will be concerned with some of the potential hazards and complications.

One immediate worry is how to extract enough material for a thermoluminescence study from expensive pieces of art. Stoneham (1983) has recently described a sampling technique whereby small cores (5 mm long, 3 mm diameter) are extracted using a diamond drill. The cores are then cut into thin (200 μm) slices for a glow-curve analysis. Fleming (1979c) describes a preparation procedure for fine grains of quartz which enhances the glow intensity and allows the use of small amounts of material. The powder is initially obtained by drilling. All extraction techniques are used on parts of the object not normally in view.

Assessing the dose rate poses several difficulties. Firstly, the environmental dose rate cannot be measured because the object is not taken from a controlled environment (e.g., buried in undisturbed soil) but is simply taken from a museum curator's shelf. Reconstructing the variations in the environmental dose over the object's lifetime is an impossible task. Secondly, the fine-grain α dose cannot easily be measured owing to the lack of available material. The α dose is not a problem if large samples are used, or if the pre-dose method is utilized (see sections 7.2.3 and 7.2.4), but the environmental dose rate has to be estimated by some means.

The cosmic ray dose rate (~ 15 mrad year^{-1}) is largely invariant and so an estimate of the upper limit to the age can be gained by using this factor only. If the age calculated is still below the supposed age, a forgery has been detected. However, for many, younger art objects this is not a sufficient test and more rigorous criteria have to be adopted. One approach is to use several similar types of artwork of known age. The age relationships given in equations (7.1) to (7.7) can be rewritten generally as

$$\text{Age} = \text{ED}/(R_i + R_e), \tag{7.19}$$

where R_i and R_e are the internal and environmental dose rates. Rearranging gives

$$R_e = \frac{ED}{Age} - R_i. \tag{7.20}$$

Using known ages estimated from art styles, glazing etc., and measured ED and R_i values, an approximation for the value of R_e can be made. This approach is possible only if the objects are all of the same type, coming from approximately the same region. This is true, for example, for Renaissance terracottas (Fleming, 1975a), the industry for which was centred in Florence. For these pieces, Fleming estimates a value for R_e of 0.18 ± 0.05 rad year^{-1}. If the fine-grain technique is to be used, a similar relationship can be used to estimate both R_e and the internal α dose.

So far these methods have evidently proved unbeatable in authenticity testing. One may imagine that a clever forger might artificially irradiate the object to give it an apparent, older age. However, he would have to know in advance the natural dose rates in order to calculate the artificial dose required to give the piece of art the right 'age'. In any case, the natural dose distribution could not be simulated. For instance, the dose to zircon grains in the pottery, with a high internal radioactivity, is largely determined by the internal α activity and is independent of the external dose rate (Zimmerman, 1979b). However, such sophistication in authenticity analysis has not yet proved necessary for routine application. A particularly good example of authenticity testing in action is provided by Aitken, Moorey & Ucko (1971), who examined 66 objects (figurines and pottery) of the Hacilar style and found 48 of them to be forgeries.

8   Geological applications

8.1   General

As noted in chapter 1, the use of thermoluminescence in geology was one of the earliest applications of the phenomenon and both terrestrial and extraterrestrial (meteorite) rocks were examined for thermoluminescence before the beginning of the twentieth century. The natural thermoluminescence from a geological specimen depends, for its character and intensity, not just upon the nature of the phosphors involved, but also upon the dose received in nature, the period over which this dose is delivered, and the temperature(s) experienced by the specimen during this time. Thus, the natural thermoluminescence contains locked within it information relating to the detailed thermal and radiation histories of the samples and unlocking this information has been the main goal of research into the natural glow from geological samples. Clearly the three parameters alluded to above, namely, dose, time and temperature, cannot be determined from a single measurement and the requisite detailed studies are not trivial. Information on kinetics, emission spectra, mineral types, radioactivity, dose response, sensitization, etc. has to be gathered if premature judgements are to be avoided.

A different 'family' of measurements which are often reported concerns the 'artificial' thermoluminescence, namely, that induced by a known dose given in a controlled environment in the laboratory. Measurements of this nature are usually carried out to derive data on the basic properties of the luminescent phosphor. Such data can be used to infer information about the source/origin of the mineral, and, for example, its degree of metamorphism, etc. Once again, detailed information on the mechanisms of thermoluminescence production needs to be available.

This chapter is devoted to a description of some of these studies on both terrestrial and extraterrestrial specimens. It will be seen that much valuable data have emerged, but on occasion the conclusions arrived at lack the support of detailed studies into the basic thermoluminescence mechanisms and properties of the materials under study.

8.2   **Meteorites**

8.2.1   *Mineralogy*[†]

Meteorites constitute the oldest material in the solar system and as such provide mankind with the potential to study its physical and chemical make-up going right back to its earliest history. Some meteorites (the so-called *carbonaceous chondrites*) have even been found to contain remnants of the material which formed the precursor to our solar system (the pre-solar nebula). By far the most common form of meteorite falling to earth is the 'stony' meteorite ($\sim 90\%$). Some 1–2% of falls are found to be pure metal (the 'irons' – 90–95% Fe :5–10% Ni alloy) whilst the remainder are classed as 'stony-irons' (consisting of Fe–Ni alloy and silicate minerals). Clearly the irons are of no interest to thermoluminescence research and apart from a brief study by Singhvi & Bhandari (1979) the silicate material from the stony-irons has not yet been examined. Thus, thermoluminescence research has concentrated almost entirely on the study of the stony meteorites.

The mineral composition of these specimens is generally understood from a consideration of the 'condensation sequence' in which the meteorites formed during condensation of the solar nebula gas. The stony meteorites can be subdivided into so-called *chondrites* and *achondrites*, the primary difference being that chondrites contain small (less than 1 mm) spherical inclusions, called *chondrules*, embedded in the matrix. The chondrules consist of high-temperature minerals, chiefly olivine, pyroxene and plagioclase feldspar. Feldspathic glass is also present. The largest group is that of the *ordinary chondrites*, for which the matrix material is also composed predominantly of olivine, pryoxene and plagioclase. Small pieces of Fe–Ni alloy are also present. Further subclassification of the ordinary chondrites is based on the amount of reduced iron and oxidized iron present, namely, H (high iron), L (low iron) and LL (low iron–low metal).

Additionally, there exist the *carbonaceous (C) chondrites*, so called because of the presence of small amounts of carbon-bearing compounds. Ca–Al rich inclusions are present in these materials and are believed to have formed at high temperatures and therefore early in the life of the solar system. Finally, there are the *enstatite (E) chondrites* for which the mineralogy is slightly different, being dominated by the presence of enstatite (Mg-rich orthopyroxene).

The achondrites divide into two classes, namely the Ca-rich and Ca-

---

[†] A brief description of all minerals referred to in this chapter can be found in Appendix A.

poor achondrites. The Ca-rich specimens are richer in plagioclase than are the ordinary chondrites, whereas the Ca-poor specimens resemble the E-chondrites, i.e., predominantly enstatite. As with the ordinary chondrites, all the stony meteorites contain small amounts of Fe–Ni alloy in varying amounts.

In some chondrites, the chondrules are seen to be physically quite distinct from the matrix, whilst in others the boundary is much less easily seen. Those meteorites with indistinct chondrules additionally show evidence of thermal metamorphism and/or shock impact alteration. These features call for further subdivision of the classes into petrologic type. Those specimens with the highest degree of metamorphism and alteration are assigned a value 6, whilst those with the least are given a value 1. Thus, chondrites are described on the basis of their class and type, i.e., H4, L3, LL5, etc. For more detailed discussions of meteorite mineralogy and petrology, the reader is referred to Sears (1978a) along with the references quoted therein.

8.2.2   *Thermoluminescence*

It has been variously claimed by different authors that the thermoluminescence of meteorites can be used to provide data on their shock/reheating history, cosmic ray exposure age, orbit, pre-atmospheric shape, ablation rate, terrestrial age and petrologic type. Some of these uses are outlined in a review by McKeever & Sears (1979). One may reasonably ask how it is that the measurement of one property can throw light on so many different aspects. Certainly the situation can be improved upon by careful sample selection; e.g., recent falls may be chosen so that terrestrial age is not a consideration; meteorites with the same cosmic ray exposure age can be found; data on the levels of shock and reheating can be gathered from other sources (e.g., $^{40}K–^{40}Ar$ measurements), etc. Additionally, measurements of natural thermoluminescence can be augmented by data on the artificial glow-curves. However, in some instances one is still left with the situation that the thermoluminescence data are a result of two or more effects acting together and, in truth, these effects cannot always be separated. Nevertheless much useful information about a meteorite's history can be gleaned from thermoluminescence if careful consideration is given to all the relevant factors, and in the sections that follow some of these cases will be discussed. Before doing so, it will be advantageous to detail some of the general characteristics of the thermoluminescence emission from these materials.

Because the character (i.e., shape, emission wavelength, intensity) of a glow-curve is critically dependent upon both the nature of the phosphor

Figure 8.1. Glow-curves from several meteorites of different classes: (*a*) Khor Temiki, a Ca-poor achondrite, in which the main phosphor is enstatite; (*b*) Ca Al-rich inclusion from Allende, a C3 chondrite; (*c*) Kapoeta, a Ca-rich achondrite, in which the phosphor is believed to be plagioclase; (*d*) Durala, an L6 chondrite, in which the phosphor is plagioclase. All ordinary chondrites have natural glow-curves similar to curve (*d*). The difference between (*c*) and (*d*) is presently unexplained. (From McKeever & Sears, 1979.)

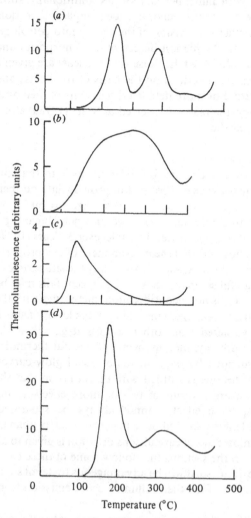

and the defects within it, it is not surprising to learn that the natural glow-curves from meteorites exhibit a wide variation, some examples of which are shown in figure 8.1. Few studies to characterize the thermoluminescence have been carried out on meteorites other than the ordinary chondrites but one example is provided by Sears & Mills (1974*a*) who examined the glow-curves produced by different mineral assemblages within the Allende C3 chondrite. Three different types of glow-curve were obtained: (i) a dominant peak at 140 °C (5 °Cs^{-1}); (ii) a dominant 200 °C peak; and (iii) a broad, ill-defined peak centred at ~ 200 °C (see figure 8.1*b*). The 140 °C peak appears to stem from the chondrules which are mainly olivine, whilst the broad-peaked glow-curve is from a Ca–Al-rich inclusion. In a detailed examination of the cathodoluminescence from these inclusions in Allende, Hutcheon *et al.* (1978) correlate the variations in the patterns of luminescence colour with mineral com-

Figure 8.2. Proton-induced luminescence spectra. (*a*) Khor Temiki (enstatite achondrite); (*b*) Khaipur (E chondrite); (*c*) Juvinas (eucrite (Ca-rich achondrite)); (*d*) Holbrook (L chondrite); (*e*) Gnadenfri (H chondrite); (*f*) Jelica (LL chondrite). Selected from Geake & Walter (1966*a*).

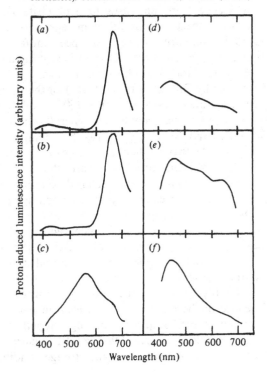

position. The brightest luminescing material is found to be plagioclase (whitish blue emission), the intensity of which is found to exhibit a negative correlation with Mg and/or Na content. Blue luminescence is also found from high-Fe spinel, whilst most of the spinel (low Fe) luminesces a bright red-orange. Deep blue to violet emission is correlated with melilite and yellow-green luminescence with grossular. A detailed examination of the spectrum of the thermoluminescence from Allende has not yet been carried out and correlations between glow-peaks and mineral type cannot confidently be made.

An examination of the thermoluminescence from achondrites, in particular a group of Ca-poor enstatite achondrites known as the *aubrites*, was carried out by Houtermans & Liener (1966) and by Grögler & Liener (1968). In addition, several reports of cathodoluminescence from these specimens have been made (e.g., Derham & Geake, 1964; Derham, Geake & Walker, 1964; Geake & Walker, 1966a, b; Reid & Cohen, 1967). The mineral enstatite is a very efficient luminescent phosphor. A typical emission spectrum reveals a weak blue emission (~ 400 nm) and a very strong red emission at ~ 670 nm, which has been related by Garlick (1964) to Mn impurities. A representative spectrum of proton-induced luminescence from an enstatite achondrite (Khor Temiki) is compared with spectra from other meteorite types in figure 8.2. In this figure, the similarity between the spectrum from the achondrite Khor Temiki and the chondrite Khaipur is striking and is due to the fact that enstatite is the main phosphor in both.

The glow-curves obtained by Grögler & Liener (1968) for the aubrites generally reveal just two natural glow peaks, at ~ 260 °C and 400 °C (heating rate not given). The artificial glow-curves generally show four peaks (~ 120, ~ 190, ~ 240 and ~ 400 °C). The two high-temperature peaks emit in the red, whilst the two low-temperature peaks emit in the blue (correlated by Grögler & Liener to low-Mn enstatite).

Apart from the above studies, by far the most detailed examination of meteorite thermoluminescence has been on ordinary chondrites, typified by the glow-curve of Durala, shown in figure 8.1d, and the glow-curve for Barwell (L5), curve A in figure 8.3. The glow-curve is characterized by an intense emission peaking around 200 °C (usually termed the low-temperature emission) followed by a less intense, broad peak at approximately 380 °C (the high-temperature peak). The exact positions of these peaks vary considerably from meteorite to meteorite, as do their shapes and sizes. The size of the low-temperature peak in particular varies enormously, and in some samples is missing altogether. An example of this is the natural glow-curve for Olivenza (LL5), curve B in figure 8.3. For these specimens the irradiation and storage conditions are such that the thermoluminescence from this region of the glow-curve is unstable and has not survived.

The variety of shape, size and position of the low- and high-temperature peaks is a clear indication that there are several overlapping peaks present in the glow-curve. This was demonstrated (McKeever (1980*a,c*)) by using the T_m–T_{STOP} analysis, described earlier in section 3.3.1, to illustrate the number and temperature range of the individual processes. The complex nature of the glow-curve precludes the use of any simple analyses and therefore a computerized curve-fitting procedure was used in the above work to analyse the kinetics of the processes and arrive at values for the trapping parameters. The second-order kinetics which resulted appear to be a characteristic of feldspars, which, as we shall see below, are the dominant luminescent phosphors in these materials. A glow-curve analysed by this method is illustrated in figure 8.4 for the Barwell meteorite. The reliability of the values obtained from a measurement such as this is related to the accuracy with which the experimental data can be collected. Thus, the fit to the experimental data is not unique and the spread of values, each of which gives an 'acceptable' fit, is an indication of the reliability of the calculation. These items are discussed in depth in the above papers and the values quoted in the caption to figure 8.4 are the 'best-fit' values only.

Figure 8.3. Natural glow-curves from ordinary chondrites: A, Barwell; B, Olivenza. Both peaks consist of several superimposed peaks. C is the black-body background signal.

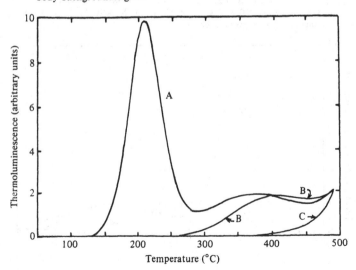

It is also true that the calculation does not take into account models which are more complex than the simple Garlick–Gibson model (cf. chapter 2) on which it is based. For instance, McKeever *et al.* (1983*a*) have shown that thermal quenching is taking place at high temperatures in these specimens. One of the simplest interpretations of thermal quenching is provided by the Schön–Klasens model (see section 2.3.4), which was treated theoretically by Bräunlich & Scharmann (1966) who indicated that under certain circumstances the glow-curve can still be described by equations of the Randall–Wilkins or Garlick–Gibson form. However, it is difficult to arrive at a proper interpretation of the meaning of the trapping parameter values calculated, assuming these simple expressions. It is clear that the kinetics of thermoluminescence production from meteorites are in need of further study, but calculations such as those described above still have value. The main purpose of the analysis is to arrive at an expression which can be used to predict the thermal stability of the meteorite glow peaks at any given temperature. McKeever (1982) has demonstrated that this can be done using the simple Garlick–Gibson model. Within the uncertainty limits of the calculation, the data can be seen to predict the actual decay of the thermoluminescence at a given temperature. Other simplified pro-

Figure 8.4. Individual glow peaks from the Barwell meteorite. The solid line is the computed glow-curve made up of the sum of the individual peaks, shown as dashed lines. The dots are experimental points. The trap depth and second-order frequency factor for each peak are: 1.35 eV, $3.0 \times 10^{13}\,\mathrm{s}^{-1}$ (216 °C); 1.55 eV, $8.0 \times 10^{14}\,\mathrm{s}^{-1}$ (228 °C); 1.20 eV, $2.7 \times 10^{10}\,\mathrm{s}^{-1}$ (261 °C); 1.81 eV, $1.7 \times 10^{15}\,\mathrm{s}^{-1}$ (296 °C); 1.22 eV, $1.2 \times 10^9\,\mathrm{s}^{-1}$ (338 °C); 2.19 eV, $3.0 \times 10^{16}\,\mathrm{s}^{-1}$ (370 °C); 1.48 eV, $9.0 \times 10^9\,\mathrm{s}^{-1}$ (406 °C); 1.7 eV, $6.0 \times 10^{10}\,\mathrm{s}^{-1}$ (425 °C). The peak temperatures are given in brackets.

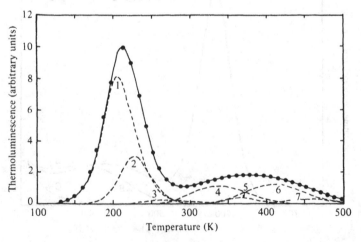

Figure 8.5. Thermoluminescence glow-curves from Wold Cottage showing how the shape of the glow-curve changes when a dose of radiation of 50 krad or less is superimposed on the natural dose, N. When doses of greater than 50 krad are added the intensity of the emission increases rapidly and the natural dose no longer has a significant effect on the shape of the glow-curve. (From Sears & McKeever, 1980.)

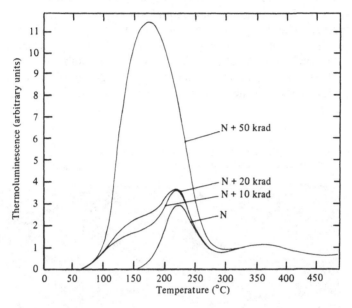

Figure 8.6. Isometric plot of thermoluminescence (TL) versus wavelength versus temperature from an X-irradiated specimen of Allegan.

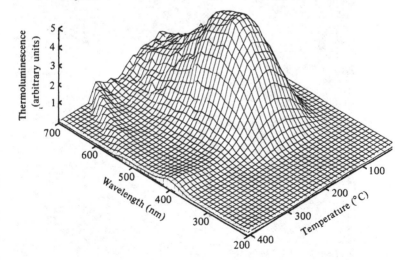

302   *Geological applications*

cedures for calculating the trap depths are unable to predict the thermoluminescence decay with the same level of accuracy (Melcher, 1981; McKeever & Townsend, 1982).

In the natural glow-curves from ordinary chondrites there is no evidence of any thermoluminescence below ~ 150 °C. However, when irradiated in the laboratory, substantial glow is seen at these lower temperatures. The changes in shapes of the glow-curve due to laboratory irradiation are shown in figure 8.5 for Wold Cottage (L6), in which it is clear that artificial irradiation is filling centres not stable enough to retain charge during natural irradiation. The effects of these low-temperature peaks on the growth curve has been discussed (Sears & McKeever 1980).

Figure 8.7. Cathodoluminescence from a chondrule extracted from the Alta'Ameem LL5 chondrite. The luminescent mineral is plagioclase.

Figure 8.6 shows an isometric plot of thermoluminescence versus wavelength versus temperature for artificial thermoluminescence from an ordinary chondrite. The thermoluminescence and radioluminescence spectra of several ordinary chondrites were examined by McKeever *et al.* (1983*a*) who observed blue ($\sim$ 450–500 nm), green ($\sim$ 500–600 nm) and red ($\sim$ 600–700 nm) emission at $\sim$ 170 °C. Cathodoluminescence (i.e., luminescence produced during electron irradiation) has proved useful in identifying the phosphors in these samples and strikingly reveals the distribution of luminescent grains (e.g., figure 8.7). From an analysis of these grains it has become established that the blue-green luminescence is from feldspar, whilst the red luminescence is from chlorapatite.

Several attempts have been made to identify the activator responsible for the luminescence in these materials. Geake, Walker and colleagues (Geake *et al.*, 1971, 1972, 1977; Telfer & Walker, 1978) deduce that Mn^{2+} substituted for Ca^{2+} acts as an emission centre in lunar and terrestrial feldspars, producing luminescence at $\sim$ 550–560 nm (green). A weaker blue luminescence ($\sim$ 450 nm) is thought to be due to a lattice defect. Fe^{3+} in Al sites gives rise to infra-red emission at $\sim$ 780 nm. All these features are discernible in the thermoluminescence spectra.

In terrestrial chlorapatite (both synthetic and natural) Lapraz & Baumer (1981) show that Ce^{3+} centres give rise to emission at $\sim$ 350 nm (not observed in meteorite spectra) whilst Mn^{2+} induces orange emission at $\sim$ 600 nm and blue emission at $\sim$ 400 nm. Sm^{3+} also induces several bands, extending well into the red – i.e., at 620 nm, 650 nm and 700 nm. There is an indication of all of these emissions in the thermoluminescence and radioluminescence spectra. Recent work by Sparks, McKimmey & Sears (1983) has indicated that the major carrier of thermoluminescence in ordinary chondrites is the plagioclase within chondrules (cf. figure 8.7)

8.2.3 *The use of the natural glow-curve*

The most obvious use of the natural thermoluminescence from meteorites is to arrive at an 'age' for the specimens, in accordance with the principles discussed in chapter 7. Sears (1978*b*) lists three events in the history of meteorites which are worth dating. These include: (i) the date at which the meteorite and its parent body formed; (ii) the date at which the meteorite parent body broke up; (iii) the date at which the meteorite became exposed to cosmic rays. Since the discovery of large numbers of meteorites in Antarctica (Cassidy & Rancitelli, 1982) we can add to the above list; (iv) the date at which the meteorite fell to earth. With the exception of this last case (see below) none of these events are amenable to dating by thermoluminescence. As discussed by McKeever & Sears (1979) and illustrated in practice by Valladas & Lalou (1973) using the Saint Severin meteorite, the glow-curve is either in equilibrium or is saturated and so the thermoluminescence will no longer be increasing with time. Event (i), as dated from Rb–Sr measurements, occurred $\sim 4.6 \times 10^9$ years ago; event (ii) has been dated by K–Ar methods to be $\sim 10^8$ years ago; and event (iii) occurred typically 10^7 years ago as determined by the accumulation of cosmogenic radio-isotopes resulting from the interaction between radionuclides already present and cosmic rays. An upper limit of 10^5 to (at most) 10^6 years applies to conventional thermoluminescence dating.

Using the orbital data for meteorites and for 'meteorite-like' meteors, one can arrive at an estimate of the variation in meteorite temperature during an orbit of the sun. Using the thermoluminescence data given in figure 8.4 it is possible to integrate the loss of thermoluminescence over one orbit and compare this with the production rate expected due to cosmic ray irradiation in space. The calculation demonstrates that whilst the high-temperature region of the glow-curve will be in saturation (cf. Saint Severin; Vallades & Lalou, 1973) the low-temperature region will have reached a pseudo-equilibrium where the thermoluminescence lost in one orbit equals that gained in the same period because of cosmic irradiation. The equilibrium level, which is reached in 10^5–10^6 years, is related to the temperature of the meteorite during irradiation. For a single trap, of depth E and second-order frequency factor s', the ratio of the equilibrium trap population to the saturation level (n/N) is given by:

$$n/N = [1 + (s'/Rr)\exp(E/kT)]^{-1} \tag{8.1}$$

where R is the probability per unit dose of filling an empty trap and r is the dose rate. (The derivation of the above equation follows the arguments of Christodoulides, Ettinger & Fremlin (1971) who derive the expression for the first-order case.) Anomalous fading is not a problem (McKeever & Durrani, 1980).

McKeever & Sears (1980) use these principles and discuss in depth the possible causes of the variation in the low-temperature levels of several ordinary chondrites. The differences, typified in figure 8.3, are quite large and can vary over several orders of magnitude. The above authors conclude that the temperature differences of the meteorites whilst in space constitute the most likely cause of this variation in the low temperature thermoluminescence, although albedo and shielding differences may be partially responsible. Because the high temperature experienced at perihelion (closest approach to the sun) is the most effective at draining thermoluminescence, it is possible to argue that the known differences in meteorite perihelia are enough to account for the observed spread in low-temperature thermoluminescence values. It is important to remember that, because of the extreme sensitivity of the calculation to the trapping parameter values, simplified methods of calculating the trap depth are inadequate and a detailed analysis such as that described in the previous section (*cf.* figure 8.4) is required.

When a meteor collides with the earth, the intense frictional heat generated during atmospheric passage is enough to melt the meteoritic material which then ablates and is lost in the atmosphere. The internal portion of the meteorite which survives the passage is covered in a thin (~ 2 mm) 'fusion crust' of blackened, melted meteorite. Obviously the high temperatures which these surface regions experienced will have been enough to drain much of the natural thermoluminescence. One would then expect that the level of the natural glow-curve would increase with depth in a way which reflects the heat penetration during passage through the atmosphere. Thus, Vaz (1971, 1972) examined the natural thermoluminescence profiles of the Lost City and Ucera chondrites and from them calculated the temperature profiles by comparison with data from laboratory annealing studies. Since Vaz's work, several other meteorites have been examined in a similar fashion (e.g., Sears, 1975; Singhvi, Pal & Bhandari, 1982) and the data have been used to test the applicability of ablation rate models for the specimens.

Calculation of the ablation rate is pertinent to an estimation of the pre-atmospheric size of the meteorites. Thermoluminescence can also provide supplementary information to this problem by examining the variation of natural thermoluminescence over greater depths than are affected by the heat of atmospheric passage. The variations over these depths (many centimetres) are believed to be the result of the depth dependence of the cosmic ray dose rate and have led to the use of thermoluminescence to estimate the pre-atmospheric size of the Estacado (Sears, 1975) and Barwell (Bagolia *et al.*, 1978) meteorites. These calculations would benefit from a clearer understanding of how the thermoluminescence varies with the depth below the pre-atmospheric surface. For instance, Vaz & Sears (1977) interpret the depth profiles as being partly due to an increase in the number of trapping centres with cosmic ray dose. Later work (Sears, 1980) rejects this proposal owing to the discovery that the thermoluminescence sensitivity is not affected by pre-irradiation – an increase in sensitivity with pre-dose would be expected from the conclusions of the earlier authors. The anomaly has yet to be cleared up.

The remaining major area in which natural thermoluminescence may be used is in the calculation of the terrestrial age of meteorites of un-known date of fall. As stated briefly earlier, this information has taken on enhanced importance in recent years due to the discovery of many thousands of meteorites in Antarctica (e.g., Cassidy & Rancitelli, 1982). These meteorites have been found in just a few locations and this, together with the large numbers found, indicates that a concentration mechanism is at work. The mechanism favoured involves radial ice flow from the centre of the continent towards the ocean where most of the ice is lost. In some locations, however, the ice becomes trapped by mountain ranges and builds up in large, stagnant ice-sheets, and it is here that the meteorites are found. Inasmuch as the ice in Antarctica may be up to $\sim 10^7$ years old, the possibility exists of finding meteorites which fell to earth a million or more years ago. This is in contrast to meteorites found in other parts of the earth: samples of such long ages have not been detected. Thus, the Antarctic meteorites are unique in that they provide specimens for research into changes throughout the last million years of the number and type of meteorites in earth-crossing orbits and in the cosmic ray flux which produces radiation effects within them. Additionally, they contribute information on the dynamics of ice flow in that the ages are relevant to the determination of the age of the ice, the flow rate and the ice ablation rate. Grouping the individual fragments into samples from the same fall will also be aided by an estimate of the terrestrial age.

Aside from thermoluminescence, the methods which have been used to determine the terrestrial age of meteorites have been based on the in-space accumulation and terrestrial decay of long-lived radionulides (e.g., ^{26}Al, ^{53}Mn, ^{36}Cl, ^{14}C) within the specimens. These nuclides are induced by cosmic rays whilst in space, and once on earth, where the cosmic ray dose rate is reduced by a factor of ~ 100, the nuclides decay at a known rate to a new equilibrium value. Measurement of the amount of each isotope and an estimate of the in-space level enables a calculation of the earth residence time to be made.

Thermoluminescence can, in principle, be used in exactly the same way, such that measurements of the present glow intensity, the in-space level and the decay rate will enable one to estimate the residence time. The principles of the technique were stated by Sears & Mills (1974*b*) and from this early work it was realized that a major difficulty lies in estimating the decay rate for the natural thermoluminescence. Sears & Durrani (1980) attempt to overcome this by arriving at an empirical decay rate estimated from a comparison of the natural thermolumines-cence with measured ^{14}C ages. However, problems here lie in the second-order nature of the decay (non-exponential, changes in peak position and shape) and in the need to calibrate the thermoluminescence data by using the ^{14}C ages of samples from the particular region of interest. Thus, it is not possible to use the ^{14}C measurements from meteorites recovered from the prairie regions of the USA as a calibration for the thermoluminescence decay from meteorites which fell in Antarctica. The vastly different mean temperatures between these regions will ensure an enormous difference in the thermoluminescence decay rates. In view of this McKeever (1982, 1983) used the detailed analysis of the natural glow-curve mentioned in the previous section to estimate typical thermoluminescence decay rates at Antarctic storage temperatures.

The natural glow-curves from ordinary chondrites in Antarctica show a wide range of glow intensities, and glow-curve shapes such as those found from Antarctic specimens can also be obtained from recent falls (see figure 8.3). Therefore it cannot be said that the variations in glow intensities observed in the Antarctic specimens are the result of thermal decay on earth only. They may equally well reflect differences in meteorite orbit, or, to a lesser extent, in shielding and albedo. These relationships cannot be separated. Thus, it would seem that only upper limits to the terrestrial age can be arrived at and this is done by assuming that the thermoluminescence was the same in all meteorites when they fell and any differences now observed are due to thermal decay during their terrestrial sojourn.

With thermoluminescence there is much potential to improve upon this state of affairs because the technique may provide a means of arriving at an absolute date for a meteorite fall. It has been noted that during fall the outer (fusion) crust of the meteorite becomes heated by atmospheric friction and in doing so loses a significant proportion of its natural glow. Thus, this event may be datable using conventional thermoluminescence dating principles (chapter 7). Since fall, the fusion crust, and the inner few millimetres, will have been accumulating thermoluminescence at a rate governed by the natural radiation dose rate. The fusion crust is blackened but sensitive measurements may be able to detect any natural glow which has accumulated since fall, or the inner few millimetres may be used instead.

8.2.4   *The use of the artificial glow-curve*

The possibility of using the thermoluminescence sensitivity to gather information about a meteorite's radiation, thermal and/or metamorphic history was first mooted by Houtermans & Liener (1966) and Liener & Geiss (1968). The sensitivity is defined as the thermoluminescence response per unit dose of artificial radiation, administered in the laboratory, per unit weight of material. Sears & McKeever (1980) examined several ways in which the sensitivity can be determined and concluded that measurements on fresh specimens should not be compared with those on samples that had been previously heated because of a colour change induced by the heating process. The measurement is complicated by the existence of rust on the specimens due to oxidation of the metal grains. The rust covering (the extent of which is variable) lowers the light output per unit dose. To obtain the true sensitivity of the material, the rust should be removed by acid washing, or by preparing the samples as fine grains precipitated on metal discs. The latter method, however, gives irreproducible results and acid washing of bulk powders is probably better. Sears & McKeever (1980) conclude that the thermoluminescence response at the maximum of the low-temperature artificial glow-curve (cf. figure 8.5) following a dose of 50 krad gives a good indication of the sensitivity.

In a study of the metamorphic history of unequilibrated (i.e., type 3)

Figure 8.8. Thermoluminescence sensitivity of several ordinary type 3 chondrites (*a*) from Sears *et al.* (1980) and (*b*) from Sears *et al.* (1982) arranged in petrological order, ranging from 3.0 to 3.9. The open and closed circles refer to fragments believed to be from two different meteorite falls.

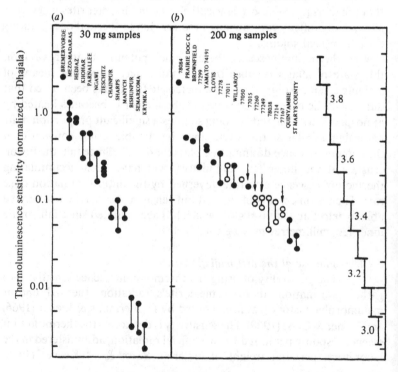

ordinary chondrites, Sears *et al.* (1980) discovered that the thermo-luminescence sensitivity of these specimens is highly dependent on their degree of metamorphism. This was further studied by Sears, Grossman & Melcher (1982) and their data are shown in figure 8.8. On the basis of the thermoluminescence sensitivity, the above authors have been able to sub-categorize the type 3 chondrites into types 3.0 to 3.9. The variation appears to be related to phase changes within the major luminescent phosphor which, as we have seen, is plagioclase feldspar, and may be related to devitrification of feldspathic glass.

The reverse of this situation, namely, the vitrification of feldspar to form a feldspathic glass, is thought to be the cause of the reduction of the sensitivity of equilibrated chondrites (types 5 and 6) which have undergone severe shock and/or reheating (i.e., the so-called 'black chondrites'; Heymann, 1967). Thus, Sears (1978*b*, 1980) believes that the relationship between thermoluminescence sensitivity and K–Ar age first noted by Liener & Geiss (1968) is a result of a shock and/or reheating event lowering both properties, the sensitivity in particular being due to heat- or shock-induced phase changes in the feldspar. In view of the spectral shifts due to shock-induced effects in the luminescence from lunar plagioclase (see next section), McKeever *et al.* (1983*a*) looked for alterations in the emission spectra which may be associated with the phase changes but were unable to identify any major effects. They concluded that trap destruction was not the cause of the sensitivity decrease but that it was more likely to be due to the introduction of non-luminescent pathways for recombination, or, more simply, to a re-duction in the amount of the phosphor (i.e., *crystalline* feldspar) present.

8.3   Lunar material

The gathering of samples of rock and regolith (surface soil) from the moon by the Apollo astronauts in the late 1960s and early 1970s stimulated a great deal of research into their thermoluminescent properties. Once again the aim was to provide information relevant to the 'thermal and radiation histories' of the samples. With hindsight, one can now say that the technique promised more than it was able to deliver owing to the immense complexity of the lunar surface history. Neverthe-less, a large list of publications exists on the thermoluminescence of lunar material and the topic calls for some discussion.

8.3.1   *Mineralogy and ages*

All of the lunar surface rocks are igneous; no sedimentary material has been found. The igneous rocks are of three major types, namely, primary or crystalline, secondary or brecciated, and regolith soil or 'fines'. The predominant igneous rocks of the lunar plains (the 'maria') are basaltic, comprising mainly plagioclase, pyroxene, ilmenite and olvine. The highland rocks are richer in feldspar (anorthosite).

The impact-generated breccias (i.e., secondary basalts formed during the collision of meteorites with the lunar surface) were fashioned by fusing together small particles of many different types of material to form a new solid rock. The breccias are widely dispersed over many areas and range from agglutinates of glass-bonded fragments to larger, consolidated deposits formed during crater-producing impacts. The regolith soil (the 'fines') consists of small (millimetre size) glassy particles and fragments of a large variety of rock and mineral types. The regolith can be several metres thick and is the result of the continuous bombardment of the surface by meteorites over millions of years.

The overall geochemical, geological and chronological evidence combines to indicate that the lunar highlands are older than the maria. Age determination using radionuclide 'clocks' suggests extensive melting, fractionation and formation of the highland crust at a time close to the condensation of the solar system (~ 4600 million years ago) and that crustal differentiation continued for another 200–300 million years. The excavation of the maria basins (by enormous meteorite impacts) was followed by flooding of the basins with basaltic magma. The last major flooding took place ~ 3200 million years ago. For further details of the mineralogy, structure and history of the moon, the reader is referred to Taylor (1975) and French (1977).

8.3.2   *Luminescence*

Most authors who have examined the luminescence emission from lunar rocks and soil agree that the dominant luminescent material is plagioclase. Examples of the cathodoluminescence emission spectra from several lunar plagioclase samples are shown in figure 8.9 where they are compared with the emission from terrestrial plagioclase. The most notable difference is the absence of the infra-red emission (related to Fe by Geake *et al.*, 1977) in the lunar samples. The main emission (green, at ~ 560 nm) is related to Mn^{2+} substituting for Ca^{2+} in the plagioclase in the same manner as for meteoritic feldspar (Geake *et al.*, 1971, 1972, 1977; Telfer & Walker, 1978). Minor amounts of cristobalite and tridymite are also found to be luminescent (Sippel & Spencer, 1970). Shock metamorphism in the breccias induces a small shift of the green peak towards the red, accompanied by an overall reduction in intensity (figure 8.9; also Sippel, 1971). The formation of maskelynite is evident during metamorphism and its luminescence spectrum is also shown in this figure.

Figure 8.9. Uncorrected cathodoluminescence spectra for: (*a*) terrestrial plagioclase; (*b*) lunar crystalline rock plagioclase; (*c*) plagioclase grain from lunar breccia; (*d*) lunar maskelynite. From Sippel & Spencer (1970).

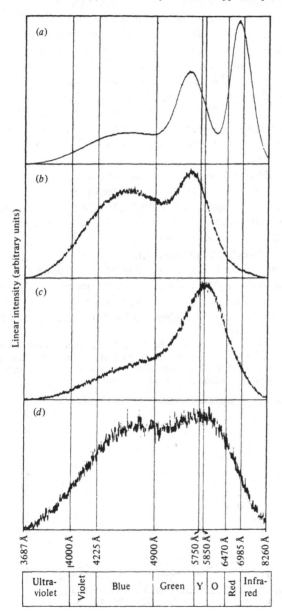

8.3.3   *Thermoluminescence*

Armed with the above background knowledge concerning the moon's surface history (information not available when the first Apollo samples were returned) and with the brief description of the luminescence emission, it is now possible to examine what can be learned from the thermoluminescence properties. The initial impetus for studying luminescence in general in lunar material stemmed from the observation of 'anomalous brightness' from the lunar surface, namely, that the dark moon as seen during eclipses was 'too bright' – see references in Greenman & Gross (1970). A possible cause was the light from luminescence but early measurements on the samples soon revealed that the luminescence intensity was far too low (e.g., Edgington & Blair, 1970). This argument applied equally well to transient lunar phenomena, the cause for which could not be thermoluminescence (Sun & Gonzales, 1966; Sidran, 1968). Nevertheless, lunar rocks and soil do emit considerable thermoluminescence when heated and many papers on the topic can be found among the proceedings of the Lunar Science conferences held each year in Houston (USA).

A set of glow-curves from unsorted lunar fines is shown in figure 8.10 from which it can be seen that the natural glow-curve shows little or no luminescence below $\sim 250\,°C$ owing to thermal drainage during irradiation. As with meteorite thermoluminescence the glow-curve is composed of several individual glow peaks. Most glow-curves from

Figure 8.10. Glow-curves from Apollo 12 lunar fines (grain diameter $\leq 106\,\mu m$; heating rate $= 5\,°Cs^{-1}$). Viewed through an Ilford broad-band blue filter (622). From Durrani *et al.* (1972). A, Natural, B, Natural $+ 35$ krad, C, natural $+ 470$ krad; D, natural $+ 960$ krad; E, natural $+ 2700$ krad.

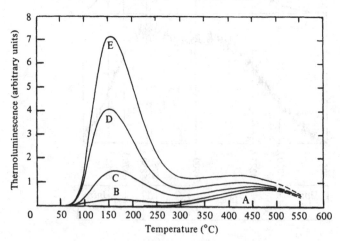

lunar material broadly resemble those of figure 8.10 although there are differences in detail. There have been very few measurements of the emission spectra of the thermoluminescence from lunar materials; certainly none as detailed as those from meteorites. Some examples were given by Durrani *et al.* (1973) and by Lalou *et al.* (1972) from which it appears that the main emission is at ~ 420–440 nm with another major emission peak near 560 nm. These emissions clearly compare closely with the luminescence seen in figure 8.9, but it is to be noted that whereas the green emission dominates in the luminescence measurements, the blue peak dominates in the thermoluminescence curves.

It becomes natural to ask, what is the main phosphor contributing to the glow? Because of the presence of a strong green emission, plagioclase is clearly an important mineral and this view was adopted by most of the early workers (e.g., Dalrymple & Doell, 1970). However, detailed mineral separation work revealed that the major proportion of the thermoluminescence was emitted by an Al–Si–K phase (with a little by a phosphate phase) and only a minor part by the plagioclase. Presumably the Al–Si–K phosphor (a potassium feldspar) is responsible for the blue luminescence (Walker, Zimmerman & Zimmerman, 1971).

There have only been some minor attempts, using simplified approaches, to evaluate the trapping parameters associated with the production of the thermoluminescence. As with most feldspars, the kinetics are non-first-order, probably second-order but this has never been confirmed. Initial rise methods (section 3.3) have been employed but a detailed examination is lacking. This is surprising considering that many of the uses to which thermoluminescence is put in the study of lunar material involve a knowledge of its thermal stability. It is unfortunate that more confidence cannot be placed upon the trap depth and frequency factor values (see chapters 2 and 3 for definitions) which are often quoted in this context.

The subject of stability leads on to the problem of anomalous fading. Garlick & Robinson (1972) presented clear evidence of athermal fading of thermoluminescence in several lunar materials. It is also to be recalled that anomalous fading is observed in many terrestrial feldspars (Wintle, 1977) but evidently not to any great extent in meteoritic plagioclase (McKeever & Durrani, 1980). The lack of knowledge concerning the kinetics and the trapping parameters, along with the existence of anomalous fading, casts severe doubt on the validity of surface temperature determinations using thermoluminescence (e.g., Durrani, Khazal & Ali, 1977c).

In consideration of the drainage of thermoluminescence especially at the lower glow-curve temperatures, several authors have expressed the

opinion that the natural thermoluminescence is likely to be in equilibrium (cf. equation (8.1); e.g., Dalrymple & Doell, 1970; Walker *et al.*,
1971). Hoyt *et al.* (1971*b*) used this principle in a study of the
thermoluminescence versus depth (beneath the lunar surface) for fines.
The natural thermoluminescence (suitably normalized to account for
sensitivity variations from sample to sample) is seen at first to increase
with depth over the first ~ 10 cm, owing to the attenuation of the diurnal
heatwave over this distance. Beyond this, the glow intensity decreases
with depth in a manner which is interpreted as being due to the outward
heatflow from the moon's interior. Hoyt *et al.* (1971*b*) estimate a
temperature gradient of (2 ± 2)K m^{-1} beyond ~ 20 cm within the
regolith but this figure must be viewed in the context of the above
discussion on kinetics and anomalous fading. An example of a
thermoluminescence versus depth profile is shown in figure 8.11.

Several authors have noted anomalous glow intensities at certain
depths which cannot be explained using the thermal gradients mentioned above (Walker *et al.*, 1971; Walker & Zimmerman, 1972;
Durrani, Bull & McKeever, 1978, 1979). These anomalies give rise to
unexpected structure in the depth profiles as can be seen in figure 8.11 at

Figure 8.11. Depth dependence of normalized natural thermoluminescence at
175 °C (crosses) and 250 °C (circles) for Apollo 12 regolith material (after Walker
et al., 1971).

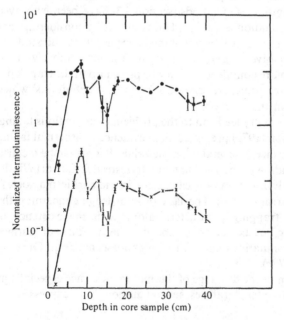

depths of ~ 15 cm and ~ 40 cm and appear to be related to variations in the internal radioactivity in those (relatively few) grains that contribute most of the glow.

Durrani et al. (1978, 1979) noted a supralinear growth of thermoluminescence from Luna 24 fines and a related increase in the sensitivity of the samples following irradiation. Earlier studies (e.g., Hoyt et al., 1971b, 1973; Walker & Zimmerman, 1972; Crozaz & Plachy, 1976) do not suggest any obvious correlations between sensitivity and radiation damage and the mechanism of sensitization remains elusive (see chapter 4).

Measurements of the glow versus depth on whole rock samples are also of interest. It is expected that a whole rock has been at the same temperature throughout its volume such that variations in the thermoluminescence with depth will reflect differences in the ionization rate rather than thermal gradients. The ionization rate from solar flares decreases with depth and initial measurements showed that the thermoluminescence also decreased with depth in the manner expected. Unfortunately, the possibility of using this to determine the solar flare spectrum (over the last 10^4 years) has never been fully exploited (Hoyt, Walker & Zimmerman, 1973; Zinner, 1980) and thermoluminescence seems less effective in this regard than solar flare particle track studies.

8.4 Terrestrial geology

The use of thermoluminescence in terrestrial geology has long been a popular area of study. The pre-1968 applications are detailed in a comprehensive book edited by McDougall (1968a), and Marfunin (1979) gives a useful treatment of the luminescence properties of several minerals. More recently, a report by Sankaran et al. (1982) gives a thorough compilation of the most recent literature. The most obvious application is in geological dating, the main principles of which were discussed in the previous chapter and no more will be added here. However, attention has been given to the use of thermoluminescence in shock detection, geothermometry, prospecting and stratigraphy and each of these topics will be discussed. Other applications will only be noted briefly, and a thorough description of the thermoluminescence properties of the main mineral phosphors is beyond the scope of the chapter. The reader is referred to Nishita, Hamilton & Haug (1974) and Sankaran et al. (1982) for information on this latter topic.

8.4.1 *Shock detection*

The evident relationship between the increase in the levels of shock experienced by meteoritic plagioclase and the reduction in thermoluminescence sensitivity as noted by Sears (1980) is a recent example of the use of thermoluminescence to study stress and strain effects in geological materials. A review of the topic is provided by Stöffler (1974) and it is evident that pressure effects in minerals can, under different circumstances, give rise to contrasting thermoluminescence properties. In general, the shock effects cause modifications to the minerals' physical, optical, electronic and structural characteristics. In certain cases of severe shock (e.g., lateral faulting, meteorite impact or nuclear detonation) vaporization, melting or structural transformation can take place (Short, 1966). The broad aim of the thermoluminescence work in this area has been to assist in the development of criteria for recognizing shock metamorphism. However, the situation is hampered by factors other than shock which can produce similar thermolumines-cence results. A coherent overview of the topic has yet to emerge.

The uncertainty as to what will happen, when and why is in part due to the wide range of materials studied in this regard, in addition to the broad range of stresses and strains utilized and the conditions under which they are applied. For example, the experiments have dealt with relatively gentle, laboratory-induced deformation, and cataclysmic, natural and manmade shock events. Measurements have been con-cerned with both the natural glow-curves and the thermoluminescence

Figure 8.12. Variation in (A) natural thermoluminescence and (B) thermolumin-escence sensitivity from rocks near the Charlevoix meteorite crater (adapted from Douglas *et al.*, 1970).

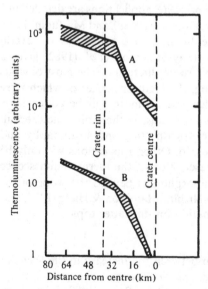

sensitivity; glow peaks both above and below room temperature have been examined. Despite these difficulties (or possibly because of them!) some general features need to be mentioned.

Douglas, Morency & McDougall (1970) summarize the effects of strain on the thermoluminescence properties of materials by suggesting that the *initial* effect is to increase the thermoluminescence emission while greater strain usually results in a decrease. By and large, this summary would appear to hold true and several examples of an increase in thermoluminescence sensitivity for small strains (and low strain-rates) can be found in the literature, although not all are concerned with geological specimens (e.g., Ausin & Alvarez Rivas, 1972). Other authors (e.g., Roach *et al.*, 1962; Manconi & McDougall, 1970; Short 1970; Douglas & McDougall, 1973) quote results consistent with the proposition of high strain decreasing the sensitivity. For example, rocks from the vicinity of the Charlevoix (Canada) meteorite crater exhibit significantly less natural glow the closer to the centre of the crater they

Figure 8.13. Intensity of the low-temperature thermoluminescence (following laboratory irradiation) of halite extracted from the vicinity of the Salmon nuclear explosion (Mississippi, USA) before (pre-shot) and after (post-shot) the detonation (830 m underground) (after Roach, 1968).

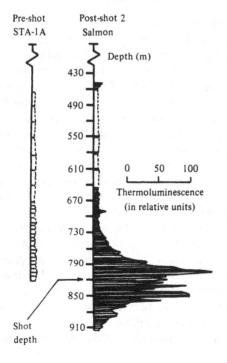

are extracted (Douglas *et al.*, 1970; see figure 8.12). Many other examples are referred to by Douglas *et al.* (1970). In addition to intensity changes, Manconi & McDougall (1970) report systematic activation energy decreases with increasing levels of shock in natural quartz.

The above picture is by no means universally observed. For example, some alkali halide crystals exhibit a *decreasing* glow sensitivity even for low deformation levels (several examples of which were discussed briefly in section 5.2.6). Roach (1968) observed a large *increase* in the thermoluminescence sensitivity of halite (NaCl) crystals, for glow peaks below room temperature, following large shock levels delivered both in the laboratory and by nuclear weapons testing. An example is shown in figure 8.13 for halite crystals adjacent to the epicentre of the 5 kton Salmon nuclear explosion. Douglas *et al.* (1970) attribute the decrease in glow intensity, whilst Roach (1968) attributes the increase in this parameter to plastic deformation. Clearly there are contrasting effects in operation and until the precise mechanisms are understood, the use of thermoluminescence to examine shock-induced metamorphism remains an uncertain area of study. A considerable increase in basic research into the fundamental thermoluminescence properties of the materials is required.

8.4.2   *Geo- and palaeothermometry*

Since the 1950s, thermoluminescence has been recognized as an investigative tool of high potential in the elucidation of the absolute and relative temperatures of formation of minerals. The technique has been used in this regard in two separate ways. One is to utilize the natural thermoluminescence of specimens in a manner which is similar to that described for meteorites and for lunar materials in previous sections – namely, to assume equilibrium and to extract information on the temperature history by examining variations in the equilibrium thermoluminescence levels. For example, Johnson (1963, 1966) examined the natural glow of rock metamorphosed by contact with volcanic magma. By monitoring the decrease in natural glow intensity with proximity to the extrusion (or intrusion), and by applying theories of heat conduction, the temperatures reached during thermal metamorphism were deduced. Similar variations (due to heat) in the natural emission from rocks near hydrothermal ores were regarded as a possible means of ore prospecting although in general the results were discouraging (MacDiarmid, 1968). More recently, the thermoluminescence from anhydrites has been used in a similar fashion to derive the palaeotemperatures of oil and gas basins (Mitin, Kononov & Pashin, 1978).

The difficulties with the interpretation of the natural thermoluminescence used in this way are that a very detailed analysis of the glow-curve to obtain the thermal stability parameters is a prerequisite, and, as stressed by McDougal (1968*b*), several other processes associated with the igneous rock or hydrothermal ore body can produce similar effects to those caused by temperature variations (e.g., impurity addition or removal, recrystallization). Indeed, the glow from rocks adjacent to hydrothermal ores has been observed to increase with proximity to the ore, or decrease, or remain constant. Establishing a single, unique cause for these effects is probably a fruitless task. As a result, the second approach, alluded to above, has been developed. This has involved the use of thermoluminescence sensitivity, following artificial irradiation. The technique has been used with apparent success to establish relative formation temperatures for fluorite (McDougall, 1970), calcite (Nambi & Mitra, 1978; Sergeyev & Ventslovayte, 1978) and quartz (David & Sunta, 1981*b*). The premise is that at high formation temperatures a greater number of defects may be incorporated into the crystal lattice than at lower temperatures. The thermoluminescence is expected to be related to the defect concentration, and so the greater the glow sensitivity, the higher the formation temperature. McDougall (1970) speculates that Frenkel defects are the defects in question in fluorite, while Sergeyev & Ventslovayte (1978) prefer Schottky defects in calcite. Neither is likely to be true. Whilst these defects are indeed more numerous at elevated temperatures they are thermodynamically unstable at low temperatures and the slow cooling conditions which will inevitably be experienced by the rocks should ensure that the non-equilibrium concentrations needed at room temperature (to explain the thermoluminescence results) will not exist. It is more likely that the sensitivity is determined by the presence of impurities. For example, incorporation of impurities can be temperature dependent and this argument is favoured by David & Sunta (1981 *b*) for natural quartz. The crystallization temperature of this mineral governs the amount of Al incorporated into the SiO_2 lattice and, as discussed in chapter 5, this impurity is intimately related to the thermoluminescence process. Alternatively, any intrinsic defects formed at high temperatures may form complexes with the impurities and thus become stabilized. Such effects have been noted recently in ion-implanted fluorite (Bangert *et al.*, 1982*b*). In general the thermoluminescence from calcite and fluorite is dependent upon impurities and either or a combination of both of the above explanations may be true.

Levy and colleagues (Pasternack, Gaines & Levy, 1977; Levy, 1979) examined another aspect of the thermoluminescence properties of natural minerals which may feasibly be used in palaeotemperature estimation – namely, the emission spectrum. In a detailed examination of the spectra from albite Levy and colleagues noted that: (i) the ratio of the 3 eV to '2.2 eV' emission bands in the artificial glow near 160 °C increased notably with increasing pre-annealing temperature; (ii) the '2.2 eV' band shifted slightly from 2.22 eV to 2.18 eV with heating time at 1050 °C. They thus suggest that emission spectra are sensitive to Al–Si disorder in this mineral. This conclusion is not supported by the recent work of Speit & Lehmann (1982*a,b*) who, from an examination of electron paramagnetic resonance and thermoluminescence of a wide variety of feldspars of different genesis, conclude that electron paramagnetic resonance and glow-curve properties (including emission spectra) are independent of Al–Si disorder. This disagreement remains to be resolved.

8.4.3   *Prospecting*
         The use of natural thermoluminescence to detect the location of hydrothermal ore deposits has been mentioned in the previous section. Whilst it appears that the glow may indeed vary near the ore body, this variation does not appear to be systematic and the use of the natural thermoluminescence in this fashion has never been developed into a useful prospecting tool. The problem is that several processes may be occurring, including thermal drainage of traps (from the heat of mineralization), recrystallization and the addition or leaching of impurities; all of these will affect the thermoluminescence (McDougall, 1970). Levy (1979) has used the artificial glow from limestone and dolomite adjacent to lead–zinc deposits to establish that there is potential in using the thermoluminescence sensitivity from these materials to prospect for these ores. As yet, this potential has remained undeveloped.
         Prospecting for *radioactive* ores using thermoluminescence is much more promising. At present, the common survey methods used to identify sites for uranium prospecting include the use of Geiger–Müller tubes, scintillation counters and radon surveys. Nielsen & Bøtter-Jensen (1973) used thermoluminescence dosemeters to evaluate the radiation levels from Greenland rocks by arranging for the geologists to carry TLD badges (cf. chapter 6) whilst in the field during mapping surveys over a period of two months. The dose rates found by this method agreed favourably with those estimated from the radioactive content of the soils, but a drawback is that the measurement cannot pinpoint the sources, but simply give an indication of the area in which the deposit may be located. Vaz & Sifontes (1978) used a similar technique and implanted TLDs in the field for a period of several months in order to survey

thorium deposits in Venezuela. Both of these examples, however, are simply applications of environmental thermoluminescence dosimetry (section 6.5) about which there has already been detailed discussion in this book.

An alternative to using TLDs is to use the natural thermoluminescence from the rocks themselves. This was investigated by Nambi, Bapat & David (1978). These authors studied the thermoluminescence from monazite deposits in India but found that although a relationship between radioactive content and natural glow was established, the technique is significantly less sensitive than conventional radiometric survey methods (e.g., radon detection).

8.4.4   *Miscellaneous applications*

Among the suggested uses for thermoluminescence in geology the study of strata was one which held a great deal of promise. Daniels (1968) summarizes many of the early studies from which it emerged that the glow-curve from sedimentary deposits could be used as a 'fingerprint' of the stratigraphical layer to which that deposit belonged. Thermoluminescence could be used to supplement information from petrographic and chemical studies and from fossil evidence, and in fact in many cases proved to be a much more sensitive tool than any of these in distinguishing the stratigraphical sequence. As a reflection of its obvious potential in this application, thermoluminescence has been used in a similar fashion in several recent studies on stratigraphy. For example, De, Rao & Kaul (1980) use the glow-curve character of deep-sea carbonate oozes in the study of deposits in the Arabian Sea. These authors use glow-curve features noted earlier by Johnson & Blanchard (1967) in an examination of the stratigraphy of fossil-bearing limestones. Similarly, the complex layering of the lava flows in India was deciphered by Bapat, Sankaran & Sunta (1980) on the basis of the glow-curve properties.

In some respects examinations of this sort are similar to the use of thermoluminescence in source identification. An example of this was given by Carriveau & Nievens (1979) for obsidian. Nevertheless, despite these apparent successes care must be taken with examinations of this sort, especially stratigraphy. It is quite possible, even likely, that the differences observed in the thermoluminescence glow-curve may also result from variations in radioactivity, shock effects (e.g., during subsequent faulting), deposition rate, etc. Such difficulties led Bergstrom (1956) to doubt the validity of using thermoluminescence in this manner. He showed that in some cases Pennsylvanian limestones showed similar glow-curve characteristics despite the fact that they were from different strata while deposits from the same layers gave differing properties. Similar difficulties have been noted by other workers.

8.5   Concluding remarks

A recurring theme in all applications of thermoluminescence to geological problems is that the observed properties of both the natural and the artificial glow-curves are dependent on several factors (mineralogy, metamorphism, impurity content, thermal history, radiation history, etc.). Any study which attempts to relate thermoluminescence to just one of the factors must be able to eliminate the effects of the remainder. In many, if not most, instances such elimination is not possible, casting a shadow of doubt over the conclusions arrived at – although in some cases such elimination can be achieved, at least partially. An example of the latter is in the thermoluminescence of meteorites, where a clearcut relationship between thermoluminescence sensitivity and petrologic type for non-equilibrated meteorites is difficult to deny, if not more difficult to explain.

The highpoint in the popularity of thermoluminescence in its application to geology appears to have been in the late 1950s and throughout the 1960s. The literature of this period abounds with papers which open by discussing the 'potential' of the technique, but then give warnings of the difficulties involved and generally conclude that more work must be done. More often than not the work in question involves fundamental studies of the thermoluminescence properties of the materials under scrutiny. An examination of chapter 5 of this book will reveal that even if the requisite detailed measurements are taken, interpretation of the data is not easy. With natural geological specimens it is exceptionally difficult.

The picture is not all gloomy. The relationship between meteorite petrology and glow sensitivity has already been noted. Add to this the detailed measurements of the fundamental solid-state properties of feldspars, particularly their thermoluminescence properties (e.g., kinetics and emission spectra; Levy, 1979; McKeever, 1980c; McKeever et al., 1983a; radiation damage effects; Speit & Lehmann, 1982a,b) and one has the possibility that an understanding of the observed effects may not be far away. Similar studies on other common minerals (e.g., quartz, fluorite) have been underway for many years and should be continued. Progress in the application of thermoluminescence in geology will only be made via this type of study, otherwise papers will continue to talk about 'potential' and not about 'exploitation'.

9   Instrumentation

9.1   Introduction

Modern thermoluminescence recording equipment can vary from the very simple to the extremely sophisticated. The nucleus of all the various designs is a light detection system, a sample heater and a temperature control unit but the designs of each of these components are many and varied. Enhanced sophistication in design is introduced if glow-curves below room temperature are required, if there is a need to record emission spectra at different glow-curve temperatures, or if simultaneous thermally stimulated current measurements are to be recorded. Extensive use of computer-controlled apparatus is becoming more and more popular and many research groups develop their own

Figure 9.1. Block diagram of the experimental arrangement for recording thermoluminescence, including emission spectra (Th/c = thermocouple).

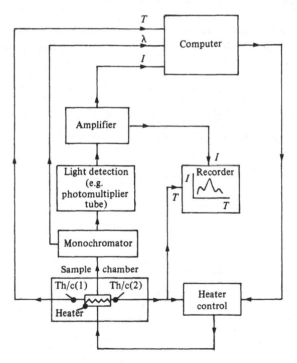

computer-based operating system. With the large numbers of micro-computers commercially available, an experimenter is faced with a wide choice of options. A typical schematic arrangement is shown in figure 9.1.

In the sections that follow, a general description of some of the necessary components for thermoluminescence detection will be given (although the exact details of the apparatus design will depend on the experimenter's individual requirements). In Appendix B, a list is given of the addresses of suppliers of commercial thermoluminescence equipment. Reference to published papers will be limited to those wherein the technique or apparatus described is particularly useful or novel, although the reader may wish to refer to two articles which deal with instrumentation for thermoluminescence in some depth – these are by Manche (1979) and by Julius (1981).

9.2   Cryostat design

9.2.1   *High temperature (> room temperature)*
The sample chamber needed for the recording of glow-curves between room temperature and, say, 600 °C has modest design requirements. The chamber houses the heater (the possible designs for which will be discussed in section 9.3) and needs to be capable of being

Figure 9.2. Typical design (schematic) for a sample chamber for recording thermoluminescence above room temperature. The reflecting cone is to reduce the signal due to radiation from the heater and aid in light collection. Lenses or light pipes may be used instead (Th/c = thermocouple.)

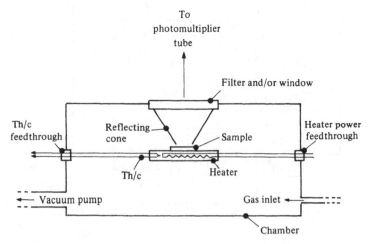

evacuated to a moderate vacuum (typically 10^{-1} to 10^{-3} torr) and/or of being flushed with inert gas. The use of a needle valve to gain fine control over the gas flow is desirable. (It is worth noting at this point that gas inlet pipes made of metal are preferable to long lengths of nylon or rubber tubing owing to oxygen outgassing from the pipes.) Furthermore, oxygen and water vapour traps are essential in the inlet line to filter these unwanted components from the inlet gas in order to reduce spurious signal levels. The gas itself may be a high-purity nitrogen or argon, or a reducing atmosphere of nitrogen/hydrogen (95:5) may be helpful.

Electrical feedthroughs for heater power and for thermocouples are required. (The thermocouple feedthrough should not be sited too close to the heater in case the junction between the feedthrough and the thermocouple becomes warmer inside the chamber than outside, with a resultant inaccuracy in temperature measurement.)

The vacuum system requires little discussion and need only consist of a simple mechanical (rotary) pump capable of reaching pressures of (at minimum) 10^{-3} torr.

The final consideration for the chamber design concerns the positioning of the light detector (usually a photomultiplier tube) with respect to the sample. Most designs have a horizontally mounted heater and a vertically mounted photomultiplier tube, although several designs exist where the opposite arrangement is used. The photomultiplier tube must be close enough to the sample for good light collection capabilities, but not so close as to damage the tube by heat from the heater. Normally, light collection devices are used to restrict the light path and aid in the elimination of large signals due to black-body radiation from the strip (e.g., reflecting cones, lenses, filters, fibre optics, light guides). Furthermore, it is often useful to be able to move the photomultiplier tube towards and away from the chamber without disturbing the vacuum and/or inert gas atmosphere. Filters and chamber windows of various types are used to seal the vacuum in order to achieve this. An example of a popular infra-red filter is a Chance Pilkington HA3, whilst broad-band blue/violet filters may also be used to eliminate further the black-body radiation. Fused silica (e.g., Suprasil) with its good transparency in the visible region is a useful material for chamber windows. A schematic diagram of a typical vacuum chamber for recording glow-curves above room temperature is shown in figure 9.2.

9.2.2 *Low temperature ($<$ room temperature)*

Low-temperature cryostats include the same basic considerations as those listed above but they incorporate some additional design features. Firstly, a higher vacuum is required ($\sim 10^{-5}$ torr) in order to prevent frosting of the chamber windows and of the sample. Suitable vacuum systems include a standard rotary and diffusion pump arrangement but care must be taken not to contaminate the chamber with diffusion pump fluid. Thus, a baffle is essential and cold traps highly desirable. In extreme cases, a mercury vapour pump or a molecular sieve pump may be needed.

The cryostat needs to have a rotating tail section so that the sample can be irradiated through one window (e.g., a beryllium or aluminium window) and subsequently viewed through another. Irradiating and viewing through the same window is not recommended because the window will inevitably phosphoresce and this will be detected by the photomultiplier tube.

9.3   Heater design and temperature control

9.3.1   *Heater design*

Perhaps the most critical part of a thermoluminescence cryostat is the heater. Several methods exist by which the temperature of the sample can be raised in a controlled fashion, and the most common of these is by passing current through a planchet or a coil (i.e., resistive heating). The use of a planchet is probably the most popular. The planchet itself is a thin metal (e.g., tantalum or nickel-chrome) strip, the typical dimensions of which may be, say, 0.025 cm $\times (1-2)$ cm $\times (4-5)$ cm. The temperature can be monitored by using a thin thermocouple (e.g., chromel-alumel, iron-constantan, copper-constantan) welded to the underside directly beneath the position of the sample. (Alternatively the thermocouple may be clamped in position but this often results in poorer thermal contact and is generally less satisfactory.) The advantage of the planchet system over other arrangements is that the heater has a low thermal mass and so fine control of the temperature is easily achieved. The low thermal inertia also enables reasonably high heating rates (tens of degrees/s) to be used and also facilitates the rapid cooling of the heater following the glow-curve readout. A disadvantage is that the planchet can often buckle and warp at high temperatures, sometimes resulting in permanent distortion. In extreme cases this can result in an alteration of the sample–photomultiplier tube distance, thereby affecting the intensity of the measured signal. Several steps can be taken to overcome this problem, including flexible clamps to allow movement of the strip during thermal expansion, ridged sides to prevent bending and a central clamp, made out of a thermal insulating material, to prevent lateral movement.

An alternative to the planchet arrangement is to have a heater block, usually of copper, which itself is heated by a coil of resistance wire electrically insulated from the block. This can be done, for example, by clamping a soldering iron heater element to the block or by wrapping a coil of wire around it. For this latter arrangement, heater wire such as Thermocoax is particularly useful. The heater block design is especially useful for low-temperature cryostats wherein the block can be maintained at low temperatures by using liquid nitrogen or liquid helium. Because of the large thermal inertia of such devices fast heating and cooling rates are not possible. In some designs, liquid coolant is piped through the block in order to aid in the cooling process when used above room temperature. The thermocouple is normally clamped in position or inserted into a suitably positioned hole near the location of the sample. This latter method enables the designer to use sheathed thermocouples which are more convenient than the loose wire versions.

The third method of using resistance heating to raise the temperature is in the use of bulb dosemeters. A typical bulb dosemeter was shown in figure 6.18 in position on the glow-curve readout unit. The dosemeter is sealed in a glass bulb and attached to a small heating filament, the electrical contacts for which extend outside the bulb. This type of heater system is used only by commercial manufacturers of thermoluminescence dosemeters and readout units.

For some applications in dosimetry, optical (infra-red) heating is proving to be useful. A commercially available thermoluminescence dosimetry readout unit (National Panasonic UD-710) uses an infra-red

Figure 9.3. Laser-heating apparatus as used by Gasiot *et al.* (1982).

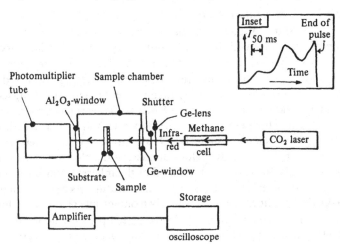

source to heat the sample to the required temperatures (several hundred degrees centigrade) in less than 1 s. Clearly such high heating rates are not suitable for simple kinetic analysis but in TLD applications this is of minor significance. The high heating rates ensure a large thermoluminescence signal and thereby allow for the detection of very small doses. This principle has recently been extended to incorporate a novel laser-heating technique (Bräunlich *et al.*, 1981; Gasiot, Bräunlich & Fillard, 1982). A schematic representation of the laser heating system is shown in figure 9.3. An 8 watt cw CO_2 laser is used to heat thin layers of TLD powders precipitated on to a glass substrate. In practice, time-averaged heating rates of up to $10^4\,\mathrm{K s}^{-1}$ can be obtained. The inset shows a typical laser heated glow-curve from TLD-100 following a dose of 1.7 R. Although the high heating rates produce intense emission for low-dose samples, they have the disadvantage of shifting the emission into a temperature region where background black-body radiation may be a problem, in accordance with the discussion of section 3.3.3. The use of infra-red filters would therefore seem necessary. Nevertheless, the technique appears to have high potential in the area of low-dose, thin film dosimetry.

Other methods of sample heating have been experimented with, including hot gas and R.F. heating. The hot gas method in particular has the advantage of reducing the temperature inhomogeneities that are almost inevitable with the other, contact-heating methods. Temperature gradients across the sample caused by heating it from one side only are discussed in the next section, but by immersing the specimen in the heating fluid (gas) these may largely be overcome.

9.3.2   Temperature control

Many applications of thermoluminescence require a linear temperature rise with respect to time. Without perfect linearity, glow-curve analysis, discussed thoroughly in chapter 3, is impossible. (An exception to this is 'reciprocal-temperature' heating, discussed briefly below.) In general, temperature management is achieved using feedback control. The control signal is the EMF of the thermocouple which, for a linear rise in temperature, is compared with a linearly increasing ramp voltage. The difference signal is then used to power the heater, usually via a thyristor drive circuit. Before discussing the electronics necessary to achieve this, the reader is reminded of the inherent non-linearity of the thermocouple EMF with respect to temperature. As can be seen in figure 9.4*a*, the non-linearity is particularly troublesome below room temperature. To overcome this, most electronic temperature controllers utilize EMF linearizing circuits, with varying degrees of success. Modern microprocessor-based controllers in particular are able to accommodate the non-linearity very well using multiple-slope approximations to the thermocouple response curve. However, recently an extremely simple method of overcoming the non-linearity has been devised by Chandler & Lilley (1981) who use a 'three-wire' sensor (figure 9.4*b*) consisting of an iron–constantan thermocouple combined with an 885:115 potentially divided copper–constantan thermocouple. The output is seen to be linear over a wide range of temperatures (figure 9.4*a*). One final reminder about thermocouples in general: it is essential to use compensating leads of the correct material if extension wires are to be used in order to prevent the formation of extra EMFs at the junctions with the extensions.

Assuming that a linear EMF–temperature response can be obtained, the electronic circuitry for the control of the temperature is quite straightforward. Several research groups have designed and built their own circuits and these have been published from time to time. Recent examples are given by Mills, Sears & Hearsey (1976) and Loggini & Sandrolini (1980). Several commercial units are available and the most sophisticated of these utilize the so-called 'three-term' or 'proportional-integral-derivative' control, along with either 'phase-angle' or 'zero-

Figure 9.4. (*a*) Thermoluminescence EMFs from − 100 °C to 0 °C compared with linear characteristics (full lines): A, Cu/Con; B, Fe/Con; C, Fe/Cu; D, combination (cf. *b*) (*b*) Circuit for combining Fe/Con and Cu/Con thermo-couples using common metal for sensor junctions. (After Chandler & Lilley, 1981.)

(*a*)
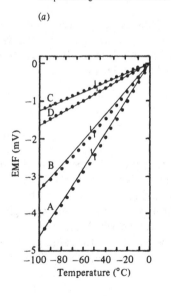
(*b*)
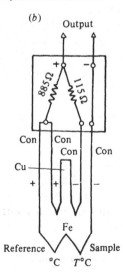

switch' firing of a silicon controlled rectifier. In the proportional mode, the power delivered to the load is proportional to the error between the desired temperature and the actual temperature. The derivative action senses the rate at which the temperature is changing and modifies the control accordingly. The band of temperatures over which the proportional control occurs is centred on the point where the heat loss from the load equals the heat input. As ambient conditions alter the proportional band will move up or down the temperature range to compensate and thus the actual temperature may drift away from the desired temperature. To correct for this, integral control is introduced: this ensures that the proportional band is always centred on the desired temperature. Three-term proportional-integral-derivative control is the most sensitive and accurate of the temperature control methods and is recommended when kinetic measurements are being attempted.

The choice between phase-angle or zero-switch firing of the silicon controlled rectifiers depends upon the type of heater being used. Phase-angle firing (i.e., power on for only a part of each half cycle) is recommended for resistive loads where the resistance is highly temperature dependent (e.g., tungsten, molybedenum). Zero-switch firing (i.e., power on and off only when the cycle crosses zero) is recommended for materials with a small temperature coefficient of resistivity (e.g., nichrome, kanthal). It is also recommended if thermally stimulated conductivity measurements are being made because the heater current switching does not then interfere with the thermally stimulated conductivity signals.

An exception to the requirement of a linear heating programme for kinetic analysis is with hyperbolic, or 'reciprocal temperature' analysis. As discussed by Stammers (1979) there are some minor advantages to be gained from using a heating scheme which varies as:

$$1/T = 1/T_0 - \beta't, \tag{9.1}$$

instead of the linear variation, namely:

$$T = T_0 + \beta t. \tag{9.2}$$

In these equations T is temperature; T_0 is initial temperature; β is the linear heating rate (in degrees/unit time), whereas β' is the hyperbolic heating rate (in/degrees time); t is time. Using equation (9.1) rather than (9.2) allows straightforward integration of the rate equations, given earlier in chapter 3 (e.g., see Kelly & Laubitz, 1966). Stammers (1979) describes two simple circuits for incorporation into the ramp generator circuit of conventional controllers in order to generate the $1/T \propto t$ law of equation (9.1).

A major problem with accurate sample temperature control is that the thermocouple is actually attached to the heater and thus the temperature being monitored is that of the heater strip or block and not that of the sample. As the sample is being heated from one side only it is inevitable that thermal gradients will exist across it. The size of these gradients will depend on several parameters (including sample thickness, sample thermal conductivity and heating rate) and unfortunately there is no guaranteed way of eliminating them. Nevertheless, several factors can be mentioned which, if properly considered, can result in the acceptable reduction of the magnitude of the gradient.

The first concerns the atmosphere in the sample chamber. If there is an inert gas atmosphere within the chamber the sample will be immersed in a thermally conducting medium so that it will, to some extent, be heated from all sides. This may prove difficult in low-temperature cryostats where frosting of the samples and windows is to be avoided at all costs. It may be possible to use extremely dry gas (by passing the gas through an efficient water trap prior to entry into the chamber) but often vacuum conditions have to be tolerated instead.

If powder samples are being used, two methods of placing the sample in contact with the heater may be employed. The first puts the powder directly on to the strip within a defined area. Small amounts of powder need to be used so that the effective thickness is only one or two grains. This method is recommended for kinetic analysis. Improving the thermal contact by mixing the powder with a thermally conducting fluid is not recommended because of absorption of light and because of the possibility of spurious glow from the fluid itself. (Especially a problem if irradiation *in situ* is taking place.) The second method involves the use of a small pan to hold the powder – for example, small grains may be precipitated on to a metal substrate (as in the fine-grain method of preparation for pottery dating, cf. chapter 7); alternatively the loose powder may be placed in a small metal container. With both of these techniques, there is then the problem of making good thermal contact between the metal pan and the heater. This may be achieved by 'gluing' the pan on to the strip with say, silicone grease or silicone spray or even with conducting (silver) paint. This latter material is very effective and does not produce spurious emission (as can the silicone materials). (Often experimenters have little choice between placing the powder directly on the strip and placing it in a metal pan. If irradiations of *the same sample* are to take place after glow-curve readout a pan has to be used.)

With larger solid samples, e.g., single crystals, thermal gradients can be an even greater problem. Large and/or irregular shaped samples which do not sit flat on the strip are to be avoided. Extremely thin, flat samples must be prepared; in the author's experience a thickness of $\leqq 0.3$ mm is an appropriate one.

Finally, whenever possible, slow heating rates ought to be used – as slow as possible. Clearly this presents difficulties in many situations. The smaller the heating rate the less intense the emission, so a heating rate commensurate with both these considerations needs to be found, depending on the application. Clearly the fast infra-red heating methods mentioned above will inevitably result in the presence of thermal gradients, but in general, unless kinetic analysis is being undertaken, this is not a consideration as long as the heating cycle is reproducible.

9.4   Light detection

9.4.1   *Photomultiplier tube: d.c. current mode*

Undoubtedly the most convenient device for measuring the intensity of the emitted luminescence is a photomultiplier tube of which there are two possible modes of operation – d.c. current and photon counting. Most photomultiplier tubes used in thermoluminescence applications have 11 or 13 dynode stages and 52 mm diameter end windows. Typical cathode sensitivities are 70–100 μA/lumen and the anode sensitivity is typically 2000 A/lumen. Thus, gains of 10^6–10^7 are often achieved. In the most usual mode of operation, the anode current is simply amplified by a suitable d.c. picoammeter, the output of which is displayed on a chart recorder or relayed via an A/D converter to a computer for data analysis.

An essential part of a photomultiplier tube design lies in the photocathode material. The photocathode is coded by an 'S' number, or by an indication of its actual composition. The most common photocathode types are the bialkalis – namely, KCs, RbCs or KNaSb. These are high-gain materials but their main drawback lies in their spectral response. Figure 9.5 illustrates typical spectral response curves for a variety of materials. The diagram displays *absolute sensitivity* (S) against wavelength (λ). Often it is more useful to determine the *quantum efficiency* (QE) and this can be calculated from:

$$QE = 123.95S/\lambda. \tag{9.3}$$

 It can be seen that the bialkalis have a sharp response peaking at ~ 420 nm. This conveniently corresponds with the maximum emission wavelength of many thermoluminescence phosphors but limits the application of the tube to the detection of glow in this narrow wavelength region. For a wider response, the S20 range of photocathodes is required, particularly if red thermoluminescence emission is being monitored. Note that this enhanced response is gained at the expense of a small loss in sensitivity. For the best spectral characteristics, a GaAs photocathode is required. However, this material may be unsuitable in many low light applications owing to its lack of sensitivity compared with some bialkali tubes. (It is to be noted that the curves shown in figure 9.5 are typical only and the response of an individual tube can be markedly different from those shown. If emission spectra are being recorded, calibration of the actual tube in use is essential.)

Also relevant to the spectral response of a photomultiplier tube is the choice of window material. Most tubes have borosilicate glass as standard, but in order to extend the photomultiplier response into the near ultra-violet, fused silica windows can be obtained.

The photocathode material has a tendency to produce electrons via thermionic emission with the result that even in the dark there is always a background 'dark current' (typically 10^{-9} A). The reduction of the dark

Figure 9.5. Spectral response of common photocathode materials expressed as absolute sensitivity. The solid lines are for borosilicate glass windows whilst the dashed lines are for fused silica windows. (Data from Thorn EMI photomultiplier tube catalogue. GaAs data from RCA photomultiplier tube catalogue.)

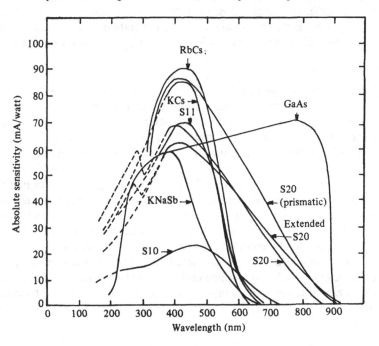

current without a loss in sensitivity is a desirable goal in most photomultiplier tube applications. Most commonly, this is achieved by cooling the tube during normal operation. Peltier Effect thermoelectric coolers can be obtained which stabilize the photomultiplier tube temperature at the manufacturer's recommended level. (Too low a temperature will induce a decrease in gain.) Pick-up from external electric and magnetic fields can also be a problem (usually alleviated with the use of mu-metal shields).

For particularly low light levels, the signal to noise ratio can be markedly increased by using a light-chopper and a lock-in amplifier. Here, the light signal is modulated at a pre-set frequency by means of a rotating shutter. A reference signal, modulated at the same frequency as the shutter rotation, is used to synchronize the phase-sensitive detector so that only signals of this frequency will be amplified. Thus, the noise, which is not modulated, is selectively reduced with respect to the signal.

A final problem to be noted with photomultiplier tubes is that of ageing, whereby the sensitivity progressively decreases with both age and usage. The only safe way to account for these changes is regularly to check the sensitivity with a reference light source. Such sources consist of a radioactive substance (e.g. ^{14}C) and a scintillator (e.g., NaI:Tl or a plastic scintillator). It should be noted that the light sources emit at different wavelengths, depending on the scintillator, and that their ageing characteristics and temperature stability are not as good as might be desired. Stray β emission can also cause difficulties (Spanne, 1973).

9.4.2   *Photon counting*

The sensitivity of the normal d.c. current mode of operation described above is not high enough for several applications of thermoluminescence. Archaeological dating is one such example where very low levels of light need to be measured. In these instances, photon counting is used instead. Photon counting involves the digital processing of current pulses from the anode corresponding to individual photons striking the cathode. Each photon incident on the cathode produces a pulse of electrons at the anode. Each pulse lasts several nanoseconds and there may be up to 10^6 arriving per second. The pulses are of different amplitudes and the baseline is not necessarily constant with time. As illustrated in figure 9.6, this signal is fed into a pulse amplifier with a gain of, typically, $\sim 10^3$. The baseline variation can be eliminated by setting a frequency threshold for the input signal before amplification will occur. After amplification, the randomly sized pulses enter a discriminator which only allows through those pulses which are above a pre-set voltage threshold. These pulses are then shaped by the discriminator into a uniform pulse height. The pulses may then be counted in pre-set counting intervals of, say, 0.1 to 10s. The total counts within these periods are thus a measure of the number of photons striking the cathode. An analogue output (for a chart recorder) can be obtained via a digital to analogue converter, whereas the digital output from the counter is ideally suited for interfacing with a computer for computerized data processing.

As yet, photon counting is not used in commercial TLD readers, but in individual research laboratories, particularly those involved with low light measurements, it is becoming the preferred mode of usage of photomultiplier tubes. Its improved signal to noise ratio and high sensitivity make it an attractive measurement technique.

Figure 9.6. Schematic view of a photon counting system showing the current pulses at various stages in the amplification/discrimination process (A,B,C). The diagram is adapted from the discussion of photon counting by Manche (1979).

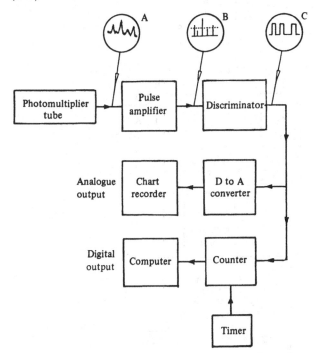

9.5   Special considerations

In several applications, it is desirable to know not just the intensity of the thermoluminescence, at a given temperature, but also the wavelength and even the region of the sample from where the emission is originating. Typical apparatus for both of these applications is described in this section. Before doing so, however, possible methods by which the black-body background can be automatically subtracted from the glow-curve will be mentioned (particularly desirable before spectral analysis).

9.5.1   *Background subtraction*

Kolberg & Prydz (1973) describe a double-beam technique for the simultaneous recording of the black-body background signal and the

Figure 9.7. Schematic view of Manche's (1978) differential thermoluminescence apparatus. f_1 and f_2 are the frequencies of the choppers; A_1, A_2 are lock-in amplifiers; A_3 is a differential amplifier.

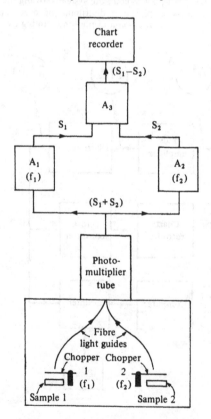

normal thermoluminescence emission during a single heating. A rotating mirror alternately reflects the light from two beams; one beam is that of the normal glow plus black-body signal whilst the other comes from a heater pan containing a pre-heated sample and is therefore the black-body signal only. The reflected beam is directed on to a photomultiplier tube which sees both signals alternately.

A more sophisticated technique than this, which also uses only one photomultiplier tube but allows for the automatic subtraction of the background signal, is described by Manche (1978). Two samples, identical except that one is pre-heated, are warmed in the glow oven. Light from each is modulated by choppers operating at two different frequencies. The modulated light outputs are fed via fibre optics into the *same* photomultiplier tube. The output from the photomultiplier tube is split between two lock-in amplifiers which are synchronized to the two different frequencies. Thus, one amplifier processes one frequency, the other processes the other frequency. The outputs from the amplifiers are demodulated and the d.c. signals fed to a differential amplifier and are thus subtracted. The difference signal is then displayed on the chart recorder and is a measure of the thermoluminescence alone, minus the black-body background. A simplified illustration of the apparatus is given in figure 9.7.

A problem with both of the above methods is that they fail to take account of any differences in the emissivities of the actual sample and that of the pre-heated sample. This can have an influence on the black-body signal. As a result, most experimenters prefer simply to take a second reading immediately after the thermoluminescence has been emitted in order to record the background signal with exactly the same specimen on the strip. Subtraction of the two signals is relatively simple using a computer, even when large amounts of data are being recorded (as is the case with spectral analysis). Even with this method, however, one can never be certain that the black-body signal is the same during the second heating as it was during the first. It is not unusual to find a mismatch of background signals.

9.5.2 *Emission spectra*

A major difficulty with the measurement of the spectrum of thermally stimulated thermoluminescence is that the light level is not constant with time, but instead changes as the sample warms up. This immediately puts a requirement on any scanning spectral recording technique, namely, that the scan be rapid. Indeed, if the scan cannot be made fast enough, it may prove necessary to use two photomultiplier tubes, one to measure the intensity from the spectrometer and one just looking at the sample to measure the light level fluctuation. Suitable normalization can then be made.

Several *rapid-scanning* spectrometers have been mentioned in the literature. Harris & Jackson (1970*b*) use a prism spectrometer, primarily for TLD work. A rotating mirror enables the device to scan the wavelength region from 200–850 nm in 0.5 s.

A difficulty with prism spectrometers is that the intensity of the light being monitored is often very low and the sensitivity of the apparatus may not be sufficient to record reliable spectra. To overcome this, Bailiff, Morris & Aitken (1977*b*) developed a rapid-scanning spectrometer using interference filters of extremely narrow bandwidth. The filters (16 in all) are revolved in order to scan from 340 to 640 nm in 125 ms. The information is stored over successive scans and an integrated spectrum is obtained.

A sophisticated version of these spectrometers, using a diffraction grating, is described by Mattern, Lengweiler & Levy (1971). This device sweeps over a wavelength range of 500 nm in 0.4 s. The data from 32 such scans are then averaged and the average is stored in computer memory. During this data processing, the temperature has risen only by 5 degrees (Celsius) so that many such scans can be taken during a single heating of the specimen. This allows for the representation of the data as an isometric plot of intensity versus temperature versus wavelength and an example of the data obtained on this apparatus was shown in figure 2.13.

Figure 9.8. Schematic diagram of the image intensifier spectrograph for emission spectra studies. The inset shows the condenser system (after Walton, 1982). PM = photomultiplier tube.

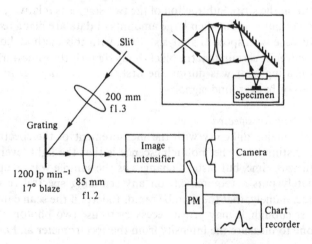

Similar equipment is described by Townsend *et al.* (1983*b*) and examples of the spectra obtained shown in figures 2.14, 6.6 and 8.6.

The above instruments (plus similar ones as referenced, for example, by Jensen & Prescott, 1982) are different forms of essentially the same type of device, namely a rapid-scanning spectrometer and with these instruments one is only looking at a small fraction of the emitted light at any one time. Clearly this is a major disadvantage in terms of the light detection efficiency which is inevitably greatly reduced. A second disadvantage relates to the size of the entrance and exit slits of the spectrometer; these must be small to gain the necessary resolution. These disadvantages may be overcome if, instead of examining a narrow wavelength range at any one time, one looks at the whole spectrum and multiplexes the data. Jensen & Prescott (1982) suggested the use of Fourier transform spectroscopy to achieve this, or alternatively it could be achieved if one were to replace the exit slit by a detector capable of spatial resolution as well as light detection. Such devices include a vidicon tube, a silicon photodiode array and photographic film.

Using the latter medium, Walton and colleagues (Walton & Debenham, 1980; Walton, 1982) developed a spectrophotometer with the aid of a high-gain image intensifier. The apparatus used by Walton and colleagues is shown in figure 9.8. The emitted spectrum is recorded on black and white film and the spectrum is then displayed graphically by measuring the intensity of the light as a function of position on the developed negative, using a densitometer. A photomultiplier tube

Figure 9.9. Spatial distribution of (*a*) natural thermoluminescence and (*b*) artificially induced thermoluminescence from a slice of calcite at different temperatures (from Walton & Debenham, 1980; reproduced with permission of

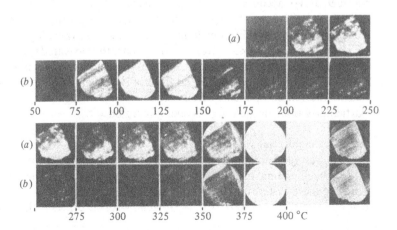

enables conventional glow-curves to be recorded. The system is calibrated using different spectral lamps (Hg, Ne, Ar and Kr) and records spectra from 390 nm to 760 nm. An added bonus is that if the spectrometer part of the apparatus is removed, the spatial distribution of the emitted thermoluminescence across the sample can be determined. The arrangement is more sensitive than the simple apparatus used by Fremlin & Srirath (1964) – cf. figure 7.1 – for similar spatial distribution studies and has the advantage of being able to photograph the distribution as a function of temperature. An example is shown in figure 9.9 and several more can be found in the proceedings of the Second and Third Specialist Seminars on Thermoluminescence Dating (Mejdahl & Aitken, 1982; Mejdahl *et al.*, 1983). The use of photographic film and a densitometer does not take the fullest advantage of a multiplexing system for recording spectra. One has to rely on reproducibility of the film sensitivity and resolution, of developing conditions, and the whole procedure is quite tedious. A major advance is not to use photographic film, but instead to use a vidicon tube or a silicon photodiode array. Intensified versions of these devices can be obtained which give one count per 5–10 photons, with a good wavelength response extending from 200 nm to ~ 800 nm. When used with a multichannel analyser these devices are among the most sensitive multiplex spectrometers currently available, spectra being recorded continuously during thermoluminescence readout. One could even use the vidicon to obtain a movie film of the changes in spatial distribution of the emission from the sample during heating.

One major consideration of all spectral recording apparatus is that the optics used (lenses, light pipes, etc.) and the detectors themselves have a non-uniform response over the wavelength range of interest (e.g., figure 9.5). To account for this the spectral response of the whole detection system has to be determined. The most straightforward way of doing this is to use a calibrated light source (e.g., a quartz–halogen bulb operated at a given temperature) and to compare the measured spectrum with the source's known spectrum over the wavelength range of interest. Calibration of this nature is essential, especially if the emission contains a substantial red component, because the response of most detectors is particularly poor in the long wavelength region.

9.6   Commercial systems

Several commercial concerns manufacture complete thermoluminescence systems, including heating and light detection apparatus. Some of these systems are of modular design: an interested purchaser can buy just one or two parts of the system, whereas others include the whole system in one package. When computing facilities are offered, various software packages can be obtained, depending on the application. The majority of available systems have been developed with dosimetry in mind and may be somewhat limited in scope for use in other research fields. Others, however, can easily be utilized for more general thermoluminescence studies.

A list of addresses of manufacturers of thermoluminescence apparatus is given in Appendix B and the interested reader is prompted to contact the manufacturers concerned for more information.

Appendix A

Minerals

A brief description of each of the minerals mentioned in chapter 8 is given below. Many closely related forms of these minerals exist and the reader is referred to any modern treatise on mineralogy for further details.

albite: see feldspar.

anhydrite: $CaSO_4$.

anorthite: see feldspar.

calcite: see limestone.

chlorapatite: member of the *apatite* family (phosphates); formula $Ca_5Cl(PO_4)_3$. (N.B. If F replaces Cl it becomes *fluorapatite*.)

cristobalite: see quartz.

dolomite: $CaCO_3$, $MgCO_3$.

enstatite: see pyroxene.

feldspar: essential constituent of most igneous rocks: chief minerals in the family are *orthoclase* ($KAl Si_3 O_8$), *albite* ($NaAlSi_3 O_8$) and *anorthite* ($CaAl_2Si_2O_8$); they are related in a tertiary phase diagram, illustrated in figure A.1 which also shows the relative compositions of the *K-feldspars* and the *plagioclases*; *oligoclase* is a plagioclase with a composition of 70–90% albite and 30–10% anorthite.

Figure A.1. Feldspar composition diagram. Region A: feldspars stable only at high temperatures. B: no feldspars stable.

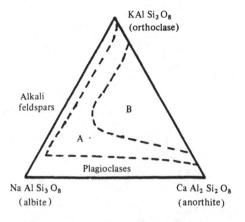

fluorite: CaF_2 (*fluorspar*).

grossular: $Ca_3Al_2(SiO_4)_3$; a type of *garnet* often found in metamorphosed rocks; general garnet formula is $(Ca,Mg,Fe^{2+},Mn)_3$ $(Al,Fe^{3+})_2(SiO_4)_3$.

halite: NaCl (rock salt).

ilmenite: $FeOTiO_2$; chief ore of titanium.

limestone: $CaCO_3$; crystalline form: *calcite*.

maskelynite: feldspathic glass; see feldspar.

melilite: family of silicate minerals ranging from $Ca_2Al_2SiO_7$ to $Ca_2MgSi_2O_7$. Found in the Ca–Al-rich inclusions in C chondrites.

obsidian: black, glassy, igneous rock of volcanic origin; a type of *rhyolite* which frequently contains porphyritic crystals of quartz and orthoclase.

oligoclase: see feldspar.

olivine: $(MgFe)_2SiO_2$; pure Mg form (Mg_2SiO_4) is called *forsterite*; pure iron form is *fayalite* (rare; Fe_2SiO_4).

orthopyroxene: see pyroxene.

plagioclase: see feldspar.

pyroxene: general formula is $(Mg, Fe, Ca)SiO_3$. If Ca $< 5\%$ of whole, known as *orthopyroxene*; if the Mg content of orthopyroxene is $> 90\%$, the mineral is *enstatite*; if $< 90\%$ Mg the mineral is *hypersthene*.

quartz: SiO_2 (rock crystal); high-temperature forms – *tridymite* (870–1470 °C) and *cristobalite* (> 1470 °C); see section 5.3.1.

spinel: family of oxide minerals which have the general formula (Mg, Fe^{2+}) $(Al,Fe^{3+},Cr)_2O_4$; the spinel found in the Ca–Al-rich inclusions of the C chondrites is $MgAl_2O_4$.

tridymite: see quartz.

Appendix B

Commercial thermoluminescence systems

1. Daybreak Nuclear and
   Medical Systems Inc.
   50 Denison Drive
   Guildford
   CT 06437
   USA

2. Harshaw Filtrol
   Partnership
   Crystal & Electronic
   Products Dept
   6801 Cochran Road
   Solon
   Ohio 44139 USA

3. Instrument AB Therados
   Dalgatan 15
   S-752-28 Uppsala
   Sweden

4. Littlemore Scientific
   Eng. Co.
   Railway Lane
   Littlemore
   Oxford OX4 4PZ England

5. National Panasonic
   Matsushita Electric Trading
   Co. Ltd
   C.P.O. Box 288
   Osaka 530–91
   Japan

6. Radiologische Dienst
   TNO
   Utrechtseweg 310
   N-Arnhem
   The Netherlands

7. Studsvik Energiteknik AB
   S-6 1182 Nykoping
   Sweden

8. Teledyne Isotopes
   50 Van Buren Avenue
   Westwood
   NJ 07675 USA

9. Victoreen Instrument Division
   10101 Woodland Avenue
   Cleveland
   Ohio 44104 USA

10. Vinten Instruments Ltd
    Jessamy Road
    Weybridge
    Surrey KT13 8LE England

References

ANSI (1978). *American Standards Criteria for Testing Dosimetry Performance*, N. 13. 11.

Aboltin, D.E., Graboviskis, V.J., Kangro, A.R., Luschik, Ch., O'Konnel-Bronin, A.A., Vitol, K.I. & Zirap, V.E. (1978). *Phys. Stat. Sol (A)*, **47**, 667.

Adam, G. & Katriel, J. (1971). *Proc. 3rd Int. Conf. Lumin. Dosim.* (Risø), 9.

Adirovitch, E. I. (1956). *J. Phys. Rad.*, **17**, 705.

Aguilar, M. Jaque. F. & Agulló-López, F. (1980). *J. de Phys.*, **41**, C6–341.

Agulló-López, F. (1981). *J. Lumin.*. **23**, 433.

Agulló-López, F., López, F.L. & Jaque, F. (1982). *Cryst. Latt. Def. & Amorph. Mat.*, **9**, 227.

Aitken, M.J. (1974). *Physics and archaeology*. Oxford University Press, Oxford.

Aitken, M.J. (1976). *Archaeometry*, **18**, 233.

Aitken, M.J. (1978a). *Phys. Rep.*, **40C**, 277.

Aitken, M.J. (ed.) (1978b). *Proc. First Specialist Seminar on Thermoluminescence Dating* (Oxford, 1978), *PACT*, **2** (1978) and **3** (1979).

Aitken, M.J. (1978c). *PACT*, **2**, 104.

Aitken, M.J. (1978d). *PACT*, **2**. 18.

Aitken, M.J. (1979). *PACT*, **3**, 319.

Aitken, M.J. (1984). *Thermoluminescence dating*. Acad. Press, London.

Aitken, M.J. & Bowman, S.G.E. (1975). *Archaeometry*, **17**, 132.

Aitken, M.J. & Bussell, G.D. (1982). *PACT*, **6**, 550.

Aitken, M.J. & Fleming, S.J. (1972). In *Topics in radiation dosimetry*. Suppl.1 (ed. Attix, F.H.), p. 2, Acad. Press, London.

Aitken, M.J., Fleming, S.J., Doell, R.R. & Tanguy, J.C. (1968a). In *Thermoluminescence of geological materials* (ed. McDougall, D.J.), p. 359. Acad. Press, London.

Aitken, M.J., Fleming, S.J., Reid, J. & Tite, M.S. (1968b). In *Thermoluminescence of geological materials* (ed. McDougall, D.J.), p. 133. Acad. Press, London.

Aitken, M.J., Moorey, P.R.S. & Ucko, P.J. (1971). *Archaeometry*, **13**, 89.

Aitken, M.J. & Murray, A.S. (1976). *The 1976 Symposium on Archaeometry and Archaeological Prospection*, Edinburgh, (ed. H. McKerrell). HMSO, London. Discussed by Aitken (1979).

Alfimov, M.V. & Nikol'skii, V.G. (1963). *Polymer Sci. USSR*, **5**, 477.

Alonso, P.J., Halliburton, L.E., Kohnke, E.E. & Bossoli, R.B. (1983). *J. Appl. Phys.* **54**, 5369.

Alvarez Rivas, J.L. (1980). *J. de Phys.*, **41**, C6–353.

Ang, T.C. & Mykura, H. (1977). *J. Phys. C: Sol. St. Phys.*, **10**, 3205.

Antinucci, M., Cevolani, M., Degli Esposti, G.C. & Petralia, S. (1977). *Lett. Nuov. Cim.*, **18**, 393.

Aramu, F., Maxia, V. & Spano, G. (1975b). *Lett. Nuov. Cim.*, **25**, 75.

Aramu, F., Maxia, V., Spano, G. & Cortese, C. (1975a). *J. Lumin.*, **11**, 197.

Arnold, G.W. (1960). *J. Phys. Chem. Sol.*, **13**, 306.

Arnold, G.W. (1973). *IEEE Trans. Nucl. Sci.*, *NS-20*, 220.

Arnold, G.W. & Compton, W.D. (1959). *Phys. Rev.*, **166**, 802.
Ascarelli, G. & Rodriguez, S. (1961). *Phys. Rev.*, **124**, 1325.
Attix, F.H. (ed.) (1965). *Proc. 1st Int. Conf. Lumin. Dosim.* (Stanford). AEC, CONF-650637.
Attix, F.H. (1974). *J. Appl. Phys.*, **46**, 81.
Ausin, V. & Alvarez Rivas, J.L. (1972a). *J. Phys. C: Sol. St. Phys.*, **5**, 82.
Ausin, V. & Alvarez Rivas, J.L. (1972b). *Phys. Rev.*, **B6**, 4828.
Ausin, V. & Alvarez Rivas, J.L. (1974). *J. Phys. C: Sol. St. Phys.*, **7**, 2255.
Auston, D.H., Shank, C.V. & Le Fur, P. (1975). *Phys. Rev. Lett.*, **35**, 1022.
Auxier, J.A., Becker, K., Robinson, E.M. (eds.) (1968). *Proc. 2nd Int. Conf. Lumin. Dosim.* (Gatlinburg). AEC, CONF-680920.
Avouris, Ph., Chang, I.F., Dove, D., Morgan, T.N. & Thefaine, Y. (1981). *J. Elect. Mat.*, **10**, 887.
Avouris, Ph. & Morgan, T.N. (1981). *J. Chem. Phys.*, **74**, 4347.
Ayyangar, K., Lakshmanan, A.R., Chandra, B. & Ramadas, K. (1974). *Phys. Med. Biol.*, **19**, 665.
Azorin, J., Salvi, R. & Moreno, A. (1980). *Nucl. Instrum. Meth.*, **175**, 81.
BCS (1977). *General criteria for laboratory approval. Provision of personal dosimetry services.* (Publ. 0803.) *Supplementary criteria for laboratory approval. Provision of personal dosimetry services using thermoluminescence dosimeters for β, γ, x and neutron irradiations.* (Publ. 0823.) British Calibration Service, London.
Bacci, C., Bernabei, R., d'Angelo, S., Furetta, C. & Laitano, R.F. (1982). *Proc. Third Int. Symp. Soc. Radiol. Protect.* (Inverness)
Bagolia, C., Doshi, N., Lal, D. & Sears, D.W. (1978). *Nucl. Track Detect.*, **2**, 29.
Bailiff, I.K. (1976). *Nature*, **264**, 531.
Bailiff, I.K. (1979). *PACT*, **3**, 345.
Bailiff, I.K. (1983). *PACT*, **9**, 207.
Bailiff, I.K., Bowman, S.G.E., Mobbs, S.F. & Aitken, M.J. (1977a). *J. Electrostat.*, **3**, 269.
Bailiff, I.K., Morris, D.A. & Aitken, M.J. (1977b). *J. Phys. E: Sci. Instrum.*, **10**, 1156.
Balarin, M. (1977). *J. Thermal Anal.*, **12**, 169.
Balarin, M. (1979a). *Phys. Stat. Sol. (A)*, **54**, K137.
Balarin, M. (1979b). *J. Thermal Anal.*, **17**, 319.
Bangert, U. & Hennig, G.J. (1979). *PACT*, **3**, 281.
Bangert, U., Thiel, K., Ahmed, K. & Townsend, P.D. (1982a) *Radiat. Eff.*, **64**, 143.
Bangert, U., Thiel, K., Ahmed, K. & Townsend, P.D. (1982b). *Radiat. Eff.*, **64**, 153.
Bapat, V.N. & Kathuria, S.P. (1981). *Phys. Stat. Sol. (A)*, **66**, K67.
Bapat, V.N., Sankaran, A.V. & Sunta, C.M. (1980). *Bull. Rad. Prot.*, **3**, 51.
Barbina, V., Contento, G., Furetta, C., Malison, M. & Padovani, R. (1981). *Radiat. Eff. Letts.*, **67**, 55.
Barbina, V., Contento, G., Furetta, C., Padovani, R. & Prokic, M. (1982). *Proc. Third Int. Symp. Soc. Radiol. Protect.* (Inverness).
Barland, M., Duval, E. & Nouailhat, A. (1980). *J. de Phys.*, **41**, 75.
Barr, L.W. & Lidiard, A.B. (1970). In *Physical chemistry–an advanced treatise*, vol. 10 (ed. Eyring, H., Henderson, D. & Jost, W.), p. 165. Acad. Press, New York.
Barthe, T., Barthe, J. & Portal, G. (1980). *Nucl. Instrum. Meth.*, **175**, 4.
Bartlett, D.T. & Sandford, D.J. (1978). *Phys. Med. Biol.*, **23**, 332.
Batrak, E.N. (1958). *Sov. Phys. – Crystall.*, **3**, 102, 633, 635.
Baur, G., Freydorf, E.V. & Koschel, W.H. (1974). *Phys. Stat. Sol. (A)*, **21**, 247.

Becker, K. (1972). *Nucl. Instrum. Meth.*, **104**, 405.
Becker, K. (1973). *Solid state dosimetry*. CRC Press, Cleveland.
Becker, K. (1974). *Health Phys.*, **27**, 321.
Becker, M., Kiessling, J. & Scharmann, A. (1973). *Phys. Stat. Sol. (A)*, **15**, 515.
Bell, W.T. (1979). *Archaeometry*, **21**, 342.
Bell, W.T. (1980). *Ancient TL*, **12**, 4.
Beltrami, M., Cappelletti, R. & Fieschi, R. (1964). *Phys. Letts.*, **10**, 279.
Bemski, G. (1958). *Proc. IRE*, **46**, 980.
Benko, L. & Koszorus, L. (1980). *Nucl. Instrum. Meth.*, **175**, 227.
Bergstrom, R.E. (1956). *Bull. Am. Ass. Petrol. Geol.*, **40**, 918.
Bhasin, B.D., Sasidharan, R. & Sunta, C.M. (1976). *Health Phys.*, **30**, 139.
Binder, W. & Cameron, J.R. (1969). *Health Phys.*, **17**, 613.
Bjarngard, B. (1967). *Proc. 1st Lumin. Dosim. Conf.* (Stanford), 195.
Björnsson, M. & Sorbe, B. (1982). *Br. J. Radiol.*, **55**, 56.
Blair, J.M., Edgington, J.A., Chen, R. & Jahn, R.A. (1972). *Proc. 3rd Lunar Sci. Conf.*, **3**, 2949.
Blak, A.R. & Watanabe, S. (1974). *Proc. 4th. Int. Conf. Lumin. Dosim.* (Krakow), 169.
Böhm, M., Peschke, W. & Scharmann, A. (1980). *Radiat. Eff.*, **53**, 67.
Böhm, M. & Scharmann, A. (1971a). *Phys. Stat. Sol. (A)*, **4**, 99.
Böhm, M. & Scharmann, A. (1971b). *Phys. Stat. Sol. (A)*, **5**, 563.
Bonfiglioli, G., Brovetto, P. & Cortese, C. (1956a). *Phys. Rev.*, **114**, 951.
Bonfiglioli, G., Brovetto, P. & Cortese, C. (1956b). *Phys. Rev.*, **114**, 956.
Booth, L.F., Johnson, T.L. & Attix, F.H. (1972). *Health Phys.*, **23**, 137.
Bosacchi, A., Bosacchi, B., Franchi, S. & Hernandez, L. (1973). *Sol. St. Commun.*, **13**, 1805.
Bosacchi, A., Fieschi, R. & Scaramelli, P. (1965). *Phys. Rev.*, **138**, 1760.
Bosacchi, A., Franchi, S. & Bosacchi, B. (1974). *Phys. Rev.*, **B10**, 5235.
Bossoli, R.B., Jani, M.G. & Halliburton, L.E. (1982). *Sol. St. Commun.*, **44**, 213.
Bøtter-Jensen, L. & Christensen, P. (1972). *Acta Radiol. Suppl.*, **313**, 247.
Bowman, S.G.E. (1979). *PACT*, **3**, 381.
Bowman, S.G.E. & Chen, R. (1979). *J. Lumin.*, **18/19**, 345.
Bowman, S.G.E. & Demetsopoullos, I.C. (1980). *Nucl. Instrum. Meth.*, **175**, 219.
Bowman, S.G.E. & Seeley, M.A. (1978). *PACT*, **2**, 151.
Bowman, S.G.E. & Sieveking, G. de G. (1983). *PACT*, **9**, 253.
Boyle, R. (1663). *Register of the Roy. Soc.*, **1663**, 213.
Bradbury, M.H. & Lilley, E. (1976). *J. Mat. Sci.*, **11**, 1849.
Bradbury, M.H. & Lilley, E. (1977a). *J. Phys. D: Appl. Phys.*, **10**, 1261.
Bradbury, M.H. & Lilley, E. (1977b). *J. Phys. D: Appl. Phys.*, **10**, 1267.
Bradbury, M.H., Nwosu, B.C.E. & Lilley, E. (1976). *J. Phys. D: Appl. Phys.*, **9**, 1009.
Brand, M.A.J.T. & Kerr, G.R. (1982). *Br. J. Radiol.*, **55**, 352.
Braner, A.A. & Israeli, M. (1963). *Phys. Rev.*, **132**, 2501.
Bräunlich, P. (ed.) (1979). *Thermally stimulated relaxation in solids*. Springer-Verlag, Berlin, Heidelberg, New York.
Bräunlich, P. & Dickinson, J.T. (1979). *Proc. VIth Int. Symp. Exoelectron Emission and Applications* (Ahrenshoop), 9.
Bräunlich, P., Gasiot, J., Fillard, J.P. & Castagne, M. (1981). *Appl. Phys. Letts*, **39**, 769.
Bräunlich, P. & Kelly, P. (1970). *Phys. Rev.*, **B1**, 1596.
Bräunlich, P. & Scharmann, A. (1966). *Phys. Stat. Sol.*, **18**, 307.
Brinck, W., Gross, K., Gels, G. & Partridge, J. (1977). *Health Phys.*, **32**, 221.

348   *References*

Brodribb, J.D., O'Colmain, D. & Hughes, D.M. (1975). *J. Phys. D: Appl. Phys.*, **8**, 856.
Broser, I. & Broser-Warminsky, R. (1954). *Br. J. Appl. Phys.*, Suppl., **4**, 90.
Brower, K.L. (1978). *Phys. Rev. Letts*, **14**, 879.
Brown, C.S. & Thomas, L.A. (1960). *J. Phys. Chem. Sol.*, **13**, 337.
Brunskill, R.T. (1968). *UKAEA PG Report* 837 (*W*), HMSO, London.
Bube, R.H. (1960). *Photoconductivity of solids.* Wiley & Sons, New York.
Buckman, W.G. (1972). *Health Phys.*, **22**, 402.
Buckman, W.G. & Payne, M.R. (1976). *Health Phys.*, **31**, 501.
Budd, T., Marshall, M., Peaple, L.H.J. & Douglas, J.A. (1979). *Phys. Med. Biol.*, **24**, 71.
Bull, C. & Mason, D.E. (1951). *J. Opt. Soc. Am.*, **41**, 718.
Burch, P.R.J. & Chesters, M.S. (1979). *Phys. Med. Biol.*, **24**, 1216.
Burgkhardt, B., Herrera, R. & Piesch, E. (1978a). *Nucl. Instrum. Meth.*, **155**, 293.
Burgkhardt, B., Herrera, R. & Piesch, E. (1978b). *Nucl. Instrum. Meth.*, **155**, 299.
Burgkhardt, B., Piesch, E. & Seguin, H. (1980). *Nucl. Instrum. Meth.*, **175**, 183.
Burlin, T.E., Chan, F.K., Zanelli, G.D. & Spiers, F.W. (1969). *Nature*, **221**, 1047.
Busuoli, G., Sermenghi, I., Rimondi, O. & Vicini, G. (1977). *Nucl. Instrum. Meth.*, **140**, 385.
CEC (1975). *Technical recommendations for the use of thermoluminescence for dosimetry in individual monitoring for photons and electrons from external sources.* CEC Rep. EUR. 5358.
Cagon, M. (1973). *J. de Phys.*, **34**, 217.
Caldas, L.V.E. & Mayhugh, M.R. (1976). *Health Phys.*, **31**, 451.
Caldas, L.V.E., Mayhugh, M.R. & Stoebe, T.G. (1983). *J. Appl. Phys.*, **54**, 3431.
Cameron, J.R. & Kenney, G.N. (1963). *Radiat. Res.*, **19**, 199.
Cameron, J.R., Suntharalingam, N. & Kenney, G.N. (1968). *Thermoluminescence dosimetry.* Univ. Wisconsin Press, Madison.
Cape, J. & Jacobs, G. (1960). *Phys. Rev.*, **118**, 9461.
Carlsson, C.A. & Alm Carlsson, G. (1970). *Radiat. Res.*, **42**, 207.
Carriveau, G.W. & Nievens, M. (1979). *PACT*, **3**, 506.
Carter, A.C. (1976). *Nature*, **260**, 133.
Cassidy, W.A. & Rancitelli, L.A. (1982). *Am. Sci.*, **70**, 156.
Catlow, C.R.A. (1976). *Chem. Phys. Letts*, **39**, 497.
Catlow, C.R.A., Corish, J., Quigley, J.M. & Jacobs, P.W.M. (1980). *J. Phys. Chem. Sol.*, **41**, 231.
Chan, F.K. & Burlin, T.E. (1970). *Health Phys.*, **18**, 325.
Chandler, P.J., Jaque, F. & Townsend, P.D. (1979). *Radiat. Eff.*, **42**, 45.
Chandler, P.J. & Lilley, E. (1981). *J. Phys. E: Sci. Instrum.*, **14**, 364.
Chandra, B., Ayyangar, K. & Lakshmanan, A.R. (1976). *Phys. Med. Biol.*, **21**, 67.
Chandra, B., Lakshmanan, A.R. & Bhatt, R.C. (1980). *Phys. Stat. Sol. (A)*, **60**, 593.
Chandra, B., Lakshmanan, A.R. & Bhatt, R.C. (1981). *Phys. Stat. Sol. (A)*, **66**, 335.
Chandra, B., Lakshmanan, A.R. & Bhatt, R.C. (1982). *J. Phys. D: Appl. Phys.*, **15**, 1803.
Chang, I.F. & Thioulouse, P. (1982). *J. Appl. Phys.*, **53**, 5873.
Chapman, M.R., Miller, A.G. & Stoebe, T.G. (1979). *Med. Phys.*, **6**, 494.
Charalambous, S. & Petridou, C. (1976). *Nucl. Instrum. Meth.*, **137**, 441.
Charles, M.W. (1977). *Proc. 5th Int. Conf. Lumin. Dosim.* (Sao Paulo), 313.
Charles, M.W., Khan, Z.U. & Mistry, H.D. (1980). *Nucl. Instrum. Meth.*, **175**, 51.
Charlesby, A. (1981). *Radiat. Phys. Chem.*, **17**, 399.
Chen, R. (1969a). *J. Appl. Phys.*, **40**, 570.
Chen, R. (1969b) *J. Electrochem. Soc.*, **116**, 1254.

Chen, R. (1969c). *J. Comput. Phys.*, **4**, 415.
Chen, R. (1971). *J. Appl. Phys.*, **42**, 5899.
Chen, R. (1976). *J. Mat. Sci.*, **11**, 1521.
Chen, R. (1979). *PACT*, **3**, 325.
Chen, R. & Fleming, R.J. (1973). *J. Appl. Phys.*, **44**, 1393.
Chen, R. & Kirsh, Y. (1981). *Analysis of thermally stimulated processes*. Pergamon Press, Oxford.
Chen, R., Kristianpoller, N., Davidson, Z. & Visocekas, R. (1981a) *J. Lumin.*, **23**, 293.
Chen, R., McKeever, S.W.S. & Durrani, S.A. (1981b) *Phys. Rev.*, **B24**, 4931.
Chen, R. & Winer, S.A. (1970). *J. Appl. Phys.*, **41**, 5227.
Chentsova, L.G. & Butuzov, V.P. (1962). In *Growth of crystals* (ed. Shubnikov, A.V. & Sheftal, N.N.) p. 336. Consultants Bureau, New York.
Chentsova, L.G., Grechushnikov, B.N. & Batrak, E.N. (1957). *Opt. Spectrosc.*, **3**, 619.
Christensen, P. (1968). *Proc. 2nd Int. Conf. Lumin. Dosim.* (Gatlinburg), 91.
Christensen, P. & Mäjborn, B. (1980). *Nucl. Instrum. Meth.*, **175**, 74.
Christodoulides, C. & Ettinger, K.V. (1971). *Mod. Geol.*, **2**, 235.
Christodoulides, C., Ettinger, K.V. & Fremlin, J.H. (1971). *Mod. Geol.*, **2**, 275.
Christodoulides, C. & Fremlin, J.H. (1971). *Nature*, **232**, 257.
Christy, R.W., Johnson, N.M. & Wilbarg, R.R. (1967). *J. Appl. Phys.*, **38**, 2099.
Cohen, A.J. (1960). *J. Phys. Chem. Sol.*, **13**, 321.
Comins, J.D. & Carragher, B.O. (1980). *J. de Phys.*, **41**, C6-166.
Cook, J.S. & Dryden, J.S. (1962). *Proc. Phys. Soc.*, **80**, 479.
Cook, J.S. & Dryden, J.S. (1981). *J. Phys. C: Sol. St. Phys.*, **14**, 1133.
Cooke D.W. (1978). *J. Appl. Phys.*, **49**, 4206.
Cooke, D.W., Payne, I.W. & Santi, R.S. (1981). *J. Appl. Phys.*, **52**, 3606.
Cooke, D.W. & Rhodes, J.F. (1981). *J. Appl. Phys.*, **52**, 4244.
Cooke, D.W., Rhodes, J.F., Santi, R.S. & Alexander, C. (1980). *J. Chem. Phys.*, **73**, 3573.
Corish, J., Quigley, J.M., Jacobs, P.W.M. & Catlow, C.R.A. (1981). *Phil. Mag. A*, **44**, 13.
Cox, F.M., Lucas, A.C. & Kaspar. N.M. (1976). *Health Phys.*, **30**, 135.
Crase, K.W. & Gammage, R.B. (1975). *Health Phys.*, **29**, 739.
Crawford, J.H. (1970). *J. Phys. Chem. Sol.*, **31**, 399.
Creswell, R.A. & Perlman, M.M. (1970). *J. Appl. Phys.*, **41**, 2365.
Crittenden, G.C., Townsend, P.D., Gilkes, J. & Wintersgill, M.C. (1974b). *J. Phys. D: Appl. Phys.*, **7**, 2410.
Crittenden, G. C., Townsend, P.D. & Townshend, S.E. (1974a) *J. Phys. D: Appl. Phys.*, **7**, 2397.
Crowell, C.R. (1976). *Appl. Phys.*, **9**, 79.
Crozaz, G. & Plachy, A.L. (1976). *Proc. 7th Lunar Sci. Conf.*, **1**, 123.
Curie, D. (1960). *Luminescence in crystals*. Methuen, London.
Curie, M. (1904). *Radioactive substances* (English translation of doctoral thesis presented to the Faculty of Science, Paris). Greenwood Press, Westpoint, 1961.
Curtis, O.L., Srour, J.R. & Chiu, K.Y. (1975). *IEEE Trans. Nucl. Sci.*, NS-22, 2174.
Cussó, F., López, F.L. & Jaque, F. (1978). *Cryst. Latt. Def*, **7**, 225.
Dalrymple, G.B. & Doell, R.R. (1970). *Science*, **167**, 713.
Daniels, F. (1968). In *Thermoluminescence of geological materials* (ed. McDougall, D.J.) p. 3. Acad. Press, London.
Daniels, F., Boyd, C.A. & Saunders, D.F. (1953). *Science*, **117**, 343.

Dauletbekova, A., Akilbekov, A. & Elango, A. (1982). *Phys. Stat. Sol. (B)*, **112**, 445.
David, M. (1981). *Ind. J. Pure Appl. Phys.*, **19**, 1048.
David, M., Kathuria, S.P. & Sunta, C.M. (1982). *Ind. J. Pure Appl. Phys.*, **20**, 519.
David, M. & Sunta, C.M. (1981a) *Ind. J. Pure Appl. Phys.*, **19**, 1041.
David, M. & Sunta, C.M. (1981b). *Ind. J. Pure Appl. Phys.*, **19**, 1054.
David, M., Sunta, C.M. & Ganguly, A.K. (1977a). *Ind. J. Pure Appl. Phys.*, **15**, 201.
David, M., Sunta, C.M. & Ganguly, A.K. (1977b). *Ind. J. Pure Appl. Phys.*, **15**, 277.
David, M., Sunta, C.M. & Ganguly, A.K. (1978). *Ind. J. Pure Appl. Phys.*, **16**, 423.
David, M., Sunta, C.M. & Ganguly, A.K. (1979). *Ind. J. Pure Appl. Phys.*, **17**, 655.
Davidson, Z. & Kristianpoller, N. (1980). *Sol. St. Commun.*, **33**, 79.
Davies, J.E. (1982). *Proc. Am. Chem. Soc. Symp. on Adsorption on and Surface Chemistry of Hydroxyapatite Surfaces* (Kansas).
Davies, J.J., Garlick, G.F.J., Richards, C.L. & Sowersby, G. (1974). *J. Lumin.*, **9**, 267.
De, R., Rao, C.N. & Kaul, I.K. (1980). *Mod. Geol.*, **7**, 231.
de Murcia, M., Bräunlich, P., Egge, M. & Mary, G. (1980). *Sol. St. Commun.*, **34**, 737.
Debenham, N.C., Driver, H.S.T. & Walton, A.J. (1982). *PACT*, **6**, 555.
Debenham, N.C. & Walton, A.J. (1983). *PACT*, **9**, 531.
Delbecq, C.J., Hartford, R., Schoemaker, D. & Yuster, P.H. (1976). *Phys. Rev.*, **B13**, 3631.
Delbecq, C.J., Toyozawa, Y. & Yuster, P.H. (1974). *Phys. Rev.*, **B9**, 4497.
Delgado, L. & Alvarez Rivas, J.L. (1979). *J. Phys. C: Sol. St. Phys.*, **12**, 3159.
Delgado, L. & Alvarez Rivas, J.L. (1981). *Phys. Rev.*, **B23**, 6699.
Delgado, L. & Alvarez Rivas, J.L. (1982). *J. Phys. C: Sol. St. Phys.*, **15**, 1591.
Dennis, J.A., Marhshall, T.O. & Shaw, K.B. (1974). *National Radiological Protection Board Report*, NRPB-R32.
Derham, C.J. & Geake, J.E. (1964). *Nature*, **201**, 62.
Derham, C.J., Geake, J.E. & Walker, G. (1964). *Nature*, **203**, 134.
Déribéré, M. (1936). *Rev. Sci.*, **76**, 382.
DeWerd, L.A. (1976). *Health Phys.*, **31**, 525.
DeWerd, L.A., Kim, T.H. & Stoebe, T.G. (1976b). *Mat. Res. Bull.*, **11**, 1413.
DeWerd, L.A. & Stoebe, T.G. (1972). *Phys. Med. Biol.*, **17**, 187.
DeWerd, L.A., White, R.P., Stang, R.G. & Stoebe, T.G. (1976a). *J. Appl. Phys.*, **47**, 4231.
Dhar, A., DeWerd, L.A. & Stoebe, T.G. (1973). *Health Phys.*, **25**, 427.
Dianov, E.M., Kornienko, L.S., Nikitin, E.P., Rybattovskii, A.O. & Chirnov, P.V. (1980). *Sov. J. Glass Phys. & Chem.*, **6**, 239.
Dieke, G.H. & Crosswhite, H.M. (1963). *Appl. Optics*, **2**, 675.
Dienes, G.J. (1960). *J. Phys. Chem. Sol.*, **13**, 272.
Dixon, R.L. & Ekstrand, K.E. (1974). *J. Lumin.*, **8**, 383.
Doheny, A.J. & Albrecht, A.C. (1977). *Can. J. Chem.*, **55**, 2065.
Doherty, S.P., Martin, J.J., Armington, A.F. & Brown, R.N. (1980). *J. Appl. Phys.*, **51**, 4164.
Dolecek, E.M. & Appleby, A. (1974). *Proc. 4th Int. Conf. Lumin. Dosim.* (Krakow), 521.
Douglas, D., Morency, M. & McDougall, D.J. (1970). *Mod. Geol.*, **1**, 211.
Douglas, G. & McDougall, D.J. (1973). In *New horizons in rock mechanics* (ed. Hardy, H.R. & Stefanko, R.), p. 121. Am. Soc. Civil. Eng., New York.
Douglas, J.A. (1981). In *Applied thermoluminescence dosimetry* (ed. Oberhofer, M. & Scharmann, A.) p. 229. Adam Hilger, Bristol.
Douglas, J.A., Baker, D.M., Marshall, M. & Budd, T. (1980). *Nucl. Instrum. Meth.*, **175**, 54.

Douglas, J.A. & Budd, T. (1981). *Phys. Med. Biol.*, **26**, 171.
Douglas, J.A. & Marshall, M. (1978). *Health Phys.*, **35**, 315.
Dreyfus, R.W. & Nowick, A.S. (1962). *Phys. Rev.*, **126**, 1367.
Driscoll, C.M.H. (1977). *National Radiological Protection Board Report*, NRPB-R68.
Driscoll, C.M.H. & McKinlay, A.F. (1981). *Phys. Med. Biol.*, **26**, 321.
Driscoll, C.M.H., Mundy, S.J. & Elliot, J.M. (1981). *Radiat. Protect. Dosim.*, **1**, 135.
Driver, H.S.T. (1979). *PACT*, **3**, 290.
Dryden, J.S. (1963). *J. Phys. Soc. Jap.*, **18**, (Suppl. III), 129.
Dryden, J.S. & Shuter, B. (1973). *J. Phys. D: Appl. Phys.*, **6**, 123.
Du Fay, C.F. (1726). *Hist. de l'Acad. Roy. de Sci. de Paris*, 1726, 56.
Du Fay, C.F. (1738). *Hist. de l'Acad. Roy. de Sci. de Paris*, 1735, 347.
Duftschmid, K.E. (1980). *Nucl. Instrum. Meth.*, **175**, 162.
Duftschmid, K.E. (1981). *Radiat. Protect. Dosim.*, **2**, 3.
Durrani, S.A., Bull, R.K. & McKeever, S.W.S. (1978). In *Mare Crisium: The view from Luna 24* (ed. Merrill, R.B. & Papike, J.J.), p. 179. Pergamon Press, New York.
Durrani, S.A., Bull, R.K. & McKeever, S.W.S. (1979). *Phil. Trans. Roy. Soc. Lond.*, **A297**, 41.
Durrani, S.A., Groom, P.J., Khazal, K.A.R. & McKeever, S.W.S. (1977b). *J. Phys. D: Appl. Phys.*, **10**, 1351.
Durrani, S.A., Khazal, K.A.R. & Ali, A. (1977c). *Nature*, **266**, 411.
Durrani, S.A., Khazal, K.A.R., McKeever, S.W.S. & Riley, R.J. (1977a). *Radiat. Eff.*, **33**, 237.
Durrani, S.A., Prachyabrued, W., Christodoulides, C., Fremlin, J.H., Edgington, J.A., Chen, R. & Blair, I.M. (1972). *Proc. 3rd Lunar Sci. Conf.*, **3**, 2955.
Durrani, S.A., Prachyabrued, W., Hwang, F.S.W., Edgington, J.A. & Blair, I.M. (1973). *Proc. 4th Lunar Sci. Conf.*, **3**, 2463.
Dussel, G.A. & Bube, R.H. (1967). *Phys. Rev.*, **155**, 764.
Edgington, J.A. & Blair, I.M. (1970). *Science*, **167**, 715.
Eernisse, E.P. (1974). *J. Appl. Phys.*, **45**, 167.
Eernisse, E.P. & Norris, C.B. (1974). *J. Appl. Phys.*, **45**, 5196.
Egemberdiev, Zh., Zazubovich, S. & Nagirnyi, V. (1982a). *Phys. Stat. Sol. (B)*, **109**, 173.
Egemberdiev, Zh., Zazubovich, S., Seeman, V. & Nagirnyi, V. (1982b). *Phys. Stat. Sol. (B)*, **109**, 473.
Eisenlohr, H.H. & Jayaraman, S. (1977). *Phys. Med. Biol.*, **22**, 18.
Ellsworth, H.V. (1932). *Can. Dept. Mines Geol. Surv. Econ. Geol. Ser.*, **11**, 55.
Endres, G.W.R., Kathren, R.L. & Kocher, L.F. (1970). *Health Phys.*, **18**, 665.
Euler, F., Liger, P., Kahan, A., Pellegrini, P., Flanagan, T.M. & Wrobel, T.F. (1978). *IEEE Trans. Nucl. Sci.*, NS-**25**, 1267.
Evans, R.C. (1966). *An introduction to crystal chemistry.* Cambridge University Press, Cambridge, England.
Ewles, J. & Lee, N. (1953). *J. Electrochem. Soc.*, **100**, 392.
Eyring, H. (1936). *J. Chem. Phys.*, **4**, 283.
Fain, J. & Monnin, M. (1977). *J. Electrostat.*, **3**, 289.
Fain, J., Montret, M. & Sahraoui, L. (1980). *Nucl. Instrum. Meth.*, **175**, 37.
Fairchild, R.G., Mattern, P.L., Lengweiler, K. & Levy, P.W. (1978a). *J. Appl. Phys.*, **49**, 4512.
Fairchild, R.G., Mattern, P.L., Lengweiler, K. & Levy, P.W. (1978b). *J. Appl. Phys.*, **49**, 4523.
Fiegl, F.J. & Anderson, J.H. (1970). *J. Phys. Chem. Sol.*, **B1**, 575.
Fiegl, F.J., Fowler, W.B. & Yip, L.K. (1974). *Sol. St. Commun.*, **14**, 225.

352   *References*

Fields, D.E. & Moran, P.R. (1974). *Phys. Rev.*, **B9**, 1836.

Fieschi, R. & Scaramelli, P. (1966). *Phys. Rev.*, **145**, 622.

Fieschi, R. & Scaramelli, P. (1968). In *Thermoluminescence of geological materials* (ed. McDougall, D.J.), p. 291. Acad. Press, London.

Fillard, J.P., Gasiot, J., Jimenez, J., Sanz, L.F. & De Sala, J.A. (1977). *J. Electrostat.*, **3**, 133.

Fillard, J.P., Gasiot, J. & Manifacier, J.C. (1978). *Phys. Rev.*, **B18**, 4497.

Fitzgerald, M., White, D.R., White, E. & Young, J. (1981). *Br. J. Radiol.*, **54**, 212.

Fleischer, R.L., Price, P.B. & Walker, R.M. (1975). *Nuclear tracks in solids.* Univ. California Press, Berkeley.

Fleming, S.J. (1966). *Archaeometry*, **9**, 170.

Fleming, S.J. (1968). *Proc. 2nd Int. Conf. Lumin. Dosim.* (Gatlinburg), 465.

Fleming, S.J. (1970). *Archaeometry*, **12**, 133.

Fleming, S.J. (1973). *Archaeometry*, **15**, 13.

Fleming, S.J. (1975a). *Authenticity in art. The scientific detection of forgery.* Inst. Phys., London.

Fleming, S.J. (1975b). *Archaeometry*, **17**, 122.

Fleming, S.J. (1976). *Dating in archaeology.* Dent and Sons, London.

Fleming, S.J. (1978). *PACT*, **2**, 125.

Fleming, S.J. (1979a). *Thermoluminescence techniques in archaeology.* Clarendon Press, Oxford.

Fleming, S.J. (1979b). *PACT*, **3**, 315.

Fleming, S.J. (1979c). *PACT*, **3**, 360.

Fleming, S.J. & Thompson, J. (1970). *Health Phys.*, **18**, 567.

Fontanella, J., Johnson, R.L., Sigel, G.H. & Andeen, C. (1979). *J. Non-Cryst. Sol.*, **31**, 401.

Fowler, J.F. (1956). *Proc. Roy. Soc.*, **A276**, 464.

Fowler, J.F. & Attix, F.H. (1966). In *Radiation dosimetry* (ed. Attix, F.H. & Roesch, W.C.), p. 241. Acad. Press, New York.

Fremlin, J.H. & Srirath, S. (1964). *Archaeometry*, **7**, 58.

French, B.M. (1977). *The moon book.* Penguin, Baltimore.

Frerichs, R. (1947). *Phys. Rev.*, **72**, 594.

Fuchs, W. & Taylor, A. (1970a). *Phys. Rev.*, **B2**, 3393.

Fuchs, W. & Taylor, A. (1970b). *Phys. Stat. Sol.*, **38**, 771.

Furetta, C. & Gennai, P. (1981). *Radiat. Eff.*, **55**, 23.

Furetta, C. & Pellegrini, C. (1981). *Radiat. Eff.*, **58**, 17.

Galkin, G.N., Kharakhorin, F.F. & Shatkovskii, E.V. (1971). *Sov. Phys.-Semicond.*, **5**, 387.

Gammage, R.B. & Cheka, J.S. (1977). *Health Phys.*, **32**, 189.

Gammage, R.B. & Garrison, A.K. (1974). *Proc. 4th Int. Conf. Lumin. Dosim.* (Krakow), 263.

Garlick, G.F.J. (1949). *Luminescent materials.* Oxford University Press, London.

Garlick, G.F.J. (1958). In *Handbüch der Physik*, Vol. 26 (ed. Flügge, S.) p. 89. Springer-Verlag, Berlin.

Garlick, G.F.J. (1964). *Nature*, **202**, 171.

Garlick, G.F.J. & Gibson, A. F. (1948). *Proc. Phys. Soc.*, **60**, 574.

Garlick, G.F.J. & Mason, D.E. (1949). *J. Electrochem. Soc.*, **96**, 90.

Garlick, G.F.J. & Richards, C.L. (1974a). *J. Lumin.*, **9**, 424.

Garlick, G.F.J. & Richards, C.L. (1974b). *J. Lumin.*, **9**, 432.

Garlick, G.F.J. & Robinson, I. (1972). In *The moon* (ed. Runcorn, S.K. & Urey, H.C.),

p. 324. International Astronomical Union; Reidel, Dordreckt.

Garlick, G.F.J. & Wilkins, M.H.F. (1945). *Proc. Roy. Soc. Lond.*, **184**, 408.

Garofano, T., Corazzari, T. & Casalini, G. (1977). *Il Nuov. Cim.*, **38B**, 133.

Gartia, R.K. (1976). *Phys. Stat. Sol. (A)*, **37**, 571.

Gartia, R.K. (1977). *Phys. Stat. Sol. (A)*, **44**, K21.

Gasiot, J., Bräunlich, P. & Fillard, J.P. (1982). *J. Appl. Phys.*, **53**, 5200.

Gasiot, J. & Fillard, J.P. (1977). *J. Appl. Phys.*, **48**, 3171.

Geake, J.E. & Walker, G. (1966a). *Geochim. Cosmochim. Acta*, **30**, 929.

Geake, J.E. & Walker, G. (1966b). *Proc. Roy. Soc. Lond.*, *A296*, 337.

Geake, J.E., Walker, G. & Mills, A.A. (1972). In *The moon* (ed. Runcorn, S.K. & Urey, H.C.), p. 279. International Astronomical Union; Reidel, Dordreckt.

Geake, J.E., Walker, G. Mills, A. A. & Garlick, G.F.J. (1971). *Proc. 2nd Lunar Sci. Conf.*, 3, 2265.

Geake, J.E., Walker, G. Telfer, D.J. & Mills, A.A. (1977). *Phil. Trans. Roy. Soc. Lond.*, *A285*, 403.

Gessel, T.F. & de Planque, G. (1980). *Nucl. Instrum. Meth.*, **175**, 186.

Gibbs, J.H. (1972). *J. Chem. Phys.*, **10**, 4473.

Gillot, P.Y., Labeyrie, J., Laj, C., Valladas, G., Guerin, G., Poupeau, G. & Delibrias, G. (1979). *Earth Planet. Sci. Letts*, **42**, 444.

Gillot, P. Y., Valladas, G. & Reyss, J.L. (1978). *PACT*, **2**, 165.

Ginther, R.J. (1954). *J. Electrochem. Soc.*, **101**, 248.

Ginther, R.J. & Kirk, R.D. (1957). *J. Electrochem. Soc.*, **104**, 365.

Glasstone, S. Laidler, K.J. & Eyring, H. (1941). *The theory of rate processes*. McGraw-Hill, New York.

Gobrecht, H. & Hofmann, D. (1966). *J. Phys. Chem. Sol.*, **27**, 509.

Goedicke, C. (1983). *PACT*, **9**, 19.

Goedicke, C., Slusallek, K. & Kubelik, M. (1983). *PACT*, **9**, 245.

Göksu, H.Y. & Fremlin, J.H. (1972). *Archaeometry*, **14**, 127.

Göksu, H.Y., Fremlin, J.H., Irwin, H.T. & Fryxell, R. (1974). *Science*, **183**, 652.

Goldstein, N. (1972). *Health Phys.*, **22**, 90.

Gorbics, S.G. & Attix, F.H. (1968). *Int. J. Appl. Radiat. Isot.*, **19**, 81.

Gorbics, S.G., Attix, F.H. & Kerris, K. (1973). *Health Phys.*, **25**, 499.

Gorbics, S.G., Nash, A.E. & Attix, F.H. (1969). *Int. J. Appl. Radiat. Isot.*, **20**, 843.

Gower, R.G., Hendee, W.R. & Ibbott, G.S. (1969). *Health Phys.*, **17**, 607.

Grant, R.M. & Cameron, J.R. (1966). *J. Appl. Phys.*, **37**, 3791.

Grant, R.M., Stowe, W.S. & Correl, J. (1968). *Proc. 2nd Int. Conf. Lumin. Dosim.* (Gatlinburg), 613.

Greaves, G.N. (1978). *Phil. Mag.*, **B37**, 447.

Greening, J.R. (1972). In *Topics in radiation dosimetry* (ed. Attix, F.H.), p. 261. Acad. Press, New York.

Greenman, N.N. & Gross, H.G. (1970). *Science*, **167**, 720.

Greitz, U. & Rudén, B.I. (1972). *Phys. Med. Biol.*, **17**, 193.

Grenet, J., Vautier, C., Carles, D. & Chabrier, J.J. (1973). *Phil. Mag.*, **28**, 1265.

Griffiths, J.H.E., Owen, J. & Ward, I.M. (1954). *Nature*, **173**, 149.

Griscom, D.L. (1978). In *The physics of SiO_2 and its interfaces* (ed. Pantelides, S.T.), p. 232. Pergamon Press, New York.

Griscom, D.L., Friebele, E.J. & Sigel, G.H. (1974). *Sol. St. Commun.*, **15**, 479.

Grögler, N., Houtermans, F.G. & Stäuffer, H. (1958). *Proc. Sec. Geneva Conf.*, **21**, 226.

Grögler, N., Houtermans, F.G. & Stäuffer, H. (1960). *Helv. Phys. Acta*, **33**, 595.

Grögler, N. & Liener, A. (1968). In *Thermoluminescence of geological materials* (ed. McDougall, D.J.), p. 569. Acad. Press, London.

Groom, P.J., Durrani, S.A., Khazal, K.A.R. & McKeever, S.W.S. (1978). *PACT*, 2, 200.

Grossweiner, L.I. (1953). *J. Appl. Phys.*, 24, 1306.

HPSSC (1977). *Proposed standard criteria for testing personal dosimetry performance*, WG/15. Health Physics Society Standards Committee. Unpublished manuscript.

Haake, C.H. (1957). *J. Opt. Soc. Am.*, 47, 649.

Haering, R.R. & Adams, E.N. (1960). *Phys. Rev.*, 117, 451.

Hagebeuk, H.J.L. & Kivits, P. (1976). *Physica*, 83B, 289.

Hagekyriakou, K. & Fleming, R.J. (1982a). *J. Phys. D: Appl. Phys.*, 15, 1795.

Hagekyriakou, K. & Fleming, R.J. (1982b). *J. Phys. D: Appl. Phys.*, 15, 163.

Hageseth, G.T. (1972). *Phys. Rev.*, B5, 4060.

Hall, T.P.P., Hughes, A.E. & Pooley, D. (1976). *J. Phys. C: Sol. St. Phys.*, 9, 439.

Halliburton, L.E., Koumvakalis, N., Markes, M.E. & Martin, J.J. (1981). *J. Appl. Phys.*, 52, 3565.

Halliburton, L.E., Markes, M., Martin, J.J., Armington, A.F. & Brown, R.N. (1979). *IEEE Trans. Nucl. Sci.*, NS-26, 4851.

Halperin, A. & Braner, A.A. (1960). *Phys. Rev.*, 117, 408.

Halperin, A., Braner, A.A., Ben-Zvi, A. & Kristianpoller, N. (1960). *Phys. Rev.*, 117, 416.

Halperin, A. & Chen, R. (1966). *Phys. Rev.*, 148, 839.

Hama, Y., Kimura, Y., Tsumura, M. & Omi, N. (1980). *J. Chem. Phys.*, 53, 115.

Harris, A.M. & Jackson, J.H. (1969). *Br. J. Appl. Phys.*, 2, 1667.

Harris, A.M. & Jackson, J.H. (1970a). *J. Phys. D: Appl. Phys.*, 3, 624.

Harris, A.M. & Jackson, J.H. (1970b). *J. Phys. E: Sci. Instrum.*, 3, 374.

Hashizume, K. (1979). *Cryst. Latt. Def.*, 8, 139.

Hashizume, T., Kato, Y., Nakajima, T., Toryu, T., Sakamoto, H., Kotera, N. & Eguchi, S. (1971). *Proc. Symp. Adv. Radiation Detectors*, IAEA-SM-143/11 (Vienna), 91.

Haskell, E.H. (1983). *PACT*, 9, 77.

Haskell, E.H., Wrenn, M.E. & Sutton, S.R. (1980). *Ancient TL*, 12, 9.

Hayashi, T., Ohata, T. & Koshino, S. (1980). *Sol. St. Commun.*, 18, 807.

Hayes, W. & Nichols, G.M. (1960). *Phys. Rev.*, 117, 993.

Hedvig, P. (1972). In *The radiation chemistry of macromolecules* (ed. Dole, M.), p. 73. Acad. Press, New York, London.

Helfrich, W., Riehl, N. & Thoma, P. (1964). *Phys. Letts*, 10, 31.

Henniger, J., Hübner, K. & Pretzsch, G. (1982). *Nucl. Instrum. Meth.*, 192, 453.

Henry, C.H. & Lang, D.V. (1975). *Phys. Rev.*, B15, 989.

Hensler, J.R. (1959). *Nature*, 183, 672.

Herreros, J.M. & Jaque, F. (1979). *J. Lumin.*, 18/19, 231.

Herschel, A.S. (1889). *Nature*, 60, 29.

Hersh, H.N. (1966). *Phys. Rev.*, 148, 928.

Heymann, D. (1967). *Icarus*, 6, 189.

Heywood, E.F. & Clarke, K.H. (1980a). *Aust. Phys. & Eng. Sci. Med.*, 3, 210.

Heywood, E.F. & Clarke, K.H. (1980b). *Aust. Phys. & Eng. Sci. Med.*, 3, 219.

Hill, J.J. & Schwed, P. (1955). *J. Chem. Phys.*, 23, 652.

Hitier, G., Curie, D. & Visocekas, R. (1981). *J. Phys.*, 42, 479.

Hobbs, L.W. (1975). In *Surface and defect properties of solids*, vol. 4, p 152. Chem. Soc., London.

Hobbs, L.W., Hughes, A.E. & Pooley, D. (1972). *Phys. Rev. Letts*, 28, 234.

References   355

Hobbs, L.W., Hughes, A.E. & Pooley, D. (1973). *Proc. Roy. Soc. Lond.*, **A332**, 167.
Hobozova, L. (1974). *Proc. 4th Int. Conf. Lumin. Dosim.* (Krakow), 1081.
Hochstrasser, G. & Antonini, J.F. (1972). *Surface Sci.*, **32**, 644.
Hoffmann, K. & Pöss, D. (1978). *Phys. Stat. Sol. (A)*, **45**, 263.
Honda, Y., Kimura, Y., Nakamura, I., Maruyama, Y., Murayama, Y. & Nakajima, T. (1980). *Nucl. Instrum. Meth.*, **175**, 169.
Hoogenstraaten, W. (1958). *Philips Res. Rep.*, **13**, 515.
Horowitz, Y.S. (1981). *Phys. Med. Biol.*, **26**, 766.
Horowitz, Y.S. (1982). *Phys. Stat. Sol. (A)*, **69**, K29.
Horowitz, Y.S., Fraier, I., Kalef-Ezra, J., Pinto, H. & Goldbart, Z. (1979a). *Nucl. Instrum. Meth.*, **165**, 27.
Horowitz, Y.S., Fraier, I., Kalef-Ezra, J., Pinto, H. & Goldbart, Z. (1979b). *Phys. Med. Biol.*, **24**, 1268.
Horowtiz, Y.S. & Freeman, S. (1978). *Nucl. Instrum. Meth.*, **157**, 393.
Horowitz, Y.S. & Kalef-Ezra, J. (1981). *Nucl. Instrum. Meth.*, **187**, 519.
Houtermans, F.G. & Liener, A. (1966). *J. Geophys. Res.*, **71**, 3387.
Howarth, J.L. (1965). *Br. J. Radiol.*, **38**, 51.
Hoyt, H.P., Miyajima, M. & Walker, R.M. (1971a). *Mod. Geol.*, **2**, 263.
Hoyt, H.P., Miyajima, M., Walker, R.M., Zimmerman, D.W., Zimmerman, J., Britton, D. & Kardos, J.L. (1971b). *Proc. Apollo 12 Lunar Sci. Conf.*, **3**, 2245.
Hoyt, H.P., Walker, R.M. & Zimmerman, D.W. (1973). *Proc. 4th Lunar Sci. Conf.*, **3**, 2489.
Huang, K. & Rhys. A. (1950). *Proc. Roy. Soc.*, **A204**, 406.
Hübner, K., Henniger, J. & Negwer, D. (1980). *Nucl. Instrum. Meth.*, **175**, 34.
Hughes, A.E. (1978). *Comments Sol. St. Phys.*, **8**, 83.
Hughes, A.E. & Jain, S.C. (1979). *Adv. Phys.*, **28**, 717.
Hughes, R.C. (1975). *Radiat. Eff.*, **26**, 225.
Huntley, D.J. & Johnson, H.P. (1976). *Can J. Earth Sci.*, **13**, 593.
Huntley, D.J. & Wintle, A.G. (1978). *PACT*, **2**, 115.
Hutcheon, I.D., Steele, I.M., Smith, J.V. & Clayton, R.N. (1978). *Proc. 9th Lunar Planet. Sci. Conf.*, 1345.
Hütt, G., Smirnov, A. & Tale, I. (1979). *PACT*, **3**, 362.
Huxtable, J. (1978). *PACT*, **2**, 7.
Huxtable, J. & Aitken, M.J. (1977). *Nature*, **265**, 40.
Huxtable, J., Aitken, M.J. & Bonhommet, N. (1978). *Nature*, **275**, 207.
Huxtable, J., Hedges, J.W., Renfrew, A.C. & Aitken, M.J. (1976). *Archaeometry*, **18**, 5.
Hwang, F.S.W. (1970). *Nature*, **227**, 940.
Hwang, F.S.W (1972). *J. Geophys. Res.*, **77**, 328.
ICRP (1977). *A system of dose limitation*, ICRP 26. International Commission on Radiological Protection. Pergamon Press, Oxford.
ICRUM (1973). *Radiation quantities and units*, Rep. 19; *Dose-equivalent*, suppl. to Rep. 19. International Commission on Radiation Units and Measurements, Washington DC, USA.
ISO (1982). *Draft standard on personal and environmental thermoluminescence dose-meters*, ISO/TC85/SC2/WG7, 10th draft revised (corrected). International Standards Organization. Unpublished manuscript.
Ichikawa, Y. (1967). *Jap. J. Appl. Phys.*, **7**, 220.
Iga, K., Yamashita, T., Takenaga, M., Yasuno, Y., Onishi, H. & Ikeda, M. (1977). *Health Phys.*, **33**, 605.
Ikeda, S. & Matsuda, K. (1979). *J. Appl. Phys.*, **50**, 6475.

Ikeya, M. (1975). *Phys. Stat. Sol. (B)*, **69**, 275.
Ikeya, M. (1978a,). *Archaeometry*, **20**, 147.
Ikeya, M. (1978b). *Naturwiss.*, **65**, 489.
Ikeya, M., Ishibashi, M. & Itoh, N. (1971). *Health Phys.*, **21**, 429.
Inabe, K. & Takeuchi, N. (1977). *J. Phys. C: Sol. St. Phys.*, **10**, 3023.
Inabe, K. & Takeuchi, N. (1978). *Jap. J. Appl. Phys.*, **17**, 1549.
Inabe, K. & Takeuchi, N. (1980). *Jap. J. Appl. Phys.*, **19**, 1165.
Ishii, T. & Rolfe, J. (1966). *Phys. Rev.*, **141**, 758.
Isoya, J., Weil, J.A. & Halliburton, L.E. (1981). *J. Appl. Phys.*, **74**, 5436.
Itoh, N. (1972). *Cryst. Latt. Def.*, **3**, 115.
Itoh, N., Royce, B.S. & Smoluchowski, R. (1965). *Phys. Rev.*, **137**, 1010.
Itoh, N. & Saidoh, M. (1969). *Phys. Stat. Sol.*, **33**, 649.
Itoh, N. & Saidoh, M. (1973). *J. de Phys.*, **34**, C9-101.
Itoh, N., Stoneham, A.M. & Harker, C.K. (1977). *J. Phys. C: Sol. St. Phys.*, **10**, 4197.
Jablonski, A. (1935). *Z. Phys.*, **94**, 38.
Jackson, J.H. & Harris, A.M. (1969). *Phys. Letts.*, **29A**, 423.
Jackson, J.H. & Harris, A.M. (1970). *J. Phys. C: Sol. St. Phys.*, **3**, 1967.
Jacobs, P.W.M. & Ong, S.H. (1980). *J. Phys. Chem. Sol.*, **41**, 437.
Jähnert, B. (1972). *Health Phys.*, **23**, 1122.
Jain, H. & Nowick, A.S. (1982). *J. Appl. Phys.*, **53**, 477.
Jain, M.P., Reddy, A.R., Nagaratnam, A. & Jain, P.C. (1979). *Br. J. Radiol.*, **52**, 499.
Jain, S.C. & Mehendru, P.C. (1970). *J. Phys. C: Sol. St. Phys.*, **3**, 1491.
Jain, V.K. (1980). *Phys. Stat. Sol. (A)*, **60**, 351.
Jain, V.K. (1981a). *Phys. Stat. Sol. (A)*, **66**, 341.
Jain, V.K. (1981b). *Health Phys.*, **41**, 363.
Jain, V.K. (1982). *Radiat. Protect. Dosim.*, **2**, 141.
Jalilian-Nosraty, M. & Martin, J.J. (1981). *J. Appl. Phys.*, **52**, 785.
Jani, M.G., Bossoli, R.B. & Halliburton, L.E. (1983). *Phys. Rev.*, **B27**, 2285.
Jaque, F. & Townsend, P.D. (1981). *Nucl. Instrum. Meth.*, **182/183**, 781.
Jasińska, M. & Niewiadomski, T. (1970). *Nature*, **227**, 1159.
Jayachandran, C.A. (1970). *Health Phys.*, **15**, 325.
Jenkins, T.R. (1978). *J. Comput. Phys.*, **29**, 302.
Jensen, H.E. & Prescott, J.R. (1982). *PACT*, **6**, 542.
Jensen, H.E. & Prescott, J.R. (1983). *PACT*, **9**, 25.
Jiménez de Castro, M. & Alvarez Rivas, J.L. (1979). *Phys. Rev.*, **B19**, 6484.
Jiménez de Castro, M. & Alvarez Rivas, J.L. (1980). *J. Phys. C: Sol. St. Phys.*, **13**, 257.
Jiménez de Castro, M. & Alvarez Rivas, J.L. (1981). *Proc. Int. Conf. Defects in Insulating Crystals* (Riga), 276.
Joelsson, I., Rudén, B.I., Costa, A., Dutreix, A. & Rosenwald, J.C. (1972). *Acta Radiol. Ther. Phys. Biol.*, **11**, 289.
Johns, T.F. (1981). *Radiat. Protect. Dosim.*, **1**, 126.
Johnson, N.M. (1963). *J. Geol.*, **71**, 596.
Johnson, N.M. (1966). *J. Geol.*, **74**, 607.
Johnson, N.M. & Blanchard, R.L. (1967). *Am. Mineral.*, **52**, 1297.
Johnson, P.D. & Williams, F.E. (1950). *J. Chem. Phys.*, **18**, 1477.
Johnson, P.D. & Williams, F.E. (1952). *J. Chem. Phys.*, **20**, 124.
Johnson, R.P. (1939). *J. Opt. Soc. Am.*, **29**, 387.
Jones, C.E. & Embree, D. (1976). *J. Appl. Phys.*, **47**, 5365.
Jones, C.E. & Embree, D. (1978). In *The physics of SiO$_2$ and its interfaces*

(ed. Pantelides, S.T.), p. 289. Pergamon Press, New York.

Jones, J.L. & Martin, J. A. (1968). *Health Phys.*, **20**, 431.

Julius, H.W. (1975). *Proc. 9th Jahrestagung Fachverband für Strahlenschutz Alpbach* (Tirol).

Julius, H.W. (1981). In *Applied thermoluminescence dosimetry* (ed. Oberhofer, M. & Scharmann, A.), p. 39. Adam Hilger, Bristol.

Jun, J.S. & Becker, K. (1975). *Health Phys.*, **28**, 459.

Kabler, M.N. (1964). *Phys. Rev.*, **136**, 1296.

Kamada, M. & Tsutsumi, K. (1981). *J. Phys. Soc. Japan*, **50**, 3370.

Kamavisdar, V.S. & Deshmukh, B.T. (1981). *J. Phys. C: Sol. St. Phys.*, **14**, 83.

Kao, J.C. & Hwang, W. (1981). *Electrical transport in solids*. Pergamon Press, Oxford.

Kao, K.J. & Perlman, M.M. (1979). *Phys. Rev.*, **B19**, 1197.

Karzmark, C.J., White, J. & Fowler, J.F. (1964). *Phys. Med. Biol.*, **9**, 273.

Kathuria, S.P. (1979). *BARC Report* I-523, cited in Jain (1982).

Kathuria, S.P. & Sunta, C.M. (1979). *J. Phys. D: Appl. Phys.*, **12**, 1573.

Kathuria, S.P. & Sunta, C.M. (1981). *Phys. Med. Biol.*, **26**, 707.

Katz, R. (1978). *Nucl. Track Detect.*, **2**, 1.

Kaul, I.K., Ganguly, D.K. & Hess, B.F.H. (1972). *Mod. Geol.*, **3**, 201.

Keating, P.N. (1961). *Proc. Phys. Soc.*, **78**, 1408.

Keller, S.P., Mapes, J.E. & Cheroff, G. (1957). *Phys. Rev.*, **108**, 663.

Kelly, J.C. (1978). *PACT*, **2**, 120.

Kelly, P.J. & Bräunlich, P. (1970). *Phys. Rev.*, **B1**, 1587.

Kelly, P.J. & Laubitz, M.J. (1966). *Can. J. Phys.*, **45**, 311.

Kelly, P.J., Laubitz, M.J. & Bräunlich, P. (1971). *Phys. Rev.*, **B4**, 1960.

Kemmey, P.J., Townsend, P.D. & Levy, P.W. (1967). *Phys. Rev.*, **115**, 917.

Kennedy, G.C. & Knopff, L. (1960). *Archaeology*, **13**, 147.

Kieffer, F., Meyer, C. & Rigaut, J. (1971). *Chem. Phys. Letts*, **11**, 359.

Kieffer, F., Meyer, C. & Rigaut, J. (1974). *Int. J. Radiat. Phys. Chem.*, **6**, 79.

Kiessling, J. & Scharmann, A. (1975). *Phys. Stat. Sol. (A)*, **32**, 459.

Kikuchi, T. (1958). *J. Phys. Soc. Japan*, **13**, 526.

King, J.C. (1959). *Bell System Tech. J.*, **38**, 573.

King, J.C. & Sander, H.H. (1972). *IEEE Trans. Nucl. Sci.*, NS-21, 390.

King, J.C. & Sander, H.H. (1975). *Radiat. Eff.*, **26**, 203.

Kirsh, Y. & Kristianpoller, N. (1977). *J. Lumin.*, **15**, 35.

Kirsh, Y., Kristianpoller, N. & Chen, R. (1977). *Phil. Mag.*, **35**, 653.

Kivits, P. (1978). *J. Lumin.*, **16**, 119.

Kivits, P. & Hagebeuk, H.J.L. (1977). *J. Lumin.*, **15**, 1.

Klasens, H.A. (1946). *Nature*, **158**, 306.

Klick, C.C., Claffy, E.W., Gorbics, S.G., Attix, F.H., Schulman, J.H. & Allard, J.G. (1967). *J. Appl. Phys.*, **38**, 3867.

Kolberg, S. & Prydz, S. (1973). *Mod. Geol.*, **4**, 137.

Kolopus, J.L., Delbecq, C.J., Schoemaker, D. & Yuster, P.H. (1967). *Bull. Am. Phys. Soc.*, **12**, 467.

Kos, H.-J. & Nink, R. (1977). *Phys. Stat. Sol. (A)*, **44**, 505.

Kos, H.-J. & Nink, R. (1979). *Phys. Stat. Sol. (A)*, **56**, 593.

Kos, H.-J. & Nink, R. (1980). *Phys. Stat. Sol. (A)*, **57**, 203.

Koumvakalis, N. (1980). *J. Appl. Phys.*, **51**, 5528.

Kreft, G.B. (1975). *Radiat. Eff.*, **26**, 249.

Kriegseis, W. & Scharmann, A. (1980). *Nucl. Instrum. Meth.*, **175**, 239.

Kristianpoller, N., Chen, R. & Israeli, M. (1974). *J. Phys. D: Appl. Phys.*, **7**, 1063.
Kristianpoller, N. & Davidson, Z. (1972). *J. Phys. C: Sol. St. Phys.*, **5**, 279.
Kristianpoller, N. & Israeli, M. (1970). *Phys. Rev.*, **B2**, 2175.
Kristianpoller, N. & Katz, I. (1970). *J. Opt. Soc. Am.*, **60**, 424.
Laksmanan, A.R. (1982). *Nucl. Tracks*, **6**, 59.
Lakshmanan, A.R. & Bhatt, R.C. (1980). *Nucl. Instrum. Meth.*, **175**, 259.
Lakshmanan, A.R. & Bhatt, R.C. (1981). *Phys. Med. Biol.*, **26**, 1157.
Lakshmanan, A.R., Bhatt, R.C. & Supe, S.J. (1981*a*). *J. Phys. D: Appl. Phys.*, **41**, 1683.
Lakshmanan, A.R., Chandra, B. & Bhatt, R.C. (1978). *Nucl. Instrum. Meth.*, **153**, 581.
Lakshmanan, A.R., Chandra, B. & Bhatt, R.C. (1981*b*). *Radiat. Protect. Dosim.*, **2**, 13.
Lakshmanan, A.R., Chandra, B. & Bhatt, R.C. (1981*c*). *Radiat. Protect. Dosim.*, **1**, 191.
Lakshmanan, A.R., Chandra, B. & Bhatt, R.C. (1982). *J. Phys. D: Appl. Phys.*, **15**, 1501.
Lakshmanan, A.R. & Vohra, K.G. (1979). *Nucl. Instrum. Meth.*, **159**, 585.
Lalou, C, Valladas, G., Brito, V., Henni, A., Geva, T. & Visocekas, R. (1972) *Proc. 3rd Lunar Sci. Conf.*, **3**, 3009.
Lampert, M.A. & Mark, P. (1970). *Current injection in solids*. Acad. Press, New York.
Land, P.L. (1969). *J. Phys. Chem. Sol.*, **30**, 1681.
Landsberg, P.T. (1970). *Phys. Stat. Sol.*, **41**, 457.
Langmead, W.A. & Wall, B.F. (1976). *Phys. Med. Biol.*, **21**, 39.
Lapraz, D. & Baumer, A. (1981). *Phys. Stat. Sol. (A)*, **68**, 309.
Lapraz, D., Baumer, A. & Iacconi, P. (1979). *Phys. Stat. Sol. (A)*, **54**, 605.
Larsson, L. & Katz, R. (1976). *Nucl. Instrum. Meth.*, **138**, 631.
Las, W.C. & Stoebe, T.G. (1981). *J. Mat. Sci.*, **16**, 1191.
Lasky, J.B. & Moran, P.R. (1977). *Phys. Med. Biol.*, **22**, 852.
Lax, M. (1959). *J. Phys. Chem. Sol.*, **8**, 66.
Lax, M. (1960). *Phys. Rev.*, **119**, 1502.
Lee, H.Y. (1969). *J. Appl. Phys.*, **38**, 4850.
Lee, T.F. & McGill, T.C. (1975). *J. Appl. Phys.*, **46**, 373.
Leiman, V.L. (1973). *Sov. Phys.-Sol. St.*, **15**, 351.
Lell, E., Kriedl, N.J. & Hensler, J.R. (1966). *Progr. Ceram. Soc.*, **4**, 3.
Lerma, I.S. & Agulló-López, F. (1973). *Phil. Mag.*, **27**, 993.
Levine, B.F. (1973). *J. Chem. Phys.*, **57**, 1479.
Levy, P.W. (1960). *J. Phys. Chem. Sol.*, **13**, 287.
Levy, P.W. (1979). *PACT*, **3**, 466.
Levy, P.W., Mattern, P.L. & Lengweiller, K. (1971). *Mod. Geol.*, **2**, 295.
Lidiard, A.B. (1957). In *Handbüch der Physik*, vol. 20 (ed Flügge, S.), p. 246. Springer-Verlag, Berlin.
Liener, A. & Geiss, J. (1968). In *Thermoluminescence of geological materials* (ed. McDougall D.J.), p. 559. Acad. Press, London.
Lilley, E. (1982). *J. Phys. D: Appl. Phys.*, **15**, L17.
Lilley, E. & McKeever, S.W.S. (1983). *J. Phys. D: Appl. Phys.*, **16**, L39.
Lilley, E. & Taylor, G.C. (1981). *J. Phys. D: Appl. Phys.*, **14**, L13.
Lind, S.C. & Bardwell, D.C. (1923). *J. Franklin Inst.*, **196**, 357.
Littlefield, T.A. & Thorley, N. (1968). *Atomic and nuclear physics: an introduction in S.I. units* (2nd edition). Van Nostrand Reinhold, London.
Liu, B.M. & Bagne, F. (1974). *Proc. 4th Int. Conf. Lumin. Dosim.* (Krakow), 937.
Loggini, M. & Sandrolini, F. (1980). L' Energia Elettrica, **27**, 1.
López, F.J., Aguilar, M. & Agulló-López, F. (1981). *Phys. Rev.*, **B23**, 3041.
López, F.J., Aguilar, M., Jaque, F. & Agulló-López, F. (1980*a*). *Sol. St. Commun.*, **34**, 869.

López, F.J., Jaque, F. & Agulló-López, F. (1980c). *J. Phys. Chem. Sol.*, **41**, 681.
López, F.J., Jaque, F., Agulló-López, F. & Aguilar, M. (1980b). *Sol. St. Commun.*, **36**, 1001.
López, F.J., Jaque, F., Fort, A.J. & Agulló-López, F. (1977). *J. Phys. Chem. Sol.*, **38**, 1101.
Lorincz, A., Puma, M., James, F.J. & Crawford, J.H. (1982). *J. Appl. Phys.*, **53**, 927.
Lóvborg, L. (1972). In *Uranium prospecting handbook* (ed. Bowie, S.H.U., Davies, M. & Ostle, D.), p. 157. Inst. Mining & Metall., London.
Lowe, D., Lakey, J.R.A. & Yorke, A.V. (1979). *Health Phys.*, **37**, 417.
Lucas, A.C. & Kaspar, B.M. (1977). *Proc. 5th Int. Conf. Lumin. Dosim.* (Sao Paulo), 131.
Lucovsky, G. (1979a). *Phil. Mag.*, **B39**, 513.
Lucovsky, G. (1979b). *Phil. Mag.*, **B39**, 531.
Lucovsky, G. (1980). *J. Non-Cryst. Sol.*, **35/36**, 825.
Lushchik, C.B. (1956). *Sov. Phys.-JETP*, **3**, 390.
Lyman, T. (1935). *Phys. Rev.*, **48**, 149.
MacDiarmid, R.A. (1968). In *Thermoluminescence of geological materials* (ed. McDougall, D.J.), p. 547. Acad. Press, London.
McCurdy, D.E., Schaiger, K.J. & Flack, E.D. (1969). *Health Phys.*, **17**, 415.
McDougall, D.J. (ed.) (1968a). *Thermoluminescence of geological materials*. Acad. Press, London.
McDougall, D.J. (1968b). In *Thermoluminescence of geological materials* (ed. McDougall, D.J.), p. 527. Acad. Press, London.
McDougall, D.J. (1970). *Econ. Geol.*, **65**, 856.
McDougall, R.S. & Rudin, S. (1970). *Health Phys.*, **19**, 281.
Mackey, J.H. (1963). *J. Chem. Phys.*, **39**, 74.
McKeever, S.W.S. (1980a). *Phys. Stat. Sol. (A)*, **62**, 331.
McKeever, S.W.S. (1980b). *Nucl. Instrum. Meth.*, **175**, 19.
McKeever, S.W.S. (1980c). *Mod. Geol.*, **7**, 105.
McKeever, S.W.S. (1982). *Earth Planet. Sci. Letts*, **58**, 419.
McKeever, S.W.S. (1983). *PACT*, **9**, 321.
McKeever, S.W.S. (ed.) (1984). *Special issue on thermoluminescence materials, Radiat. Protect. Dosim.*, **8**.
McKeever, S.W.S., Ahmed, K., Chandler, P.J., Strain. J.A., Rendell, H.M. & Townsend, P.D. (1983a). *PACT*, **9**, 187.
McKeever, S.W.S., Chen, R., Groom, P.J. & Durrani, S.A. (1980). *Nucl. Instrum. Meth.*, **175**, 43.
McKeever, S.W.S. & Durrani. S.A. (1980). *Mod. Geol.*, **7**, 75.
McKeever, S.W.S. & Hughes, D.M. (1975). *J. Phys. D: Appl. Phys.*, **8**, 1520.
McKeever, S.W.S. & Lilley, E. (1981). *J. Phys. C: Sol. Phys.*, **14**, 3547.
McKeever, S.W.S. & Lilley, E. (1982). *J. Phys. Chem. Sol.*, **43**, 885.
McKeever, S.W.S. & Sears, D.W. (1979). *Meteoritics*, **14**, 29.
McKeever, S.W.S. & Sears, D.W. (1980). *Mod. Geol.*, **7**, 137.
McKeever, S.W.S., Strain, J.A., Townsend, P.D. & Udval, P. (1983b). *PACT*, **9**, 123.
McKeever, S.W.S. & Townsend, P.D. (1982). *Geochim. Cosmochim. Acta*, **46**, 1997.
McKeogh, K.J., Rendell, H.M. & Townsend, P.D. (1983). *Proc. 1983 Archaeometry Symposium (Naples)*.
McKinlay, A.F. (1981). *Thermoluminescence dosimetry*. Adam Hilger, Bristol.
McKlveen, J.W. (1980a). *Nucl. Instrum. Meth.*, **175**, 201.
McKlveen, J.W. (1980b). *Health Phys.*, **39**, 211.
McMorris, D.W. (1971). *J. Geophys. Res.*, **76**, 7875.

Malik, D.M., Kohnke, E.E. & Sibley, W.A. (1981). *J. Appl. Phys.*, **52**, 3600.

Malik, S.R., Durrani, S.A. & Fremlin, J.H. (1973). *Archaeometry*, **15**, 249.

Manche, E.P. (1978). *Rev. Sci. Instrum.*, **49**, 715.

Manche, E.P. (1979). *J. Chem. Educ.*, **56**, 273.

Manconi, J.W. & McDougall, D.J. (1970). *Am. Mineral.*, **55**, 398.

Mangini, A., Pernicka, E. & Wagner, G.A. (1983). *PACT*, **9**, 49.

Marat-Mendes, J.N. & Comins, J.D. (1975). *Cryst. Latt. Def.*, **6**, 141.

Marat-Mendes, J.N. & Comins, J.D. (1977). *J. Phys. C: Sol. St. Phys.*, **10**, 4425.

Marfunin, A.S. (1979). *Spectroscopy, luminescence and radiation centres in minerals.* Springer-Verlag, Berlin.

Mariani, D.F. & Alvarez Rivas, J.L. (1978). *J. Phys. C: Sol. St. Phys.*, **11**, 3499.

Marinello, G., Barret, C. & Le Bourgeois, J.P. (1980). *Nucl. Instrum. Meth.*, **175**, 198.

Martensson, B.K. (1969). *Phys. Med. Biol.*, **14**, 119.

Martin, A. & Harbison, S.A. (1979). *An introduction to radiation protection.* Chapman & Hall, New York, London.

Martini, M. (1981). *Ancient TL*, **1**, 4.

Martini, M., Piccinini, G. & Spinolo, G. (1983b). *PACT*, **9**, 87.

Martini, M., Spinolo, G. & Dominici, G. (1983c). *PACT*, **9**, 9.

Martini, M., Spinolo, G. & Vedda, A. (1983a). *PACT*, **19**, 205.

Maschmeyer, D., Niemann, K., Hake, H., Lehmann, G. & Rauber, A. (1980). *Phys. Chem. Mineral.*, **6**, 145.

Mason, E.W., McKinlay, A.F. & Clark, I. (1976). *Phys. Med. Biol.*, **21**, 60.

Mason, E.W., McKinlay, A.F., Clark, I. & Saunders, D. (1974). *Proc. 4th Int. Conf. Lumin. Dosim.* (Krakow), 219.

Mason, E.W., McKinlay, A.F. & Saunders, D. (1977). *Phys. Med. Biol.*, **22**, 29.

Mataré, H.F. (1970). *Defect electronics in semiconductors.* Wiley-Interscience, New York.

Mattern, P.L., Lengweiler, K. & Levy, P.W. (1971). *Mod. Geol.*, **2**, 293.

Mattern, P.L., Lengweiler, K. & Levy, P.W. (1975). *Radiat. Eff.*, **26**, 237.

Mattern, P.L., Lengweiler, K., Levy, P.W. & Esser, P.D. (1970). *Phys. Rev. Letts*, **24**, 1287.

Maxia, V. (1977). *Lett. Nuov. Cim.*, **20**, 443.

Maxia, V. (1978). *Phys. Rev.*, **B17**, 3262.

Maxia, V. (1979). *Lett. Nuov. Cim.*, **24**, 89.

Maxia, V. (1980). *Phys. Rev.*, **B21**, 749.

May, C.E. & Partridge, J.A. (1964). *J. Chem. Phys.*, **40**, 1401.

Mayhugh, M.R. (1970). *J. Appl. Phys.*, **41**, 4776.

Mayhugh, M.R. & Christy, R.W. (1972). *J. Appl. Phys.*, **41**, 2968.

Mayhugh, M.R., Christy, R.W. & Johnson, N.M. (1970). *J. Appl. Phys.*, **41**, 2968.

Mayhugh, M.R. & Fullerton, G.D. (1974). *Med. Phys.*, **1**, 275.

Mayhugh, M.R. & Fullerton, G.D. (1975). *Health Phys.*, **28**, 297.

Meakins, R.L., Dickson, B.L. & Kelly, J.C. (1978). *PACT*, **2**, 97.

Medlin, W.L. (1961). *Phys. Rev.*, **123**, 502.

Medlin, W.L. (1963). *J. Chem. Phys.*, **38**, 1132.

Medlin, W.L. (1968). In *Thermoluminescence of geological materials* (ed. McDougall, D.J.), p. 193. Acad. Press, London.

Mehendru, P.C. (1970a). *Phys. Rev.*, **B1**, 809.

Mehendru, P.C. (1970b). *J. Phys. C: Sol. St. Phys.*, **3**, 530.

Mehendru, P.C. & Radhakrishna, S. (1969). *J. Phys. C: Sol. St. Phys.*, **2**, 796.

Mehta, S.K., Merklin, J.F. & Donnert, H.J. (1977). *Phys. Stat. Sol. (A)*, **44**, 679.

Mehta, S.K. & Sengupta, S. (1976a). *Health Phys.*, **31**, 176.
Mehta, S.K. & Sengupta, S. (1976b). *Phys. Med. Biol.*, **21**, 955.
Mehta, S.K. & Sengupta, S. (1977). *Phys. Med. Biol.*, **22**, 863.
Mehta, S.K. & Sengupta, S. (1978). *Phys. Med. Biol.*, **23**, 471.
Mehta, S.K. & Sengupta, S. (1981). *Nucl. Instrum. Meth.*, **187**, 515.
Mejdahl, V. (ed.) (1971). *Proc. 3rd Int. Conf. Lumin. Dosim.* (Risø), Rep. 249. Danish Atomic Energy Commission.
Mejdahl, V. (1978a). *PACT*, **2**, 35.
Mejdahl, V. (1978b). *PACT*, **2**, 70.
Mejdahl, V. (1983). *PACT*, **9**, 351.
Mejdahl, V. & Aitken, M.J. (ed.) (1982). *Proc. Second Specialist Seminar on Thermoluminescence Dating* (Oxford, 1980), *PACT*, **6**.
Mejdahl, V., Bowman, S.G.E., Wintle, A.G. & Aitken, M.J. (ed.) (1983). *Proc. Third Specialist Seminar on Thermoluminescence Dating and Electron Spin Resonance* (Helsingør, 1982), *PACT*, **9**.
Mejdahl, V. & Winther-Nielsen, M. (1982). *PACT*, **6**, 426.
Melcher, C.L. (1981). *Geochim. Cosmochim. Acta*, **45**, 615.
Melcher, C.L. & Zimmerman, D.W. (1977). *Science*, **197**, 1359.
Miallier, D., Sanzelle, S. & Fain, J. (1983). *Ancient TL*, **1**, 5.
Michael, C.T., Andronikos, P. & Haikalis, D. (1983). *PACT*, **9**, 95.
Mieke, S. & Nink, R. (1979). *J. Lumin.*, **18/19**, 411.
Mieke, S. & Nink, R. (1980). *Nucl. Instrum. Meth.*, **175**, 10.
Mikhailov, A.J. (1971). *Dokl. Phys. Chem.*, **197**, 223.
Miki, T. & Ikeya, M. (1978). *Jap. J. Appl. Phys.*, **17**, 1703.
Miller, L.D. & Bube, R.H. (1970). *J. Appl. Phys.*, **41**, 3687.
Mills, A.A., Sears, D.W. & Hearsey, R. (1976). *J. Phys. E: Sci. Instrum.*, **10**, 51.
Milne, A.G. (1973). *Deep impurities in semiconductors.* Wiley-Interscience, New York.
Mitchell, E.W.J. & Paige, E.G.S. (1956). *Phil. Mag.*, **1**, 1085.
Mitin, S.M., Kononov, O.V. & Pashin, V.N. (1978). *Dokl. Acad. Nauk. SSR. Earth Sci.*, **243**, 169.
Mohan, N.S. & Chen, R. (1970). *J. Phys. D: Appl. Phys.*, **3**, 243.
Moharil, S.V. (1980). *Sol. St. Commun.*, **33**, 697.
Moharil, S.V. (1981). *J. Phys. D: Appl. Phys.*, **14**, 1677.
Moharil, S.V. (1982). *J. Phys. D: Appl. Phys.*, **15**, L18.
Moharil, S.V., Kamavisdar, V.S. & Deshmukh, B.T. (1979). *Phys. Stat. Sol. (A)*, **55**, K167.
Moharil, S.V. & Kathuria, S.P. (1983). *J. Phys. D: Appl. Phys.*, **16**, 425.
Mooney, G.H. (1981). *Br. J. Radiol.*, **54**, 861.
Moos, W.S. (1979). *Int. J. Appl. Radiat. Isot.*, **30**, 499.
Moran, P.R. & Podgoršak, E.B. (1971). *USAEC Rep.* COO-1105-164.
Morato, S.P., Gordon, A.M.P., Dos Santos, E.N., Gomes, L., Campos, L.L., Prado, L., Vieira, M.M.F. & Bapat, V.N. (1982). *Nucl. Instrum. Meth.*, **200**, 449.
Morozov, G.V. (1969). *US Geol. Surv. Cat.* No. 208 M8280 (in Russian). Cited by Wintle & Huntley (1979a).
Morse, H.W. (1905). *Astrophys. J.*, **21**, 410.
Mort, J. (1965). *Sol. St. Commun.*, **3**, 263.
Mort, J. (1966). *Phys. Letts*, **21**, 124.
Mott, N.F. (1977). *Contemp. Phys.*, **18**, 225.
Mott, N.F. & Davies, E.A. (1979). *Electronic processes in non-crystalline materials* (2nd edition). Clarendon Press, Oxford.

Mott, N.F. & Gurney, R.W. (1948). *Electronic processes in ionic crystals* (2nd edition). Oxford University Press, London.

Muccillo, R. & Rolfe, J. (1974). *Phys. Stat. Sol.* (*b*), **61**, 579.

Mukherjee, B. (1981). *Nucl. Instrum. Meth.*, **190**, 207.

Murray, A.S. (1982). *PACT*, **6**, 53.

Murray, A.S., Bowman, S.G.E. & Aitken, M.J. (1978). *PACT*, **2**, 84.

Murti, Y.V.G.S. & Sucheta, N. (1982). *Phys. Stat. Sol.* (*B*), **109**, 325.

Nagpal, J.S. (1979). *Radiat. Eff. Letts*, **43**, 173.

Nahum, J. & Halperin, A. (1963). *J. Phys. Chem. Sol.*, **24**, 823.

Nakajima, T. (1969). *Int. J. Appl. Radiat. Isot.*, **19**, 789.

Nakajima, T. (1970). *J. Phys. D: Appl. Phys.*, **3**, 300.

Nakajima, T. (1972). *Health Phys.*, **23**, 133.

Nakajima, T., Kato, Y., Kotera, N. & Eguchi, S. (1971). *Health Phys.*, **21**, 118.

Nakajima, T., Murayama, Y., Matsuzawa, T. & Koyano, A. (1978). *Nucl. Instrum. Meth.*, **157**, 155.

Nakamura, S., Inabe, K., Takeuchi, N. & Kos, H.-J. (1981). *Phys. Stat. Sol.* (*B*), **104**, K139.

Nambi, K.S.V. (1979). *PACT*, **3**, 299.

Nambi, K.S.V. (1982). *Nucl. Instrum. Meth.*, **197**, 453.

Nambi, K.S.V., Bapat, V.N. & David, M. (1978). *Ind. J. Earth Sci.*, **5**, 154.

Nambi, K.S.V., Bapat, V.N. & Ganguly, A.K. (1974). *J. Phys. C: Sol. St. Phys.*, **7**, 4403.

Nambi, K.S.V. & Mitra, S. (1978). *Thermochim. Acta*, **27**, 61.

Nash, A.E., Attix, F.H. & Schulman, J.H. (1965). *Proc. Int. Conf. Lumin. Dosim.* (Stanford), 244.

Nash, A.E. & Johnson, T.L. (1977). *Proc. 5th Int. Conf. Lumin. Dosim.* (Sao Paulo), 393.

Nassau, K. & Prescott, B.E. (1975). *Phys. Stat. Sol.* (*a*), **29**, 659.

Nelson, C.M. & Crawford, J.H. (1960). *J. Phys. Chem. Sol.*, **13**, 296.

Nepomnyachikh, A.I. & Radyabov, E.A. (1980). *Optica i Spektroskopiya*, **48**, 273.

Nicholas, K.H. & Woods, J. (1964). *Br. J. Appl. Phys.*, **15**, 783.

Nielsen, B.L. & Bøtter-Jensen, L. (1973). *Mod. Geol.*, **4**, 119.

Niewiadomski, T. (ed.) (1974). *Proc. 4th Int. Conf. Lumin. Dosim.* (Krakow), Inst. Phys.

Niewiadomski, T. (1976a). *Health Phys.*, **31**, 373.

Niewiadomski, T. (1976b). *Nucleonica*, **21**, 1097.

Nikol'skii, V.G. & Buben, N.Ya. (1960). *Proc. Acad. Sci. USSR Phys. Chem.*, **134**, 827.

Nink, R. & Kos, H.-J. (1976). *Phys. Stat. Sol.* (*A*), **35**, 121.

Nink, R. & Kos, H.-J. (1980). *Nucl. Instrum. Meth.*, **175**, 15.

Nishimaki, N. & Shimanuki, S. (1980). *Phys. Stat. Sol.* (*B*), **97**, K119.

Nishimura, S. (1970). *Earth Planet. Sci. Letts*, **8**, 293.

Nishita, N., Hamilton, M. & Haug, R.M. (1974). *Soil Sci.*, **117**, 211.

Nuttal, R.H.D. & Weil, J.A. (1980). *Sol. St. Commun.*, **33**, 99.

Nuttal, R.H.D., Weil, J.A. & Claridge, R.F.C. (1976). *Sol. St. Commun.*, **19**, 141.

Oatley, D., Hudson, G. & Plato, P. (1981). *Health Phys.*, **41**, 513.

Oberhofer, M. & Scharmann, A. (ed.) (1981). *Applied thermoluminescence dosimetry*. Adam Hilger, Bristol.

O'Brien, K. (1977). *Phys. Med. Biol.*, **22**, 836.

O'Brien, M.C.M. (1955). *Proc. Roy. Soc. Lond.*, **A231**, 404.

Oldenburg, H. (1676). *Phil. Trans. Roy. Soc. Lond.*, **11**, 867.

Oliveri, E., Fiorella, O. & Mangia, M. (1979). *Nucl. Instrum. Meth.*, **163**, 569.

Olsen, L., Thoresen, M. & Mejdahl, V. (1983). *PACT*, **9**, 56.

Ong, S.H. & Jacobs, P.W.M. (1979). *Can. J. Phys.*, **57**, 1031.

Opyrchal, H. & Nierzewski, K.D. (1979). *Phys. Stat. Sol. (B)*, **35**, 251.

Osvay, M. & Biró, T. (1980). *Nucl. Instrum. Meth.*, **175**, 60.

Owen, G.P. & Charlesby, A. (1974). *J. Phys. C: Sol. St. Phys.*, **7**, L400.

PTB (1976). *Requirements for the type approval of TLD systems for radiation protection measurement.* Physikalisch Technische Bundesanstalt.

Panizza, E. (1964). *Phys. Letts*, **10**, 37.

Pantelides, S.T. (ed.) (1978). *The physics of SiO₂ and its interfaces.* Pergamon Press, New York.

Paradysz, R.E. & Smith, W.L. (1975). *Radiat. Eff.*, **26**, 213.

Partridge, R.H. (1965). *J. Polymer Sci.*, **A3**, 2817.

Partridge, R.H. (1972). In *The radiation chemistry of macromolecules* (ed. Dole, M.), p. 194. Acad. Press, New York, London.

Pässler, R. (1976). *Phys. Stat. Sol. (B)*, **78**, 625.

Pässler, R. (1978). *Phys. Stat. Sol. (B)*, **85**, 203.

Pasternack, E.S., Gaines, A.M. & Levy, P.W. (1977). *Proc. Int. Conf. Defects in Insul. Crystals* (Oak Ridge), 333.

Paus, H.J. & Scheu, W. (1980). *J. de Phys.*, **41**, C6-116.

Paus, H.J. & Strohm, K.M. (1979). *Z. Physik*, **B24**, 263.

Paus, H.J. & Strohm, K.M. (1980). *J. Phys. C: Sol. Phys.*, **13**, 57.

Pender, L.F. & Fleming, R.J. (1975). *Thin Sol. Films*, **30**, L11.

Pender, L.F. & Fleming, R.J. (1977a). *J. Phys. C: Sol. St. Phys.*, **10**, 1561.

Pender, L.F. & Fleming, R.J. (1977b). *J. Phys. C. Sol. St. Phys.*, **10**, 1571.

Petralia, S. & Gnani, G. (1972). *Lett. Nuov. Cim.*, **4**, 483.

Petridiou, Ch., Christodoulides, C. & Charalambous, S. (1978). *Nucl. Instrum. Meth.*, **150**, 247.

Pfannhauser, J., Hohenau, W. & Gross, F. (1978). *Acta Phys. Austriaca*, **49**, 75.

Pick, H. (1939). *Ann. Phys.*, **35**, 73.

Piesch, E. (1977). *Health Phys.*, **145**, 613.

Piesch, E. (1981a). In *Applied thermoluminescence dosimetry* (ed. Oberhofer, M. & Scharmann, A.), p. 167. Adam Hilger, Bristol.

Piesch, E. (1981b). In *Applied thermoluminescence dosimetry* (ed. Oberhofer, M. & Scharmann, A.), p. 197. Adam Hilger, Bristol.

Piesch, E., Burgkhardt, B. & Hofmann, J. (1976). *Nucl. Instrum. Meth.*, **138**, 157.

Piesch, E., Burgkhardt, B. & Sayed, A.M. (1978). *Nucl. Instrum. Meth.*, **157**, 179.

Podgoršak, E.B., Moran, P.R. & Cameron, J.R. (1971). *J. Appl. Phys.*, **42**, 2761.

Pooley, D. (1966). *Proc. Phys. Soc.*, **87**, 245.

Pooley, D. & Runciman. W.A. (1970). *J. Phys. C: Sol. St. Phys.*, **3**, 1815.

Portal, G. (ed.) (1980). *Proc. 6th Int. Conf. Lumin. Dosim.* (Toulouse), *Nucl. Instrum. Meth.*, **175**.

Portal, G. (1981). In *Applied thermoluminescence dosimetry*, (ed. Oberhofer, M. & Scharmann, A.), p. 97. Adam Hilger, Bristol.

Portal, G., Bermann, F., Blanchard, Ph. & Pringent, R. (1971). *Proc. 3rd Int. Conf. Lumin. Dosim.* (Risø), 410.

Portal, G., Lorrain, S. & Valladas, G. (1980). *Nucl. Instrum. Meth.*, **175**, 12.

Pradhan, A.S. (1981). *Radiat. Protect. Dosim.*, **1**, 153.

Pradhan, A.S. & Bhatt, R.C. (1977). *Phys. Med. Biol*, **22**, 873.

Pradhan, A.S. & Bhatt, R.C. (1979). *Nucl. Instrum. Meth.*, **166**, 497.

Pradhan, A.S. & Bhatt, R.C. (1981). *Phys. Stat. Sol. (A)*, **68**, 405.

Pradhan, A.S. & Bhatt, R.C. (1982). *Radiat. Protect. Dosim.*, **2**, 23.

364   *References*

Pradhan, A.S. & Bhatt, R.C. (1983). *J. Phys. D: Appl. Phys.*, **16**, 53.
Pradhan, A.S., Bhatt, R.C., Lakshmanan, A.R., Chandra, B. & Shinde, S.S. (1978). *Phys. Med. Biol.*, **23**, 723.
Pradhan, A.S., Bhatt, R.C. & Vohra, K.G. (1982). *Bull. Rad. Protect.*, **5**, 13.
Prokic, M. (1978). *Nucl. Instrum. Meth.*, **151**, 603.
Prokic, M. (1980). *Nucl. Instrum. Meth.*, **175**, 83.
Puma, M., Laredo, E., Galavis, M.E. & Figueroa, D.R. (1980). *Phys. Rev.*, **B22**, 5791.
Purdy, A.E. & Murray, R.B. (1975). *Sol. St. Commun.*, **16**, 1293.
Racz, Z. (1981). *J. Lumin.*, **23**, 255.
Radyabov, E.A. & Nepomnyachikh, A.I. (1981). *Phys. Stat. Sol. (A)*, **68**, 77.
Randall, J.T. & Wilkins, M.H.F. (1945a). *Proc. Roy. Soc. Lond.*, **184**, 366.
Randall, J.T. & Wilkins, M.H.F. (1945b). *Proc. Roy. Soc. Lond.*, **184**, 390.
Rao, S.M.D. (1974). *Phys. Stat. Sol. (A)*, **24**, 519.
Rascón, A. & Alvarez Rivas, J.L. (1978). *J. Phys. C: Sol. St. Phys.*, **11**, 1239.
Rascón, A. & Alvarez Rivas, J.L. (1981). *J. Phys. C: Sol. St. Phys.*, **14**, L741.
Reddy, K.N., Rao, M.L. & Babu, V.H. (1982). *Phys. Stat. Sol. (A)*, **70**, 335.
Reid, A.M. & Cohen, A.J. (1967). *Geochim. Cosmochim. Acta*, **31**, 661.
Rendell, H.M. (1983). *PACT*, **9**, 523.
Richayzen, G. (1957). *Proc. Roy. Soc. Lond.*, **A241**, 480.
Riehl, N. (1970). *J. Lumin.*, **1**, 1.
Riehl, N. & Schön, M. (1939). *Z. Physik*, **114**, 682.
Rieke, J.R. & Daniels, F. (1957). *J. Phys. Chem.*, **61**, 629.
Roach, C.H. (1968). In *Thermoluminescence of geological materials* (ed. McDougall, D.J.), p. 591. Acad. Press, London.
Roach, C.H., Johnson, C.R., McGrath, J.G. & Sterrett, T.S. (1962). *US Geol. Surv. Prof. Paper* 450-D, 98.
Roberts, G.G., Apsley, N. & Munn, R.W. (1980). *Phys. Rep.*, **60**, 59.
Rodine, E.T. & Land, P.L. (1971). *Phys. Rev.*, **B4**, 2701.
Rose, A. (1963). *Concepts in photoconductivity and allied problems*. Interscience, New York.
Rossister, M.J., Rees-Evans, D.B. & Ellis, S.C. (1970). *J. Phys. D: Appl. Phys.*, **3**, 1816.
Roth, M. & Halperin, A. (1982). *J. Phys. Chem. Sol.*, **43**, 609.
Rubio, J.O. Aguilar, M.G., López, F.J., Galan, M., Garcia-Solé, J. & Murrieta, H.S. (1982). *J. Phys. C: Sol. St. Phys.*, **15**, 6113.
Rudlof, G., Becherer, J. & Glaefeke, H. (1978). *Phys. Stat. Sol. (A)*, **49**, K121.
Rudlof, G., Becherer, J. & Glaefeke, H. (1979). *Phys. Stat. Sol (A)*, **52**, K137.
Ruffa, A.R. (1973/74). *J. Non-Cryst. Sol.*, **13**, 37.
Rutherford, E. (1913). *Radioactive substances and their radiations*. Cambridge University Press, Cambridge, England.
Rutherford, E., Chadwick, J. & Ellis, C.D. (1957). *Radiations from radioactive substances*. Cambridge University Press, Cambridge, England.
Rzyski, B.M. & Morato, S.P. (1980). *Nucl. Instrum. Meth.*, **175**, 62.
Sagastibelza, F. & Alvarez Rivas, J.L. (1981). *J. Phys. C: Sol. St. Phys.*, **14**, 1873.
Saidoh, M. & Townsend, P.D. (1975). *Radiat. Eff.*, **27**, 1.
Sankaran, A.V., Nambi, K.S.V. & Sunta, C.M. (1982). *BARC Rep.* 1156.
Sasidharan, R., Sunta, C.M. & Nambi, K.S.V. (1979). *PACT*, **3**, 401.
Sastry, S.B.S. & Sapru, S. (1981). *J. Lumin.*, **23**, 281.
Saunders, I.J. (1969). *J. Phys. C: Sol. St. Phys.*, **2**, 2181.
Scarpa, G. (1970). *Health Phys.*, **19**, 91.
Scharmann, A. (ed.) (1977). *Proc. 5th Int. Conf. Lumin. Dosim.* (Sao Paulo). Physikalischer Institut, Giessen.

Schayes, R., Brooke, C., Kozlowitz, I. & Lheureux, M. (1967). *Proc. 1st Int. Conf. Lumin. Dosim.* (Stanford), 138.

Schirmer, O.F. (1976). *Sol. St. Commun.*, **18**, 1349.

Schlesinger, M. (1964). *Phys. Letts*, **10**, 49.

Schlesinger, M. (1965). *J. Phys. Chem. Sol.*, **26**, 1761.

Schmidt, K., Linemann, H. & Giessing, R. (1974). *Proc. 4th Int. Conf. Lumin. Dosim.* (Krakow).

Schön, M. (1942). *Z. Physik.*, **119**, 463.

Schön, M. (1951a). *Z. Naturforsch*, **6a**, 251.

Schön, M. (1951b). *Naturwiss*, **38**, 235.

Schulman, J.H., Attix, F.H., West, E.J. & Ginther, R.J. (1960). *Rev. Sci. Instrum.*, **31**, 1263.

Schulman, J.H., Ginther, R.J., Gorbics, S.G., Nash, A.E., West, E.J. & Attix, F.H. (1969). *J. Appl. Rad. Isot.*, **20**, 253.

Schulman, J.H., Kirk, R.D. & West, E.J. (1967). *Proc. 1st Int. Conf. Lumin. Dosim.* (Stanford), 113.

Sears, D.W. (1975). *Mod. Geol.*, **5**, 155.

Sears, D.W. (1978a). *The nature and origin of meteorites*. Adam Hilger, Bristol.

Sears, D.W. (1978b). *PACT*, **2**, 231.

Sears, D.W. (1980). *Icarus*, **44**, 190.

Sears, D.W. & Durrani, S.A. (1980). *Earth Planet. Sci. Letts*, **46**, 159.

Sears, D.W., Grossman, J.N. & Melcher, C.L. (1982). *Geochim. Cosmochim. Acta*, **46**, 247.

Sears, D.W., Grossman, J.N., Melcher, C.L., Ross, L.M. & Mills, A.A. (1980). *Nature*, **287**, 791.

Sears, D.W. & McKeever, S.W.S. (1980). *Mod. Geol.*, **7**, 202.

Sears, D.W. & Mills, A.A. (1974a). *Earth Planet. Sci. Letts*, **22**, 391.

Sears, D.W. & Mills, A.A. (1974b). *Meteoritics*, **9**, 47.

Seeley, M.A. (1975). *J. Archaeol. Sci.*, **2**, 17.

Seitz, F. (1940). *Modern theory of solids*. McGraw-Hill, New York.

Seitz, F. (1952). *Imperfections in nearly perfect crystals*. Wiley and Sons, New York.

Sergeyev, V.M. & Ventslovayte, Y.J. (1978). *Dokl. Akad. Nauk. SSR Earth Sci.*, **241**, 195.

Shekhmametev, R.I. (1973). *Opt. Spectrosc.*, **34**, 288.

Shelkoplyas, V.N. (1971). In *Chronology of the glacial age*. Geogr. Soc. USSR (Leningrad). Cited by Wintle & Huntley (1979a).

Shenker, D. & Chen, R. (1971). *J. Phys. D: Appl. Phys.*, **4**, 287.

Shenker, D. & Chen, R. (1972). *J. Comput. Phys.*, **10**, 272.

Shinde, S.S. & Sastry, S.S. (1979). *Int. J. Appl. Radiat. Isot.*, **30**, 501.

Shockley, W. & Read, W.T. (1952). *Phys. Rev.*, **87**, 835.

Short, N.M. (1966). *J. Geol. Educ.*, **14**, 149.

Short, N.M. (1970). *J. Geol.*, **78**, 705.

Shrimpton, P.C., Wall, B.F. & Fisher, E.S. (1981). *Phys. Med. Biol.*, **26**, 133.

Sibley, W.A., Martin, J.J., Wintersgill, M.C. & Brown, J.D. (1979). *J. Appl. Phys.*, **50**, 5449.

Sidran, M. (1968). *J. Geophys. Res.*, **73**, 5196.

Sigel, G.H. (1973/74). *J. Non-Cryst. Sol.*, **13**, 372.

Sigel, G.H., Friebele, E.J., Marrone, M.J. & Gingerich, M.E. (1981). *IEEE Trans. Nucl. Sci.*, **NS-28**, 4095.

Sigel, G.H. & Marrone, M.J. (1981). *J. Non-Cryst. Sol.*, **45**, 235.

Silsbee, R.H. (1961). *J. Appl. Phys.*, **32**, 1459.

Simmons, J.G. & Taylor, G.W. (1971). *Phys. Rev.*, **B4**, 302.
Simmons, J.G. & Taylor, G.W. (1972). *Phys. Rev.*, **B5**, 1619.
Singhvi, A.K. & Bhandari, N. (1979). *Meteoritics*, **14**, 536.
Singhvi, A.K., Pal, S. & Bhandari, N. (1982). *PACT*, **6**, 404.
Singhvi, A.K., Sharma, Y.P. & Agarwal, D.P. (1982). *Nature*, **295**, 313.
Singhvi, A.K. & Zimmerman, D.W. (1978). *PACT*, **2**, 12.
Sippel, R.F. (1971). *Proc. 2nd Lunar. Sci. Conf.*, **1**, 247.
Sippel, R.F. & Spencer, A.B. (1970). *Proc. Apollo 11 Lunar Sci. Conf.*, **3**, 2413.
Sonder, E. & Sibley, W.A. (1972). In *Point defects in solids* (ed. Crawford, J.H. & Slifkin, L.M.), p. 1. Plenum Press, New York.
Song, K.S., Stoneham, A.M. & Harker, A.H. (1975). *J. Phys. C: Sol. St. Phys.*, **8**, 1125.
Spanne, P. (1973). *Health Phys.*, **24**, 568.
Sparks, M.H., McKimmey, P.M. & Sears, D.W. (1983). *J. Geophys. Res.*, **88**, (Suppl.), A773.
Speit, B. & Lehmann, G. (1982a) *J. Lumin.*, **27**, 127.
Speit, B. & Lehmann, G. (1982b). *Phys. Chem. Miner.*, **8**, 77.
Spurný, Z. (1980). *Nucl. Instrum. Meth.*, **175**, 71.
Spurný, Z., Milu, C. & Racoveanu, N. (1973). *Phys. Med. Biol*, **18**, 276.
Sridaran, P., Gartia, R.K., Bhuniya, R.C. & Ratnam, V.V. (1981). *Phys. Stat. Sol. (A)*, **64**, 127.
Srinivasan, M. & DeWerd, L.A. (1973). *J. Phys. D: Appl. Phys.*, **6**, 2142.
Srivastava, J.K. & Supe, S.J. (1979). *Nucl. Instrum. Meth.*, **160**, 529.
Srour, J.R., Curtis, O.L. & Chiu, K.Y. (1974). *IEEE Trans. Nucl. Sci.*, **NS-21**, 73.
Srour, J.R., Othmer, S., Curtis, O.L. & Chiu, K.Y. (1976). *IEEE Trans. Nucl. Sci.*, **NS-23**, 1513.
Stadler, A. & Wagner, G.A. (1979). *PACT*, **3**, 448.
Stammers, K. (1979). *J. Phys. E: Sci. Instrum.*, **12**, 637.
Stapelbroek, M., Griscom, D.L., Friebele, E.J. & Sigel, G.H. (1979). *J. Non-Cryst. Sol.*, **32**, 313.
Stevels, J.M. & Volger, J. (1962). *Philips Res. Rep.*, **7**, 283.
Stoddard, A.E. (1960). *Phys. Rev.*, **120**, 114.
Stoebe, T.G. (ed.) (1984). *Proc. 7th Int. Conf. Lumin. Dosim.* (Ottawa, 1984). *Radiat. Protect. Dosim.*, **6**.
Stoebe, T.G. & Watanabe, S. (1975). *Phys. Stat. Sol. (A)*, **20**, 11.
Stoebe, T.G., Wolfenstine, J.B. & Las, W.C. (1980). *J. de Phys.*, **41**, C6-265.
Stöffler, D. (1974). *Fortscher. Miner.*, **51**, 256.
Stoneham, D. (1983). *PACT*, **9**, 227.
Stott, J.P. & Crawford, J.H. (1971). *Phys. Rev.*, **4**, 639.
Strohm, K.M. & Paus, H.J. (1980). *J. de Phys.*, **41**, C6-119.
Strutt, J.E. & Lilley, E. (1976). *Phys. Stat. Sol. (A)*, **33**, 229.
Suhr, N.H. & Ingamells, C.D. (1966). *Anal. Chem.*, **36**, 730.
Sun, K.H. & Gonzales, J.L. (1966). *Nature*, **212**, 23.
Sunta, C.M. (1970). *Phys. Stat. Sol. (A)*, **37**, K81.
Sunta, C.M. (1979). *Phys. Stat. Sol. (A)*, **53**, 127.
Sunta, C.M. & Bapat, V.N. (1982). *PACT*, **6**, 252.
Sunta, C.M. & Kathuria, M. (1978). *PACT*, **2**, 185.
Sunta, C.M. & Watanabe, S. (1976). *J. Phys. D: Appl. Phys.*, **9**, 1271.
Suntharalingam, N. & Cameron, J.R. (1969). *Phys. Med. Biol.*, **14**, 397.
Sutton, S.R. & Zimmerman, D.W. (1976). *Archaeometry*, **18**, 125.
Suzuki, K. (1961). *J. Phys. Soc. Japan*, **16**, 67.

Sweet, M.A.S. & Urquhart, D. (1980). *Phys. Stat. Sol. (A)*, **59**, 223.
Sweet, M.A.S. & Urquhart, D. (1981). *J. Phys. D: Appl.*, Phys., **14**, 773.
Swyler, K.J. & Levy, P.W. (1976). *Proc. 3rd Conf. Partially Ionized and Uranium Plasmas* (Princeton), 1.
Szabó, P.P., Félszerfalvi, J. & Lénárt, A. (1980). *Nucl. Instrum. Meth.*, **175**, 45.
Takenaga, M., Yamamoto, O. & Yamashita, T. (1977). *Proc. 5th Int. Conf. Lumin. Dosim.* (Sao Paulo), 148.
Takenaga, M., Yamamoto, O. & Yamashita, T. (1980). *Nucl. Instrum. Meth.*, **175**, 77.
Takeuchi, N., Inabe, K., Kido, H. & Yamashita, I. (1978). *J. Phys. C: Sol. St. Phys.*, **11**, L147.
Takeuchi, N., Inabe, K. & Nanto, H. (1975). *J. Mat. Sci.*, **10**, 1.
Takeuchi, N., Inabe, K. & Okung, T. (1982). *Phys. Stat. Sol. (A)*, **73**, K47.
Tale, I.A. (1981). *Phys. Stat. Sol. (A)*, **66**, 65.
Tanaka, S.I. & Furuta, Y. (1974). *Proc. 4th Int. Conf. Lumin. Dosim.* (Krakow), 1213.
Tanaka, S.I. & Furuta, Y. (1977). *Nucl. Instrum. Meth.*, **140**, 395.
Tanimura, K., Fujiwara, M., Okada, T. & Hagihara, T. (1978). *Sol. St. Commun.*, **25**, 315.
Tanimura, K., Fujiwara, M., Okada, T. & Suita, T. (1974a). *Phys. Letts*, **50A**, 301.
Tanimura, K. & Okada, T. (1976). *Phys. Rev.*, **B13**, 1811.
Tanimura, K. & Okada, T. (1980). *Phys. Rev.*, **B21**, 1690.
Tanimura, K., Okada, T. & Suita, T. (1974b). *Sol. St. Commun.*, **14**, 107.
Tatake, V.G. (1975). *Proc. Natn. Symp. on Thermoluminescence* (Madras), 262.
Taylor, G.C. & Lilley, E. (1978). *J. Phys. D: Appl. Phys.*, **11**, 567.
Taylor, G.C. & Lilley, E. (1982a). *J. Phys. D: Appl Phys.*, **15**, 1243.
Taylor, G.C. & Lilley, E. (1982b). *J. Phys. D: Appl. Phys.*, **15**, 1253.
Taylor, G.C. & Lilley, E. (1982c). *J. Phys. D: Appl. Phys.*, **15**, 2053.
Taylor, S.R. (1975). *Lunar science: a post-Apollo view.* Pergamon Press, New York.
Telfer, D.J. & Walker, G. (1978). *Mod. Geol.*, **6**, 199.
Templar, R.H. & Walton, A.J. (1983). *PACT*, **9**, 299.
Thomas, J.H. & Fiegl, F.J. (1970). *Sol. St. Commun.*, **8**, 1669.
Thomasz, E., D'Alotto, V., Nollmann, C.E. & Kunst, J.J. (1980). *Nucl. Instrum. Meth.*, **175**, 196.
Thompson, J.J. & Ziener, P.L. (1972). *Health Phys.*, **22**, 399.
Thompson, J.J. & Ziener, P.L. (1973). *Health Phys.*, **25**, 435.
Tochilin, E. & Goldstein, N. (1966). *Health Phys.*, **12**, 1705.
Tochilin, E., Goldstein, N. & Miller, W.G. (1969). *Health Phys.*, **16**, 1.
Tochin, V.A. & Nikol'skii, V.G. (1969). *High Energy Chem.*, **3**, 256.
Tournon, J. & Bergé, P. (1964). *Phys. Stat. Sol.*, **5**, 117.
Townsend, P.D. (1968). In *Thermoluminescence of geological materials* (ed. McDougall, D.J.), p. 51. Acad. Press, London.
Townsend, P.D. (1976). *J. Phys. C: Sol. St. Phys.*, **9**, 1871.
Townsend, P.D. (1982). *Nucl. Instrum. Meth. Phys. Res.*, **197**, 9.
Townsend, P.D. & Agulló-López, F. (1980). *J. de Phys.*, **41**, C6-279.
Townsend, P.D., Ahmed, K., Chandler, P.J., McKeever, S.W.S. & Whitlow, H.J. (1983b). *Radiat. Eff.*, **72**, 245.
Townsend, P.D., Blanke, J.H., Price, E.W. & Rendell, H.M. (1983a). *Radiat. Eff.*, **72** 103.
Townsend, P.D., Clarke, C.D. & Levy, P.W. (1967). *Phys. Rev.*, **155**, 908.
Townsend, P.D. & Kelly, J.C. (1973). *Colour centres and imperfections in insulators and semiconductors.* Sussex University Press, London.

Townsend, P.D., Mahjoobi, A., Michael, A.J. & Saidoh, M. (1976). *J. Phys. C: Sol. St. Phys.*, **9**, 4203.

Townsend, P.D., Taylor, G.C. & Wintersgill, M.C. (1978). *Radiat. Eff.*, **41**, 11.

Toyotomi, Y. & Onaka, R. (1979a). *J. Phys. Soc. Japan*, **46**, 1861.

Toyotomi, Y. & Onaka, R. (1979b). *J. Phys. Soc. Japan*, **46**, 1869.

Treadaway, M.J., Passenheim, B.C. & Kitterer, B.D. (1975). *IEEE Trans. Nucl. Sci.*, *NS-22*, 2253.

Trousil, J. & Kokta, L. (1982). *Radiat. Protect. Dosim.*, **2**, 169.

Trowbridge, J. & Burbank, J.E. (1898). *Am. J. Sci.*, Ser. 4, **5**, 55.

Trukhin, A.N. & Plaudis, A.E. (1979). *Sov. Phys. Sol. St.*, **21**, 644.

Turner, R.C., Radley, J.M. & Mayneord, W.V. (1958). *Br. J. Radiol.*, **31**, 397.

Turner, W.H. & Lee, H.A. (1966). *J. Chem. Phys.*, **44**, 1428.

Unger, S. & Perlman, M.M. (1974). *Phys. Rev.*, **B10**, 3692.

Unger, S. & Perlman, M.M. (1977). *Phys. Rev.*, **B15**, 4105.

Urbach, F. (1930). *Wiener Ber.*, **139**, 363.

Valladas, G. & Ferreira, J. (1980). *Nucl. Instrum. Meth.*, **175**, 216.

Valladas, G. & Gillot, P.Y. (1978). *PACT*, **2**, 141.

Valladas, G., Gillot, P.V. & Guerin, G. (1979). *PACT*, **3**, 251.

Valladas, G. & Lalou, C. (1973). *Earth Planet. Sci. Letts*, **18**, 168.

Valladas, G. & Valladas, H. (1983). *PACT*, **9**, 73.

Valladas, H. (1978). *PACT*, **2**, 180.

van den Brom, W.E. & Volger, J. (1974). *Physica*, **75**, 245.

van Puymbroeck, W. & Schoemaker, D. (1981). *Phys. Rev.*, **23**, 1670.

van Roosbroeck, W. (1973). *J. Non-Cryst. Sol.*, **12**, 232.

van Roosbroeck, W. & Casey, H.C. (1972). *Phys. Rev.*, **B5**, 2154.

van Roosbroeck, W. & Shockley, W. (1954). *Phys. Rev.*, **94**, 1558.

van Turnhout, J. & van Rheenen, A.H. (1977). *J. Electrostat.*, **3**, 213.

van Utiert, L.G. (1960). *J. Electrochem.*, **107**, 803.

Vaz, J.E. (1971). *Meteoritics*, **6**, 207.

Vaz, J.E. (1972). *Meteoritics*, **7**, 77.

Vaz, J.E., Kemmey, P.J. & Levy, P.W. (1968). In *Thermoluminescence of geological materials* (ed. McDougall, D.J.), p. 111. Acad. Press, London.

Vaz, J.E. & Sears, D.W. (1977). *Meteoritics*, **12**, 47.

Vaz, J.E. & Sifontes, R.S. (1978). *Mod. Geol.*, **6**, 147.

Velkley, D.E., Cunningham, D.E. & Strockbine, M.F. (1980). *Int. J. Radiat. Oncol. Biol. Phys.*, **6**, 1739.

Vignolo, J. & Alvarez Rivas, J.L. (1980). *J. Phys. C: Sol. St. Phys.*, **13**, 5291.

Visocekas, R. (1979). *PACT*, **3**, 258.

Visocekas, R. (1981). *Int. Conf. Defects in Insul. Crystals* (Riga), 4

Visocekas, R., Ceva, T., Marti, C., Lefaucheux, F. & Robert, M.C. (1976). *Phys. Stat. Sol. (A)*, **35**, 315.

Visocekas, R. & Geoffroy, A. (1977). *Phys. Stat. Sol. (A)*, **41**, 499.

Vora, H., DeWerd, L.A. & Stoebe, T.G. (1974). *Proc. 4th Int. Conf. Lumin. Dosim.* (Krakow), 143.

Wachter, W. (1982). *J. Appl. Phys.*, **53**, 5210.

Wachter, W., Vana, N.J. & Aiginger, H. (1980). *Nucl. Instrum. Meth.*, **175**, 21.

Wald, J., DeWerd, L.A. & Stoebe, T.G. (1977). *Health Phys.*, **33**, 303.

Waligórski, M.P.R. & Katz, R. (1980). *Nucl. Instrum. Meth.*, **172**, 463.

Walker, R.M. & Zimmerman, D.W. (1972). *Earth Planet. Sci. Letts*, **13**, 419.

Walker, R.M., Zimmerman, D.W. & Zimmerman, J. (1971). *Proc. Conf. Lunar Geophys.* (Houston), 308.

Wall, B.F., Driscoll, C.M.H., Strong, J.C. & Fisher, E.S. (1982). *Phys. Med. Biol.*, 27, 1023.

Wall, B.F., Fisher, E.S., Paynter, R., Hudsen, A. & Bird, P.D. (1979b). *Br. J. Radiol.*, 52, 727.

Wall, B.F., Green, D.A.C. & Veerapan, R. (1979a). *Br. J. Radiol.*, 52, 189.

Wall, B.F., Rae, S., Darby, S.C. & Kendall, G.M. (1981). *Br. J. Radiol.*, 54, 719.

Walton, A.J. (1977). *Adv. Phys.*, 26, 887.

Walton, A.J. (1982). *PACT*, 6, 524.

Walton, A.J. & Debenham, N.C. (1980). *Nature*, 284, 42.

Walton, A.J. & Debenham, N.C. (1982). *PACT*, 6, 202.

Watson, P. (1977). *Br. J. Radiol.*, 50, 745.

Watterich, A., Foldvari, I. & Voszka, R. (1980). *J. de Phys.*, 41, C6-159.

Watterich, A. & Voszka, R. (1979). *Phys. Stat. Sol. (B)*, 93, K161.

Weeks, R.A. (1963). *Phys. Rev.*, 130, 570.

Weeks, R.A. & Nelson, C.M. (1960). *J. Amer. Ceram. Soc.*, 43, 399.

Weil, J.A. (1975). *Radiat. Eff.*, 26, 261.

West, R.H. & Carter, A.R. (1980). *Radiat. Eff. Letts*, 57, 129.

Wick, F.G. (1924). *Phys. Rev.*, 24, 272.

Wick, F.G. (1927). *J. Opt. Soc. Am.*, 14, 33.

Wick, F.G. & Gleason, J.M. (1924). *J. Opt. Soc. Am.*, 9, 639.

Wick, F.G. & Slattery, M.K. (1927). *J. Opt. Soc. Am.*, 14, 125.

Wick, F.G. & Slattery, M.K. (1928). *J. Opt. Soc. Am.*, 16, 398.

Wiedemann, E. & Schmidt, G.C. (1895). *Ann. Phys. Chem. Neue Folge*, 54, 604.

Williams, E.W. & Hall, R. (1978). *Luminescence and the light emitting diode.* Pergamon Press, Oxford.

Williams, F.E. (1949). *J. Opt. Soc. Am.*, 39, 648.

Williams, F.E. & Eyring, H. (1947). *J. Chem. Phys.*, 15, 289.

Williams, M.L., Landel, R.F. & Ferry, J.D. (1955). *J. Am. Chem. Soc.*, 77, 4701.

Williams, R.T. (1978). *Semicond. Insul.*, 3, 251.

Wintersgill, M.C. & Townsend, P.D. (1978a). *Radiat. Eff.*, 38, 113.

Wintersgill, M.C. & Townsend, P.D. (1978b). *Phys. Stat. Sol. (A)*, 47, K67.

Wintersgill, M.C., Townsend, P.D. & Cusso-Perez, F. (1977). *J. de Phys.*, C7, 123.

Wintle, A.G. (1973). *Nature*, 245, 143.

Wintle, A.G. (1975a). *Geophys. J. Roy. Astr. Soc.*, 41, 107.

Wintle, A.G. (1975b). *Mod. Geol*, 5, 165.

Wintle, A.G. (1977). *J. Lumin.*, 15, 385.

Wintle, A.G. (1980). *Archaeometry*, 22, 113.

Wintle, A.G. (1981). *Nature*, 289, 478.

Wintle, A.G. & Aitken, M.J. (1977). *Archaeometry*, 19, 11.

Wintle, A.G. & Huntley, D.J. (1979a). *Nature*, 279, 710.

Wintle, A.G. & Huntley, D.J. (1979b) *PACT*, 3, 374.

Wintle, A.G. & Huntley, D.J. (1980). *Can. J. Earth Sci.*, 17, 348.

Wochos, J.F., DeWerd, L.A., Hilko, R., Meyer, J.A., Stovall, M., Spearman, D., Thomason, C. & Dubuque, G.L. (1982). *Med. Phys.*, 9, 920.

Wood, D.L. (1960). *J. Phys. Chem. Sol.*, 13, 326.

Wu, S. & Kendall, F.H. (1982). *PACT*, 6, 34.

Yamaoka, Y (1978). *Health Phys.*, 35, 708.

Yamashita, T. Nada, N., Onishi, H. & Kitamura, S. (1968). *Proc. 2nd Int. Conf. Lumin. Dosim.* (Gatlinburg), 4.

Yamashita, T., Nada, N., Onishi, H. & Kitamura, S. (1971). *Health Phys.*, **21**, 295.

Yokota, K. (1953). *Phys. Rev.*, **91**, 1013.

Zanelli, G.D. (1968). *Phys. Med. Biol.*, **13**, 393.

Zeller, E.J. (1968). In *Thermoluminescence of geological materials* (ed. McDougall, D.J.), p. 311. Acad. Press, London.

Zimmerman, D.W. (1967). *Archaeometry*, **10**, 26.

Zimmerman, D.W. (1971c). *Archaeometry*, **13**, 29.

Zimmerman, D.W. (1971d). *Science*, **74**, 818.

Zimmerman, D.W. (1972). *Radiat. Eff.*, **14**, 81.

Zimmerman, D.W. (1977a). *Abstr. Symp. Archaeometry and Archaeological Prospection* (Philadelphia), 42.

Zimmerman, D.W. (1977b). *J. Electrostat.*, **3**, 257.

Zimmerman, D.W. (1978). *PACT*, **2**, 1.

Zimmerman, D.W. (1979a) *PACT*, **3**, 257.

Zimmerman, D.W. (1979b). *PACT*, **3**, 458.

Zimmerman, D.W. & Cameron, J.R. (1968). In *Thermoluminescence of geological materials* (ed. McDougall, D.J.), p. 485. Acad. Press, London.

Zimmerman, D.W., Rhyner, C.R. & Cameron, J.R. (1966). *Health Phys.*, **12**, 526.

Zimmerman, J. (1971a). *J. Phys. C: Sol. St. Phys.*, **4**, 3265.

Zimmerman, J. (1971b). *J. Phys. C: Sol. St. Phys.*, **4**, 3277.

Ziniker, Z.M., Rusin, J.M. & Stoebe, T.G. (1973). *J. Mat. Sci.*, **8**, 407.

Zinner, E. (1980). In *The ancient sun: Fossil record in the earth, moon and meteorites* (ed. Pepin, R.O., Eddy, J.A. & Merrill, R.B.), p. 201. Pergamon Press, New York.

Index

Index

V-centres 111, 139–40, 143, 154–87
vidicon tube 339–40

Williams –Landel–Ferry 113, 116
windows (for cryostats) 325–6, 333

X-ray fluorescence 280

Z-centre model 179–81

Z-centres 112, 129, 133, 163, 179–81, 184–5
ZnO 143
ZnS 40, 81, 110, 143–5, 148–9, 279
ZnSe 36, 38, 63, 84, 143, 179
$ZnSiO_4$ 32, 100, 108, 144
zero dose 207, 234, 245
zeroing mechanism (in sediment dating) 282–3, 287
zircon 144, 256–7, 266, 271, 292
zoning 258

Printed in the United States
By Bookmasters